T0138006

Learning Analytics in R with SNA, LSA, and MPIA

Fridolin Wild

Learning Analytics in R with SNA, LSA, and MPIA

Fridolin Wild
Performance Augmentation Lab
Department of Computing and
Communication Technologies
Oxford Brookes University
Oxford
UK

The present work has been accepted as a doctoral thesis at the University of Regensburg, Faculty for Languages, Literature, and Cultures under the title "Learning from Meaningful, Purposive Interaction: Representing and Analysing Competence Development with Network Analysis and Natural Language Processing".

Cover photo by Sonia Bernac (http://www.bernac.org)

ISBN 978-3-319-80425-5 ISBN 978-3-319-28791-1 (eBook)
DOI 10.1007/978-3-319-28791-1

Softcover reprint of the hardcover 1st edition 2016

Printed on acid-free paper

This Springer imprint is published by Springer Nature
The registered company is Springer International Publishing AG Switzerland

*To my father Kurt Wild (*1943–†2010).*

Acknowledgement

This work would not have been possible without the support of my family, my son Maximilian, my brothers Benjamin, Johannes, and Sebastian, and my parents Brunhilde and Kurt, only one of whom survived its finalisation (Dad, you are missed!).

I am indebted to Prof. Dr. Peter Scott, at the time Director of the Knowledge Media Institute of The Open University and Chair for Knowledge Media, now with the University of Technology in Sydney as Assistant Deputy Vice-Chancellor, and Prof. Dr. Christian Wolff, Chair for Media Informatics of the University of Regensburg (in alphabetical order), who both helped improve this book and saved me from my worst mistakes. Without you and our engaging discussions, this book would simply not have happened.

There are many people who contributed to this work, some of them unknowingly. I'd like to thank the coordinators of the EU research projects that led to this work, also in substitution for the countless researchers and developers with whom I had the pleasure to work in these projects over the years: Wolfgang Nejdl for Prolearn, Barbara Kieslinger for iCamp, Ralf Klamma and Martin Wolpers for ROLE, Peter van Rosmalen and Wolfgang Greller for LTfLL, Peter Scott for STELLAR, Vana Kamtsiou and Lampros Stergioulas for TELmap, and Paul Lefrere, Claudia Guglielmina, Eva Coscia, and Sergio Gusmeroli for TELLME.

I'd also like to thank Prof. Dr. Gustaf Neumann, Chair for Information Systems and New Media of the Vienna University of Economics and Business (WU), for whom I worked in the years 2005–2009, before I joined Peter Scott's team at the Knowledge Media Institute of The Open University. Special thanks go to Prof. Dr. Kurt Hornik of the Vienna University of Economics and Business, who first got me hooked on R (and always had an open ear for my tricky questions). I'd also like to express my gratitude to Dr. Ambjoern Naeve of the KTH Royal Institute of Technology of Sweden for awakening my passion for mathemagics more than a decade ago and believing in me over years that I have it in me.

I'd like to particularly thank Debra Haley and Katja Buelow for the work in the LTfLL project, particularly the data collection for the experiment further analysed in Sects. 10.3.2 and 10.3.3.

I owe my thanks to my current and former colleagues at the Open University and at the Vienna University of Economics and Business, only some of which I can explicitly mention here: Thomas Ullmann, Giuseppe Scavo, Anna De Liddo, Debra Haley, Kevin Quick, Katja Buelow, Chris Valentine, Lucas Anastasiou, Simon Buckingham Shum, Rebecca Ferguson, Paul Hogan, David Meyer, Ingo Feinerer, Stefan Sobernig, Christina Stahl, Felix Moedritscher, Steinn Sigurdarson, and Robert Koblischke.

Not least, I'd like to stress that I am glad and proud to be a citizen in the European Union that has brought the inspiring research framework programmes and Horizon 2020 on the way and provided funding to those projects listed above.

Fridolin Wild

The original version of the book was revised: Acknowledgement text "Cover photo by Sonia Bernac (http://www.bernac.org)" has been included in the copyright page. The Erratum to the book is available at DOI 10.1007/978-3-319-28791-1_12

Contents

1 Introduction 1
1.1 The Learning Software Market 2
1.2 Unresolved Challenges: Problem Statement 3
1.3 The Research Area of Technology-Enhanced Learning 5
1.4 Epistemic Roots 7
1.5 Algorithmic Roots 9
1.6 Application Area: Learning Analytics 10
1.7 Research Objectives and Organisation of This Book 14
References 17

2 Learning Theory and Algorithmic Quality Characteristics 23
2.1 Learning from Purposive, Meaningful Interaction 25
2.2 Introducing Information 27
2.3 Foundations of Culturalist Information Theory 28
2.4 Introducing Competence 30
2.5 On Competence and Performance Demonstrations 32
2.6 Competence Development as Information Purpose of Learning 32
2.7 Performance Collections as Purposive Sets 34
2.8 Filtering by Purpose 35
2.9 Expertise Clusters 35
2.10 Assessment by Equivalence 36
2.11 About Disambiguation 37
2.12 A Note on Texts, Sets, and Vector Spaces 38
2.13 About Proximity as a Supplement for Equivalence 39
2.14 Feature Analysis as Introspection 39
2.15 Algorithmic Quality Characteristics 40
2.16 A Note on Educational Practice 41
2.17 Limitations 41
2.18 Summary 42
References 43

3 Representing and Analysing Purposiveness with SNA 45
3.1 A Brief History and Standard Use Cases 45
3.2 A Foundational Example 48
3.3 Extended Social Network Analysis Example 56
3.4 Limitations 68
References 69

4 Representing and Analysing Meaning with LSA 71
4.1 Mathematical Foundations 73
4.2 Analysis Workflow with the R Package 'lsa' 77
4.3 Foundational Example 80
4.4 State of the Art in the Application of LSA for TEL 91
4.5 Extended Application Example: Automated Essay Scoring 95
4.6 Limitations of Latent Semantic Analysis 101
References 101

5 Meaningful, Purposive Interaction Analysis 107
5.1 Fundamental Matrix Theorem on Orthogonality 108
5.2 Solving the Eigenvalue Problem 112
5.3 Example with Incidence Matrices 114
5.4 Singular Value Decomposition 117
5.5 Stretch-Dependent Truncation 122
5.6 Updating Using Ex Post Projection 125
5.7 Proximity and Identity 126
5.8 A Note on Compositionality: The Difference of Point, Centroid, and Pathway 128
5.9 Performance Collections and Expertise Clusters 129
5.10 Summary 130
References 131

6 Visual Analytics Using Vector Maps as Projection Surfaces 133
6.1 Proximity-Driven Link Erosion 135
6.2 Planar Projection With Monotonic Convergence 136
6.3 Kernel Smoothing 141
6.4 Spline Tile Colouring With Hypsometric Tints 142
6.5 Location, Position, and Pathway Revisited 145
6.6 Summary 146
References 147

7 Calibrating for Specific Domains 149
7.1 Sampling Model 150
7.2 Investigation 152
7.3 Results 155
7.4 Discussion 158
7.5 Summary 162
References 163

8 Implementation: The MPIA Package 165
8.1 Use Cases for the Analyst 165
8.2 Analysis Workflow 167
8.3 Implementation: Classes of the *mpia* Package 170
8.3.1 The *DomainManager* 174
8.3.2 The *Domain* 175
8.3.3 The *Visualiser* 175
8.3.4 The *HumanResourceManager* 176
8.3.5 The *Person* 176
8.3.6 The *Performance* 177
8.3.7 The Generic Functions 177
8.4 Summary 181
References 181

9 MPIA in Action: Example Learning Analytics 183
9.1 Brief Review of the State of the Art in Learning Analytics 184
9.2 The Foundational SNA and LSA Examples Revisited 185
9.3 Revisiting Automated Essay Scoring: Positioning 196
9.4 Learner Trajectories in an Essay Space 212
9.5 Summary 221
References 222

10 Evaluation 223
10.1 Insights from Earlier Prototypes 225
10.2 Verification 230
10.3 Validation 231
10.3.1 Scoring Accuracy 232
10.3.2 Structural Integrity of Spaces 234
10.3.3 Annotation Accuracy 238
10.3.4 Visual (in-)Accuracy 240
10.3.5 Performance Gains 241
10.4 Summary and Limitations 242
References 243

11 Conclusion and Outlook 247
11.1 Achievement of Research Objectives 248
11.1.1 Objective 1: Represent Learning 249
11.1.2 Objective 2: Provide Instruments for Analysis 250
11.1.3 Objective 3: Re-represent to the User 252
11.1.4 Summary 256
11.2 Open Points for Future Research 258
11.3 Connections to Other Areas in TEL Research 260
11.4 Concluding Remarks 263
References 263

Erratum to: Learning Analytics in R with SNA, LSA, and MPIA E1

Annex A: Classes and Methods of the *mpia* Package 265

Application examples included

Social Network Analysis

3.2 A foundational example
3.3 Extended social network analysis example

Latent Semantic Analysis

4.3 Foundational latent semantic analysis example
4.5 Extended application example: automated essay scoring

Meaningful, Purposive Interaction Analysis

(8.3 Implementation: classes of the mpia package)
9.2 The foundational SNA and LSA examples revisited
9.3 Revisiting Automated Essay Scoring: Positioning
9.4 Learner trajectories in an essay space

Chapter 1
Introduction

Learning can be seen as the co-construction of knowledge in isolated and collaborative activity—and as such poses an eternal riddle to the investigation of human affairs. Scholars in multiple disciplines have been conducting research to understand and enrich learning with new techniques for centuries, if not millennia. With the advent of digital technologies, however, this quest has undoubtedly been accelerated and has brought forward a rich set of support technologies that can render learning more efficient and effective, when applied successfully.

This work contributes to the advancement knowledge about such learning technology by first providing a sound theoretical foundation, to then develop a novel type of algorithm for representing and analysing semantic appropriation from meaningful, purposive interaction. Moreover, the algorithmic work will be challenged in application and through evaluation with respect to its fitness to serve as analytics for studying the competence development of learners.

This work is not in academic isolation. Over years, recurring interest in research and development lead to the formation of a new, interdisciplinary research field in between computer science and humanities, often titled 'technology-enhanced learning' (TEL). With respect to development, this repeated interest in TEL is reflected in the growth of educational software and learning tools for the web (Wild and Sobernig 2007). Research in this area has seen a steadily growing body of literature, as collected and built up by the European Networks of Excellence Prolearn (Wolpers and Grohmann 2005), Kaleidoscope (Balacheff et al. 2009), and Stellar (Gillet et al. 2009; Fisichella et al. 2010; Wild et al. 2009).

Both research and development have created many new opportunities to improve learning in one way or the other. Only for some of them, however, research was able to provide evidence substantiating the claims on such improvement (for reviews see Schmoller and Noss 2010, p. 6ff; Means et al. 2010, pp. xiv ff. and pp. 17ff.; Dror 2008; Campuzano et al. 2009, pp. xxiv).

F. Wild, *Learning Analytics in R with SNA, LSA, and MPIA*,
DOI 10.1007/978-3-319-28791-1_1

1.1 The Learning Software Market

Already early empirical studies in technology-enhanced learning report large variety in web-based software used to support learning activity in Higher Education: Paulsen and Keegan (2002) and Paulsen (2003), for example, reports 87 distinct learning content management systems in use in 167 installations from the interviews with 113 European experts in 17 countries.

Wild and Sobernig (2007) report on results from a survey on learning tools offered at 100 European universities in 27 countries. The survey finds 182 distinct learning tools in 290 installations at the universities investigated. Figure 1.1 gives an overview of the variety of these identified tools: block width indicates the number of distinct tools, whereas block height reveals the number of installations of these tools at the universities studied. The majority of tools reported are classified as all-rounders, i.e., multi-purpose learning management and learning content management systems with broad functionality. Special purpose tools such as those *restricted* to content management (CMS), registration and enrolment management (MS), authoring (AUT), sharing learning objects (LOR), assessment (ASS), and collaboration (COLL) were found less frequent and with less variety in addition to these all-rounders.

Among the 100 universities screened in Wild and Sobernig (2007), functionality of the tools offered covers a wide range of learning activities: text-based communication and assessment is supported in almost all cases and quality assurance/ evaluation plus collaborative publishing still in the majority of cases. Additional functionalities offered still frequently at more than a quarter of the universities at that time were individual publishing tools, tools for social networking, authoring tools for learning designs, and audio-video conferencing and broadcasting tools.

Today, the Centre for Learning and Performance Technologies (2014) lists over 2000 tools in the 13 categories of its directory of learning and performance tools, although little is known about how and which of these are promoted among the European universities.

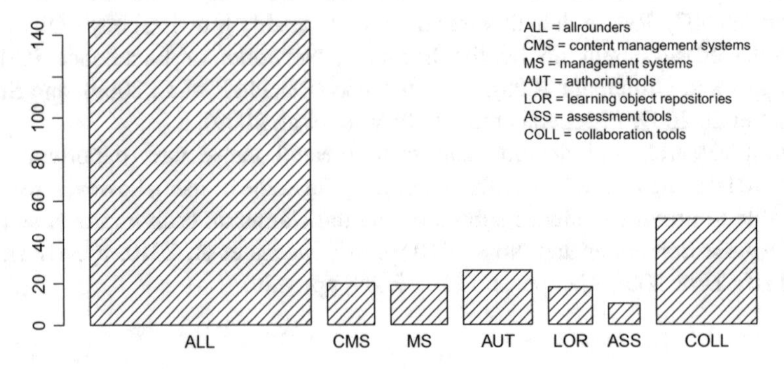

Fig. 1.1 Learning tools: types and installations according to Wild and Sobernig (2007, new graphic from raw data)

This variety of learning tools available online, however, is only in parts a blessing to learners. Choice can be a burden and when it involves processes of negotiation, as often required in collaborative situations, it can actually get into the way of knowledge work with content. The plethora of tools definitely forms an opportunity, but at the same time it affords new learning competences and literacies in its utilisation in practice.

On the other hand, this plethora of tools is evidence of the existence of a large and growing market of learning technology. And indeed, the market-research firm Ambient Insight assesses the global market for "self-paced eLearning products and services" to have reached a revenues volume of "$32.1 billion in 2010" and predicts a further annual growth of 9.2 % over the coming years till 2015, with the big buyers to be found in North America and Western Europe (Adkins 2011, p. 6).

1.2 Unresolved Challenges: Problem Statement

Although there is an abundance of learning technology, this does not mean that enhancing learning with technology is free of any challenges—quite the contrary is found to be the case, when asking researchers and professionals in the field. The STELLAR network of excellence[1] in technology-enhanced learning of the European Union (Gillet et al. 2009) identifies several Grand Challenges of relevance that pose significant unresolved problems.

Grand Challenges (in STELLAR's terminology) refer to problems, "whose solutions are likely to enhance learning and education in Europe, providing clear directions for stakeholders, policy makers and funders" (Sutherland et al. 2012, p. 32). STELLAR and its capacity building instruments lifted proposals for all in all 32 Grand Challenge problems.

Following several iterations of refinement, these problem statements were rated by stakeholders and researchers along the following indicators: how likely they are to leave lasting educational (EDI), technological (TECI), social (SOCI), and economic impact (ECI); and how well defined the problem descriptions they are (assessing the aspects of clarity (CLAR), feasibility (FEAS), measures of success (SUCC), and their potential for scalability (SCAL)). This rating exercise lead to an expert ranking of the most achievable and most impacting challenges, see Fig. 1.2 for the ratings of the top ten ranked challenges.

The top ranked challenges circled around the following focus points: fostering engagement, early years technology, networked and seamless learning, improved teacher training, and—last but not least—learning analytics (cf. Sutherland et al. 2012, p. 32; cf. Fischer et al. 2014, for full problem descriptions[2]).

[1] http://teleurope.eu and http://www.stellarnet.eu/

[2] The Grand Challenge descriptions extracted from the STELLAR Alpine Rendezvous.

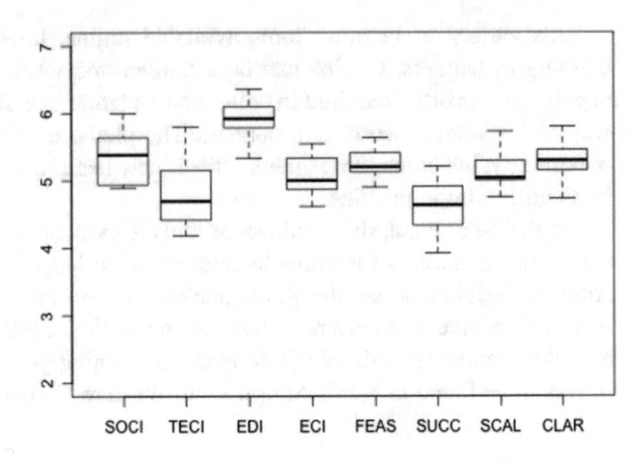

Fig. 1.2 Expert rating of expected success of the top ten Grand Challenges (new figure from raw data)

Looking just at the latter, i.e. 'learning analytics', three Grand Challenge proposals related to it that scored among the top ten. The first of these, entitled 'New forms of assessment for social TEL environments', derives the requirement to adapt the prevalent, individual-centric assessment model in education to changed circumstances of a more open and more social educational environment that emerged over the past decade (Whitelock 2014a,[3] p. 54). With the possibility to harvest 'big data' from the learning tools involved, this creates the opportunity to "to support assessment of learning processes and learning outcomes" (Whitelock 2012a, p. 1). This particularly involves, so Whitelock (2014a, p. 54), "learning network analysis" for "assessing networks and driving the development of groups and networks that provide effective support for learners" as well as "learning content analysis" for "assessing the resources available to learners, and using this information to recommend appropriate materials, groups and experts". Moreover, this involves the "development of visualisations and dashboards that present these assessments to learners and teachers in accessible and meaningful ways" (Whitelock 2012a, p. 1).

The second related Grand Challenge problem is entitled "Assessment and automated feedback" (Whitelock 2014b, p. 22). Though overlapping with the previous, it further stresses the importance of "timely feedback" and the potential that "automatic marking and feedback systems for formative assessment" can play not least with respect to scaling up facilitating support (Whitelock 2012b, p. 1). For Whitelock (2014b, p. 23), this includes "the development and evaluation of technologies that make for example intensive use of text- and data mining or natural language processing approaches".

The third related Grand Challenge deals with "Making use and sense of data for improving teaching and learning" (Plesch et al. 2012). By "generating data as a

[3] For the appearance in print in Fischer et al. (2014), the Grand Challenge problem descriptions had to be shortened in order to fit the book into the SpringerBriefs in Education format: this means that Whitelock (2014a) is a shortened version of Whitelock (2012a); and Whitelock (2014b) is a shortened version of Whitelock (2012b).

side-product of learning activities", i.e. more specifically "real time and outcome data", teachers as well as students can be provided feedback "about their [...] progress and success" (Plesch et al. 2012). For Plesch et al. (2012), this particularly involves to define, what data is required and "how can this data be collected and presented in an efficient and useful way".

In a nutshell, one key challenge contained in these problem descriptions, which at present still seems unresolved, can be formulated as follows:

> To automatically represent conceptual development evident from interaction of learners with more knowledgeable others and resourceful content artefacts; to provide the instruments required for further analysis; and to re-represent this back to the users in order to provide guidance and support decision-making about and during learning.

This work tries to provide a potential answer to this single challenge, integrating the three Stellar Grand Challenges described above: it elaborates means to understand and model knowledge co-construction, detached from the media in which it takes place.

In order to facilitate this analysis of knowledge co-construction, a representation algorithm is developed, used for mapping and analysing conceptual development from the textual artefacts created and consumed in learning. This novel algorithm and software package is called meaningful, purposive interaction analysis (short 'mpia'), providing one possible solution for shallow social semantic network analysis amongst the actors of a learning ecosystem. The name *mpia* refers to the software package and model, which significantly differs from its predecessors *lsa*, *sna*, and *network*,[4] adding, for example, functionality for identifying competences and extracting learning paths.

As will be shown, with the help of this novel method, support for competence development of learners can be extended with new software to foster analysis of and decision making in learning.

As evident in the use of signal words such as 'learning' and 'software' already in this rough outline, the study of this problem and its solution is an interdisciplinary one.

1.3 The Research Area of Technology-Enhanced Learning

Growth in research and development activity around learning with technology in recent years has lead to a new interdisciplinary research area titled 'technology-enhanced learning' (TEL).

Technology-enhanced learning is directed at human creation of knowledge and the human development of competence and its codification in media as heterogeneous as, e.g., courses, books, or instant messages. Technology-enhanced learning means supporting human activity needed for knowledge creation and competence

[4] See also Chaps. 3–7 for details and Chap. 2 for conceptual clarification.

development with tools that afford isolated or collaborative endeavours in formal and informal situations.

Technology-enhanced learning is often contrasted with a set of different notions: e-learning, online learning, learning technology, computer-based training and computer based learning, computer aided instruction, or computer supported collaborative learning. While these notions are often used interchangeably, slight contextual differences can be identified: As Tchounikine (2011, p. 2) remarks, "using one or another of these terms may contextually denote a particular perspective such as when emphasizing the on-line or the teaching dimension".

Technology-enhanced learning is rooted in many disciplines, bridging between engineering and social sciences. Meyer (2011) and Meyer et al. (2013) report on the state of affairs in interdisciplinarity in a study conducted amongst the members of the research community platform of STELLAR, the European Network of Excellence in technology-enhanced learning (see Fig. 1.3). They find, that the community

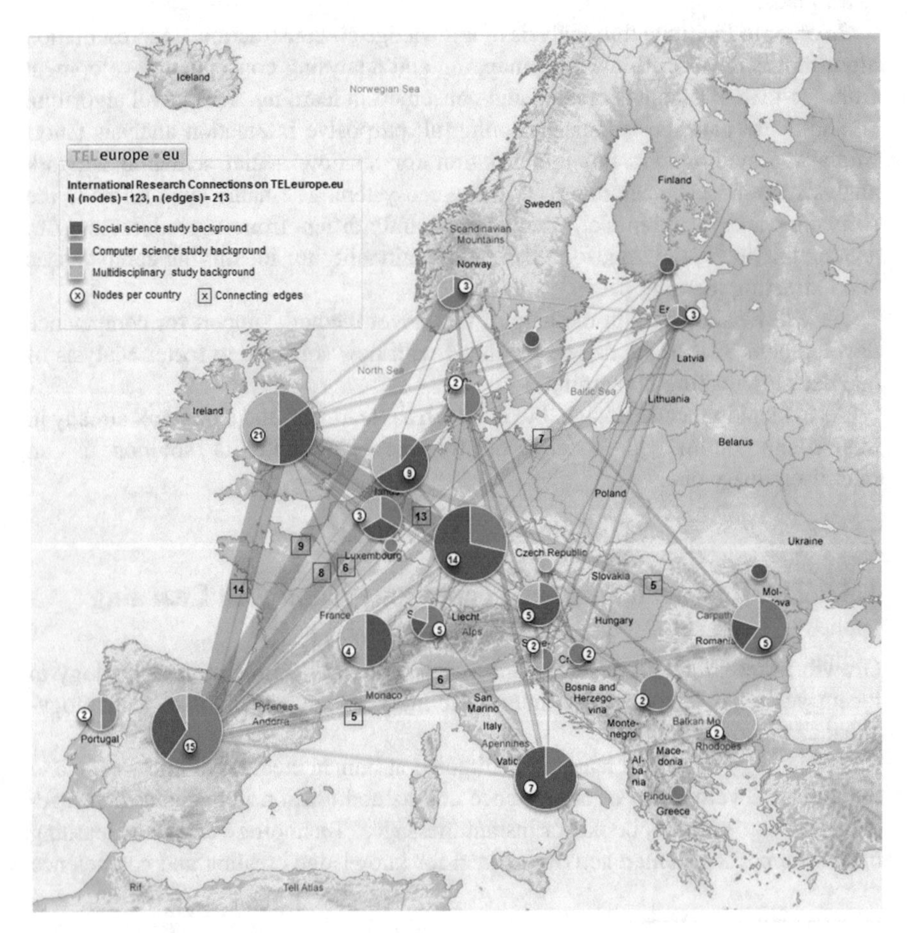

Fig. 1.3 Study backgrounds and connectedness in Europe among the members of TELeurope (Meyer 2011, p. 92)

integrates researchers from disciplines such as psychology, pedagogy, social science, cognitive science, but also computer science and engineering.

By looking at their disciplinary orientation and their level of engagement in TEL, Meyer et al. (2013, p. 7ff) identify eight different ideal types of researchers in this field. Thereby, engagement in TEL is—in the study—reflected in involvement in core TEL research areas (Spada et al. 2011), use and knowledge of methodology and theory, audience targeted, publication output, project participation, and platform connectedness. Disciplinary orientation is composed out of the interdisciplinarity of study background, work, attitudes, terminology, methodology, theory, audience targeted, publication habits, and collaboration.

Six of the eight ideal types found in the research community investigated are at least multidisciplinary—if not interdisciplinary—oriented and the largest group of 21 % of the researchers surveyed is classified as 'TEL interdisciplinarians'.

1.4 Epistemic Roots

This book is an interdisciplinary work, rooted in this emerging field of technology-enhanced learning. It bases on strands of Epistemic Theory rooted in the learning sciences and humanities as well as strands of Knowledge Representation and Algorithm Theory rooted in information and computer sciences. In its application to learning, it has its roots in the emerging research field of Learning Analytics.

The book is meant to be well-formed. It derives a new learning theory foundation and proposes a complementary, computational model and representation algorithm. The implementation of a real system is then used to verify and validate the model with realistic pilots in an applied setting in Higher Education—for Learning Analytics.

Any TEL research can implicitly or explicitly be based on schools of thought of epistemic theory such as contemporary derivatives of constructivism (like culturalism, socio-constructivism, or activity theory), cognitivism (like the unified theory of cognition, knowledge transfer theories, or neuro-engagement theories), or—though today rather limited—behaviourism (such as connectivism or knowledge transaction theory). A bit of a boundary demarcation of the school of thinking applied in this work will be set out in Chap. 2.

Implicit linking of the state of the art to these perspectives is often visible from the epistemological stance vested into the terminology used: the notion 'competence acquisition', for example, relates to cognitivist ideas of transfer, whereas 'competence development' refers to the constructivist ideas of autopoiesis. More often than basing work explicitly in the epistemological strands of analytical philosophy, researchers in the field of TEL refer to certain models of learning: they utilise 'smaller' theories such as communities of practice, situated cognition, or cognitive apprenticeship.

It should be noted, that the learning theory presented is not an instructional-design theory (as promoted by Reigeluth 2007; Reigeluth and Carr-Chellman 2009;

or Mandl et al. 2004)—it is void of instructional aspects such as organisation form, establishment of situational context, or effective pacing. Though it can serve as a component in models of instructional-design.

This work is mainly connected to one epistemic theory strand: it integrates and extends the perspective of socio-constructivism, specifically the viewpoint taken by methodical culturalism (Janich 1998, 1999; Hesse et al. 2008; which itself is rooted in methodical constructivism, cf. Kamlah and Lorenzen 1967).

Methodical culturalism sees human action as constitutive for cognition. The theory of action at its centre encompasses communicative acts. Learning and informing happens from communicative acts in successful, cooperative conversation with others. Conversations can be 'direct' using oral language or 'mediated' by artefacts and other communication tools, as proposed in the theory of mediated action (Wertsch 1994).

Where communicative action is mediated, processes of internalisation construct mental activity from social interaction; at the same time, in processes of objectification, mental activity can produce new or can adapt existing tools and content artefacts (see also Engestroem 1999; Engestroem and Miettinen 1999).

Learning is usefully thought of as competence development from and with the help of an ecosystem, in which people interact with each other and with learning contents with the help of learning tools in more or less planful learning activity. Together and over time, they develop and enact a culture of shared professional language.

This shared language culture can be captured from communication via the textual traces left behind in the activity conduced in tools. Tools hereby denote software applications; today they are often web-based in nature. Textual artefacts consumed by and exchanged among learners in their interaction with tools and with each other in these social semantic networks enable the automated calculation of a representation of such 'shared meaning'. They also allow for identifying, where the use of professional language differs from the norm, such as expressed in model solutions (created by experts) or crowd solutions (aggregated from a number of peers). This representation can be used in deep analysis such as required for formative or summative assessments, serving as foundation for planning, or utilised when reasoning on past and future competence development.

The theoretical foundation mapped out in Chap. 2 about how learning takes place in social interaction provides a frame grounded in epistemic theory, against which the computational representation model (further substantiated in Chaps. 5 and 6) is developed and evaluated. This Chap. 2 elaborates the theory frame of competence development from information exchange, while it at the same time provides the conceptual space explicating all essential definitions required for understanding the work presented subsequently.

1.5 Algorithmic Roots

TEL research on social network representation and knowledge representation is based in the traditions of algorithm theory. Mathematically, and without going too much into detail, the class of the algorithms referred to and extended in this work are rooted in algebraic graph theory, particularly a subarea called linear algebra (cf. Hartmann 2003, pp. 5ff.). These algorithms are latent semantic analysis (Deerwester et al. 1990) and social network analysis (cf. Monge and Contractor 2003). Foundational mathematical working principles will be detailed where required in the Chaps. 3–5.

Latent semantic analysis (LSA) is an indexing technique that uses bags of words to capture meaning and association by inspecting word context across a larger number of typically paragraph-sized units of text with the help of a two-mode factor analysis (Wild et al. 2005). LSA calculates a small number of factors to represent meaning of terms and documents in a lower dimensional vector space. LSA originated as an information retrieval technique, but has been adapted to a wide range of application scenarios, amongst others information filtering, essay scoring, tutoring, and analogical reasoning (Dumais 2005).

LSA is sometimes referred to as a knowledge representation theory (Landauer and Dumais 1997). This status as a representation theory, however, has been ever so often challenged (e.g. in Wandmacher 2005) as it was upheld in the past. Problems challenging its status as a representation theory of human cognition are attributed to difficulties in calibration that can result in too low performance (Dumais 2005), in-principle wrong-kind representation failures (Perfetti 1998), its ignorance of syntactic and rhetoric structure (Perfetti 1998), and in its shortcomings with respect to directly distinguishing typed lexical semantic relationships (Wandmacher 2005). Chapter 4 will introduce in depth to LSA and its prevailing applications.

Still, LSA is part of a class of related natural language processing techniques, which over the last quarter of a century have been quite commonly accepted as a new type of "empirical distributional semantics" (Cohen and Widdows 2009). Turney and Pantel (2010, p. 148) classify LSA in their framework for vector space models of semantics as "word-context matrix". It may be a bit far fetched to speak of a "new scientific paradigm" that will lead to a "semantic general theory of everything" (as claimed by Samsonovich et al. 2009, p. 1). It seems, however, to be established that LSA is an accepted semantic representation technique—even if only so for being able to perform (Perfetti 1998, p. 374) and not as a theory of human cognition or memory.

The advent of social software (Hippner and Wilde 2005) and social media (Kaplan and Haenlein 2010) in learning technology (Grodecka et al. 2008; Klamma et al. 2007; Li et al. 2012) has sparked interest in Social Network Analysis (SNA).

SNA has emerged from the sociometrics movement in the late nineteenth and early twentieth century, with the notion 'social network analysis' starting to be systematically used in the 1950s (Freeman 1996, 2004). It is an analysis technique, with which social relations captured in graph structures can be further analysed for their structural properties.

Networks are typically modelled as nodes (also known as vertices or actors) that have links (also titled edges, ties, or relations). A rich body of research has brought forward calculation methods and metrics to investigate and predict network, group, or individual characteristics and structural relations (cf. Wassermann and Faust 1994: Part III, p. 167ff).

SNA has been successfully applied in technology-enhanced learning to investigate the structure and characteristics of learning networks. For example, Song et al. (2011), use social network analysis to support 160,000 European teachers in the eTwinnings network with monitoring tools to inspect their competence development. Law and Nguyen-Ngoc (2007) deployed SNA to evaluate self-direction in online collaborative learning with social software. Harrer et al. (2005) use SNA to explore the complex communication patterns emerging in a blended learning university course. These are just three examples of a wide range of contributions picking up social network analysis as a method to evaluate, monitor, assess, or predict social behaviour relevant to learning. Chapter 3 will introduce in depth to SNA and its key applications.

It will be argued in this work, that the social semantics communicated in learning conversations can be captured with a clever fusion algorithm of social network analysis and latent semantic analysis.

The motivation for merging these two different algorithms can be found in their shortcomings applying when used separately. Social Network Analysis (and even the broader Network Analysis) does not provide means to deal with semantics: it does not provide means to create semantic representations, nor helps with the disambiguation of unstructured texts.

Latent Semantic Analysis, on the other hand, does not provide methods and methodology, how to work with and visualise the resulting latent semantic graph structures. It lacks the instruments to, for example, deal with aggregation, component manipulation, and any other manipulation and measurements beyond the evaluation of fundamental proximity relations.

By engrafting network analysis onto latent semantic graphs, thereby retaining provenance data, a novel technology emerges: meaningful, purposive interaction analysis, used to analyse, visualise, and manipulate social, semantic networks. The foundations of this novel fusion algorithm will be documented in Chaps. 5 and 6.

1.6 Application Area: Learning Analytics

Fitness of a representation can best be judged in its application, not least because “representation and reasoning are inextricably intertwined” (Davis et al. 1993, p. 29).

The application area, for which meaningful interaction analysis is proposed, is the area of Learning Analytics. Learning Analytics is defined as the scientific study of gathering and analysing usage data of learners, with the aim to "observe and understand learning behaviors in order to enable appropriate interventions" (Brown 2011, p. 1). The analytics bring together "large data sets, statistical techniques, and predictive modeling" (Campbell et al. 2007, p. 42).

One of the more influential forecasting initiatives on emerging technologies in teaching and learning is run by the New Media Consortium: the so-called 'Horizon Project' and its connected reports, which in 2012 celebrated its 10th anniversary with a strategic retreat. The three annual Horizon Reports "cumulatively have some 1.25 million readers and hundreds of thousands of downloads per year" (Larry Johnson, CEO of the New Media Consortium, personal communication, 19. 12. 2011). Readers are based "in over 100 countries" and the report series has undergone "27 translations in the last ten years" (Larry Johnson, personal communication, 18. 11. 2011).

In its 2011 edition, the Horizon Report predicts for 'Learning Analytics' to have an estimated time to adoption of 4–5 years (Johnson et al. 2011). Quite similarly, Sharples et al. (2012) predict in a forecasting report on pedagogical innovation for Learning Analytics to have a "medium timescale" of adoption of 2–5 years. In the subsequent years, learning analytics shall return as a key trend identified by the annual Horizon Reports (Johnson et al. 2012, p. 22ff); Johnson et al. 2013, p. 24ff) to finally, in 2014, list it as key development with an estimated time-of-adoption of 1 year or less (Johnson et al. 2014, p. 38ff).

Analytics have been successful in other areas such as Web, Business, or Visual Analytics, with some of them having a long-standing scientific discourse. Jansen (2009, p. 24), for example, traces the idea of web analytics back to the 1960s, where first transaction log studies were conducted. Web Analytics deal "with Internet customer interaction data from Web systems" (Jansen 2009, p. 2), assuming that this user-generated trace data "can provide insights to understanding these users better or point to needed changes or improvements to existing Web systems" (ibid, p. 2).

Different from that, Visual Analytics sets the focus on re-representation of data in visual form with the aim of supporting analytical reasoning: "Visual analytics is the science of analytical reasoning facilitated by interactive visual interfaces" (Thomas and Cook 2005, p. 4). Initially, Visual Analytics has been sponsored as a new research field by the US department of homeland security as means to improve terrorism threat detection (Thomas and Kielman 2009, p. 310). As Thomas and Kielman (2009, p. 310), however, denote, research and development has "almost immediately" broadened out to other domains. Visual Analytics encompasses four areas: the study of "analytical reasoning techniques", "visual representation and interaction techniques", "data representations and transformations", and techniques for "presentation, production, and dissemination" (Thomas and Cook 2005, p. 4; Thomas and Kielman 2009, p. 310).

Business Analytics again has very historic roots, with the term itself arising somewhere following the turn of the millennium. Google's ngram viewer,[5] for

[5] http://bit.ly/R4V6KN

example, shows the phrase "Business Analytics" to start to appear with rising frequency only in the years 1999/2001 in the indexed English books (for the year 2000 no book indexed contained the phrase).

Business Analytics is about "giving business users better insights, particularly from operational data stored in transactional systems" (Kohavi et al. 2002). It is applied in areas as rich as marketing, customer relationship management, supply chain management, price optimisation, or work force analysis (cf. Kohavi et al. 2002, p. 47; Davenport et al. 2001, p. 5; Trkman et al. 2010, p. 318).

Sometimes 'business analytics' is used synonymously with 'business intelligence'. Authoritative sources, however, define it as the subset of business intelligence that deals with "statistics, prediction and optimization" (Davenport and Henschen 2010).

A very closely related area to Learning Analytics is Educational Data Mining, which is defined as "a field that exploits statistical, machine-learning, and data-mining (DM) algorithms over the different types of educational data" (Romero and Ventura 2010, p. 601; Romero et al. 2011). According to Duval (2012), the difference may be found mainly therein, that Learning Analytics is about "collecting traces that learners leave behind and using those traces to improve learning", whereas educational data mining merely "can process the traces algorithmically and point out patterns or compute indicators". Following this argument, Learning Analytics aims at improving learning, whereas educational data mining aims at detecting interesting patterns in educational data. In Siemens and Baker (2012), two key protagonists of the disjunct research communities analyse in a joint publication the differences between Learning Analytics and Educational Data Mining. Though many of the points unveiled therein appear as a mere difference in weight assigned, the most striking ones can be found in the techniques and methods studied: while Learning Analytics focus on areas such as social network analysis, discourse analysis, "concept analysis", or "influence analytics", so Siemens and Baker (2012, p. 253), Educational Data Mining is seen to focus rather areas such as classification, clustering, relationship mining, and "discovery with models". While the first is seen to focus more holistically on "systems as wholes", the latter serves the development of applicable components (ibid). Still, even though differences can be assessed, it remains to be shown in the future, whether Learning Analytics and educational data mining are in fact two independent (merely overlapping) research strands.

All three related analytics areas,—Web, Business, and Visual Analytics–, as well as Educational Data Mining share with Learning Analytics big data, statistical methods for processing, and predictive modelling. They, however, differ in their application domain, the types of data and data traces available, and objective of the analysis (see Table 1.1).

Learning Analytics has two direct precursors that seem to have been absorbed in it or abandoned by now. Academic Analytics serves the administration/institution as its main stakeholder and focuses particularly on supporting enrolment management and facilitating drop out prevention (Goldstein 2005; Campbell and Oblinger

Table 1.1 Related analytics approaches

	Domain	Data	Objective
Web analytics	Web use	Operative systems	Website improvement
Visual analytics	Decision making	Anything	Analytical reasoning
Business analytics	Business	Operative systems	Performance improvement
Educational data mining	Education	Operative systems	Pattern detection
Academic analytics	Education	Mainly LMS, SIS	Organisational performance
Action analytics	Education	Mainly LMS, SIS	Organisational performance
Learning analytics	Learning	Operative systems, user generated	Learner performance

2007). This area has brought forward already a rich set of success stories (see below).

Lifting this to the next level, Norris et al. (2008) postulate the investment into learner-centric Action Analytics, that focus not only on mere measurement, but move "from data to reporting to analysis to action" (p. 46) in order to help organisations improve cognitive and pedagogical performance of their learners. Action Analytics, however, all the same retain an organisation-centric perspective (Chatti et al. 2013), supporting institutions rather than individuals.

The Society of Learning Analytics Research (SOLAR) describes Learning Analytics as "the measurement, collection, analysis and reporting of data about learners and their contexts, for purposes of understanding and optimising learning and the environments in which it occurs" (Siemens et al. 2011, p. 4; Long and Siemens 2011, p. 34).

More precisely, it can be defined as the scientific study of the data left behind in knowledge co-construction by learners in their interaction with information systems using statistical, predictive modelling techniques with the aim of improving competence and thus building up potential for future performance.

Learning Analytics can serve analysis on a cross-institutional (macro) level, an institutional (meso) level, or individual (micro) level, as the Unesco policy brief on Learning Analytics outlines (Buckingham Shum (2012, p. 3). The report further emphasises, that Learning Analytics are "never neutral", as "they unavoidably embody and thus perpetuate particular pedagogy and assessment regimes" (ibid, p. 9).

Several proposals for frameworks exist, ranging from empirically founded dimensional models (Greller and Drachsler 2012; Chatti et al. 2013) to classification of the subsumed fields by strands (Ferguson 2013; Buckingham Shum and Ferguson 2011). While Greller and Drachsler (2012) as well as Chatti et al. (2013) focus on the broad picture, looking also at limitations and constraints, types of stakeholders, and objectives, the most elaborate technical canon can be found in Buckingham Shum and Ferguson (2011) and Ferguson (2013).

Buckingham Shum and Ferguson (2011, p. 13) and Ferguson (2013, p. 312) offer a classification of learning analytics by focus, differentiating five distinct types of analyses studied in Learning Analytics: social network analysis, discourse analysis, content analysis, disposition analysis, and context analysis.

While social network analysis has already been introduced in this work (see above, Sect. 1.5), content analysis is here defined as the upper class of latent semantic analysis, containing also non-bag-of-words oriented approaches to "examine, index, and filter online media assets" (Buckingham Shum and Ferguson 2011, p. 17).

Discourse analysis relates to the wider "analysis of series of communicative events" (ibid, p. 14). The analysis of learning dispositions refers to means to "assess and characterise the complex mixture of experience, motivation and intelligences that a learning opportunity evokes for a specific learner" (ibid, p. 19). Finally, the analysis of context aims to "expose, make use of or seek to understand" contexts as different as found in formal settings, as well as informal, or even mobile learning situations.

This work will show, how the proposed novel representation technique,—meaningful, purposive interaction analysis—, extends the state of the art in Learning Analytics in its application by bringing together the two areas of social network analytics and content analytics. It will demonstrate the usefulness of such integrated method with the help of practical learning analytics application examples. These evaluation examples (Chap. 9) and the subsequent evaluation trials (Chap. 10) take place in realistic settings and will provide evidence that the proposed novel technique works in this chosen application area.

1.7 Research Objectives and Organisation of This Book

As already touched upon above, this work commits to three intertwined, overarching objectives. The work underlying is presented here in such a way that these objectives form the common thread leading through the book.

The first two objectives are *to represent learning* and *to provide the instruments required for further analysis*. As already mentioned above, 'learning' is to be understood as knowledge co-construction in isolated or collaborative activity of actors performed in (online) tools while using and creating learning artefacts (see also Chap. 2). According to Davis et al. (1993), knowledge representations have to satisfy five different and sometimes conflicting criteria: 'representing' means to develop a computation algorithm and notation that serves as a human-readable, but machine-manipulable surrogate for intelligent reasoning, with clearly expressed ontological commitments. The objective is about capturing 'learning' from the textual interaction in a digital learning ecosystem populated by learners and their conversation partners, who exchange textual artefacts via tools in the fulfilment of their learning activity.

The third objective is *to re-represent learning*, i.e. to provide (visual) interfaces to these representations that support assessment, planning, and reasoning about learning. The user interface, especially the visualisations, will have to work naturally for the users in affording internal, cognitive representations inline with what has been intended to convey. According to Scaife and Rogers (1996, p. 189), the relation between internal, cognitive and external, graphical representations is controlled by three elements: their capability for "computational offloading highlights the cognitive benefits of graphical representations, re-representation relates to their structural properties and graphical constraining to possible processing mechanisms". This objective is about (visually) presenting learning to end-users so as to provide accurate and visual learning analytics for decision making about and during learning.

This book is structured as follows. Within this introduction, the scope of the work was further refined. The three roots introduced above thereby also define the three areas to which this work contributes: epistemic theory, algorithms for representation, visualisation, and analysis, and the application area of Learning Analytics, see Table 1.2.

The three roots provide the brackets for this work. Epistemic foundations scope the concept space of this book and provide definitions of as well as explicate relations between key concepts. With respect to algorithms, the foundational roots as well as the full derivation in matrix theory will be elaborated in depth. Since the best way to challenge a specific representation algorithm for its fitness is to subject it to application, Learning Analytics provide a concrete context for adaptation, deployment, and testing.

The remainder of this book is organised as follows. Chapter 2 defines the most central concepts and relates them to each other into a novel theoretical foundation, clarifying and defining how learning takes place in social environments. This

Table 1.2 Structural organization of the book and key contributions

	Epistemology	Algorithms	Application area
Aim	• Define key concepts and their relations	• Derive and develop computational model for representation and analysis	• Adapt, deploy, and subject to testing in application area
Contribution	• Learning from a methodo-culturalist perspective	• Meaningful, purposive interaction analysis (MPIA)	• Learning Analytics with MPIA
Chapters	• Learning theory and requirements (Chap. 2)	• Algorithmic roots in SNA (Chap. 3) and LSA (Chap. 4); • MPIA (Chaps. 5 and 6) • Calibration for specific domains (Chap. 7) • Package implementation (Chap. 8)	• Application examples in Learning Analytics (Chap. 9) • Evaluation (Chap. 10)

theory, in itself an extension of methodo-culturalist information theory, serves as basis for the subsequent chapters.

Chapter 3 reviews the state of affairs in social network analysis, while Chap. 4 reviews latent semantic analysis, including a brief documentation of the software implementation by the author in the package 'lsa'. A foundational and an extended example each, to work out advantages and shortcomings of the two methods, support both Chaps. 3 and 4 (and they will be revisited in Chap. 9).

Chapter 5 develops the math for meaningful, purposive interaction analysis and Chap. 6 adds the complementary visualisation techniques. The Chapter 5 relates to the first two objectives of creating a representation of learning, while providing the required analysis instruments, whereas Chap. 6 relates to the third objective of visualising such representation and analysis in convenient and aesthetic form. Chapter 7 reports on trends identified from a calibration experiment on how to prepare corpora with respect to sampling and sanitising for specific domains.

Chapter 8 introduces to the implementation in the package 'mpia'.

The foundational example from Chaps. 3 and 4 guides through the algorithm development and is revisited again in Chap. 9, where two additional enhanced examples illustrate the use of the novel algorithm.

Evaluation, as presented in Chap. 10, looks back on the validity of the approaches proposed to achieve the top-level objectives. Thereby strong emphasis is given to the representation and analysis aspects.

Finally, Chap. 11 rounds up the book with a conclusion and an outlook on open research questions. The conclusion draws a resume on the three top-level objectives stated in this introduction.

The code examples provided throughout the publication aim to foster *re-executable* (same code, same data) and *reproducible* research (same code, different data) in order to push rigour of the work conducted, while at the same time lowering barriers to uptake and reuse. Therefore, the application examples are carefully selected and written in a tutorial-style, striving for completeness, even if that—in cases—may involve adding seemingly trivial routines for, e.g., class instantiation, initialisation, or data storage.

Often these minimalist instantiations abstract a more complex process of analysis that is encapsulated in the package implementations. Together, the listings cover the full flow of analysis. All lines of code provided in the core body of text, however, have been reduced to their bare minimum, stripping them of additional, but optional configuration parameters, where not relevant (for example, when adding axis headings to depicted visualisations).

Inline with tradition of the R community, the full application examples are included as commented demos in the packages released as part of this work on R-Forge (Theußl and Zeileis 2009), allowing for direct execution and adaptation to new data sources.

The book follows in its presentation the cycle from idea elaboration, to implementation, to evaluation. The underlying work, however, took place in iterations, with evaluation results repeatedly informing ideation and implementation.

Advances from and differences in these iterations will be indicated wherever it is deemed relevant for subsequent chapters.

References

Adkins, S.: The Worldwide Market for Self-paced eLearning Products and Services: 2010-2015 Forecast and Analysis. Ambient Insight Comprehensive Report. http://www.ambientinsight.com/Resources/Documents/Ambient-Insight-2010-2015-Worldwide-eLearning-Market-Executive-Overview.pdf (2011)

Balacheff, N., Ludvigsen, S., de Jong, T., Lazonder, A., Barnes, S.: Technology-Enhanced Learning: A Kaleidoscopic View, pp. v–xvi. Springer, Berlin (2009)

Brown, M.: Learning analytics: the coming third wave. In: ELI Briefs ELIB1101, EDUCAUSE Learning Initiative (ELI). http://educause.edu/Resources/LearningAnalyticsTheComingThir/227287 (2011)

Buckingham Shum, S.: Learning Analytics, Policy Brief, UNESCO Institute for Information Technologies in Education. Russian Federation, Moscow (2012)

Buckingham Shum, S., Ferguson, R. Social Learning Analytics. Technical Report KMI-11-01, Knowledge Media Institute, The Open University, UK. http://kmi.open.ac.uk/publications/pdf/kmi-11-01.pdf (2011). Accessed March 4, 2016

Campbell, J.P., Oblinger, D.G.: Academic Analytics. Educause White Paper. http://net.educause.edu/ir/library/pdf/PUB6101.pdf (2007)

Campbell, J.P., DeBlois, P.B., Oblinger, D.G.: Academic analytics: a new tool for a new era. EDUCAUSE Rev. **42**(4), 40–57 (2007)

Campuzano, L., Dynarski, M., Agodini, R., Rall, K.: Effectiveness of Reading and Mathematics Software Products: Findings from Two Student Cohorts (NCEE 2009-4041). National Center for Education Evaluation and Regional Assistance, Institute of Education Sciences, U.S. Department of Education, Washington, DC (2009)

Centre for Learning & Performance Technologies: Directory of Learning & Performance Tools. http://c4lpt.co.uk/directory-of-learning-performance-tools/ (2014). Accessed 8 Aug 2014

Chatti, A., Dyckhoff, A.L., Schroeder, U., Thüs, H.: A reference model for learning analytics. Int. J. Technol. Enhanc. Learn. **4**(5–6), 318–331 (2013)

Cohen, T., Widdows, D.: Empirical distributional semantics: methods and biomedical applications. J. Biomed. Inform. **42**(2), 390–405 (2009)

Davenport, T., Henschen, D.: Analytics at Work: Q&A with Tom Davenport. Interview with Thomas Davenport, InformationWeek. http://www.informationweek.com/news/software/bi/222200096 (2010)

Davenport, T.H., Harris, J.G., De Long, D.W., Jacobson, A.L.: Data to Knowledge to Results: Building an Analytic Capability. Working Paper, Accenture (2001)

Davis, R., Shrobe, H., Szolovits, P.: What is a knowledge representation? AI Mag. **14**(1), 17–33 (1993)

Deerwester, S., Dumais, S., Furnas, G., Landauer, T.K., Harshman, R.: Indexing by latent semantic analysis. J. Am. Soc. Inf. Sci. **41**(6), 391–407 (1990)

Dror, I.E.: Technology enhanced learning: the good, the bad, and the ugly. Pragmat. Cogn. **16**(2), 215–223 (2008)

Dumais, S.: Latent semantic analysis. Ann. Rev. Inf. Sci. Technol. **38**(1), 188–230 (2005). Chapter 4

Duval, E.: Learning Analytics and Educational Data Mining. Blog post. http://erikduval.wordpress.com/2012/01/30/learning-analytics-and-educational-data-mining/ (2012). Accessed March 4, 2016

Engestroem, Y.: Activity theory and individual and social transformation. In: Engestroem, Y., Miettinen, R., Punamaeki, R.-L. (eds.) Perspectives on Activity Theory. Cambridge University Press, New York (1999)

Engestroem, Y., Miettinen, R.: Introduction. In: Engestroem, Y., Miettinen, R., Punamaeki, R.-L. (eds.) Perspectives on activity theory, pp. 1–16. Cambridge University Press, Cambridge (1999)

Ferguson, R.: Learning analytics: drivers, developments and challenges. Int. J. Technol. Enhanc. Learn. **4**(5–6), 304–317 (2013)

Fischer, F., Wild, F., Sutherland, R., Zirn, L.: Grand Challenges in Technology Enhanced Learning. Springer, New York (2014)

Fisichella, M., Herder, E., Marenzi, I., Nejdl., W.: Who are you working with? Visualizing TEL Research Communities. In: Proceedings of the World Conference on Educational Multimedia, Hypermedia & Telecommunications (ED-MEDIA 2010), 28 June–2 July 2010, AACE, Toronto, Canada (2010)

Freeman, L.C.: Some antecedents of social network analysis. Connections **19**(1), 39–42 (1996). INSNA

Freeman, L.C.: The Development of Social Network Analysis: A Study in the Sociology of Science. Empirical Press, Vancouver (2004)

Gillet, D., Scott, P., Sutherland, R.: STELLAR European research network of excellence in technology enhanced learning. In: International Conference on Engineering Education & Research, 23–28 August 2009, Seoul, Korea (2009)

Goldstein, P.: Academic Analytics: The Uses of Management Information and Technology in Higher Education. In: ECAR Key Findings, Educause (2005)

Greller, W., Drachsler, H.: Translating learning into numbers: a generic framework for learning analytics. Educ. Technol. Soc. **15**(3), 42–57 (2012)

Grodecka, K., Wild, F., Kieslinger, B. (eds.): How to Use Social Software in Higher Education. Wydawnictwo Naukowe Akapit, Kraków (2008)

Harrer, A., Zeini, S., Pinkwart, N.: The effects of electronic communication support on presence learning scenarios. In: Koschmann, T., Suthers, D., Chan, T.-W. (eds.) Computer supported collaborative learning 2005: the next 10 years! pp. 190–194. Lawrence Erlbaum Associates, Mahwah, NJ (2005)

Hartmann, P.: Mathematik fuer Informatiker. Vieweg, Braunschweig/Wiesbaden (2003)

Hesse, W., Mueller, D., Ruß, A.: Information, information systems, information society: interpretations and implications. Poiesis Prax. **5**, 159–183 (2008)

Hippner, H., Wilde, T.: Social Software. Wirtschaftsinformatik **47**(6), 441–444 (2005)

Janich, P.: Informationsbegriff und methodisch-kulturalistische Philosophie. EuS **9**(2), 169–182 (1998)

Janich, P.: Die Naturalisierung der Information. Franz Steiner Verlag, Stuttgart (1999)

Jansen, B.J.: Understanding User–Web Interactions via Web Analytics. Morgan & Claypool, San Rafael, CA (2009)

Johnson, L., Smith, R., Willis, H., Levine, A., Haywood, K.: The 2011 Horizon Report. The New Media Consortium, Austin, TX (2011)

Johnson, L., Adams, S., Cummins, M.: The NMC Horizon Report: 2012 Higher Education Edition. The New Media Consortium, Austin, TX (2012)

Johnson, L., Adams Becker, S., Cummins, M., Estrada, V., Freeman, A., Ludgate, H.: NMC Horizon Report: 2013 Higher Education Edition. The New Media Consortium, Austin, TX (2013)

Johnson, L., Adams Becker, S., Estrada, V., Freeman, A.: NMC Horizon Report: 2014 Higher Education Edition. The New Media Consortium, Austin, TX (2014)

Kamlah, W., Lorenzen, P.: Logische Propaedeutik. Bibliographisches Institut, Mannheim (1967)

Kaplan, A.M., Haenlein, M.: Users of the world, unite! The challenges and opportunities of Social Media. Bus. Horiz. **53**(1), 59–68 (2010)

Klamma, R., Chatti, M.A., Duval, E., Hummel, H., Hvannberg, E.T., Kravcik, M., Law, E., Naeve, A., Scott, P.: Social Software for Life-long Learning. Educ. Technol. Soc. **10**(3), 72–83 (2007)

Kohavi, R., Rothleder, N.J., Simoudis, E.: Emerging trends in business analytics. Commun. ACM **45**(8), 45–48 (2002)

Landauer, T.K., Dumais, S.: A solution to Plato's problem: the latent semantic analysis theory of acquisition, induction and representation of knowledge. Psychol. Rev. **104**(2), 211–240 (1997)

Law, E., Nguyen-Ngoc, A.V.: Fostering self-directed learning with social software: social network analysis and content analysis. In: Dillenbourg, P., Specht, M. (eds.) Times of Convergence: Technologies Across Learning Contexts. Lecture Notes in Computer Science, vol. 5192, pp. 203–215. (2007)

Li, N., El Helou, S., Gillet, D.: Using social media for collaborative learning in higher education: a case study. In: Proceedings of the 5th International Conference on Advances in Computer-Human Interactions, 30 January–4 February 2012, Valencia, Spain (2012)

Long, P., Siemens, G.: Penetrating the fog. EDUCAUSE Rev. **46**(5), 31–40 (2011)

Mandl, H., Kopp, B., Dvorak, S.: Aktuelle theoretische Ansätze und empirische Befunde im Bereich der Lehr-Lern-Forschung: Schwerpunkt Erwachsenenbildung. Deutsches Institut für Erwachsenenbildung, Bonn (2004)

Means, B., Toyama, Y., Murphy, R., Bakia, M., Jones, K.: Evaluation of Evidence-Based Practices in Online Learning: A Meta-Analysis and Review of Online Learning Studies, U.S. Department of Education. http://www2.ed.gov/rschstat/eval/tech/evidence-based-practices/finalreport.pdf (2010). Revised 2010

Meyer, P.: 'Technology-enhanced learning' as an interdisciplinary epistemic community. Master Thesis, University of Augsburg (2011)

Meyer, P., Kelle, S., Ullmann, T.D., Scott, P., Wild, F.: Interdisciplinary cohesion of tel: an account of multiple perspectives. In: Hernández-Leo, D., Ley, T., Klamma, R., Harrer, A. (eds.) Scaling Up Learning for Sustained Impact. Lecture Notes in Computer Science, vol. 8095. Springer, Berlin (2013)

Monge, P., Contractor, N.: Theories of Communication Networks. Oxford University Press, New York, NY (2003)

Norris, D., Baer, L., Leonard, J., Pugliese, L., Lefrere, P.: Action analytics: measuring and improving performance that matters in higher education. Educ. Rev. **43**(1), 42–67 (2008)

Paulsen, M.F.: Experiences with learning management systems in 113 European Institutions. Educ. Technol. Soc. **6**(4), 134–148 (2003). http://ifets.ieee.org/periodical/6_4/13.pdf

Paulsen, M. F., Keegan, D.: European experiences with learning management systems. In: ZIFF PAPIERE 118: Web-Education Systems in Europe, Zentrales Institut für Fernstudienforschung, FernUniversitaet Haagen, pp. 1–22 (2002)

Perfetti, C.: The limits of co-occurrence. Discourse Process. **25**(2–3), 363–377 (1998)

Plesch, C., Wiedmann, M., Spada, H.: Making Use and Sense of Data for Improving Teaching and Learning (DELPHI GCP 10). http://www.teleurope.eu/pg/challenges/view/147663 (2012). Accessed March 4, 2016

Reigeluth, C.: Instructional-Design Theories and Models: Volume II. Routledge, London (2007)

Reigeluth, C., Carr-Chellman, A.: Instructional-Design Theories and Models: Volume III: Building a Common Knowledge Base. Routledge, London (2009)

Romero, C., Ventura, S.: Educational data mining: a review of the state-of-the-art. IEEE Trans. Syst. Man Cybern. Part C Appl. Rev. **40**(6), 601–618 (2010)

Romero, C., Ventura, S., Pechenizkiy, M., Baker, R.: Introduction. In: Romero, C., Ventura, S., Pechenizkiy, M., Baker, R. (eds.) Handbook of Educational Data Mining. Chapman & Hall, Boca Raton, FL (2011)

Samsonovich, A., Goldin, R.F., Ascoli, G.A.: Toward a Semantic General Theory of Everything. Complexity **15**(4), 12–18 (2009)

Scaife, M., Rogers, Y.: External cognition: how do graphical representations work? Int. J.Hum. Comput. Stud. **45**(2), 185–213 (1996)

Schmoller, S., Noss, R. (eds.): Technology in Learning. Working Paper, Association for Learning Technology and Technology Enhanced Learning Research Programme. http://www.esrc.ac.uk/my-esrc/grants/RES-139-34-0001/outputs/read/e2b8e1b2-d766-45ab-a605-5c9776e7925c (2010)

Sharples, M., McAndrew, P., Weller, M., Ferguson, R., FitzGerald, E., Hirst, T., Mor, Y., Gaved, M., Whitelock, D.: Innovating Pedagogy 2012. Open University Innovation Report 1. http://www.open.ac.uk/iet/main/sites/www.open.ac.uk.iet.main/files/files/ecms/web-content/Innovating_Pedagogy_report_July_2012.pdf (2012). Last Access: March 4, 2016

Siemens, G., Gasevic, D., Haythornthwaite, C., Dawson, S., Buckingham Shum, S., Ferguson, R., Duval, E., Verbert, K., Baker, R.: Open Learning Analytics: an integrated & modularized platform. SOLAR: Society of Learning Analytics Research. http://www.solaresearch.org/OpenLearningAnalytics.pdf (2011). Accessed March 4, 2016

Siemens, G., Baker, R.: Learning analytics and educational data mining: towards communication and collaboration. In: Proceedings of the 2nd International Conference on Learning Analytics and Knowledge (LAK'12), pp. 252–254. ACM, New York (2012)

Song, E., Petrushyna, Z., Cao, Y., Klamma, R.: Learning analytics at large: the lifelong learning network of 160,000 European Teachers. In: Kloos, C.D., Gillet, D., Crespo, R.M., Wild, F., Wolpers, M. (eds.) Towards Ubiquitous Learning. Lecture Notes in Computer Science, vol. 6964/2011, pp. 398–411. Springer, Berlin (2011)

Spada, H., Plesch, C., Kaendler, C., Deiglmayr, A., Mullins, D., Rummel, N., Kuebler, S., Lindau, B.: Intermediate Report on the Delphi Study—Findings from the 2nd and 3rd STELLAR Delphi rounds: Areas of Tension and Core Research Areas (STELLAR deliverable 1.3A) (2011)

Sutherland, R., Eagle, S., Joubert, M.: A Vision and Strategy for Technology Enhanced Learning: Report from the STELLAR Network of Excellence, STELLAR Consortium. http://www.teleurope.eu/mod/file/download.php?file_guid=152343 (2012)

Tchounikine, P.: Computer Science and Educational Software Design: A Resource for Multidisciplinary Work in Technology Enhanced Learning. Springer, Berlin (2011)

Theußl, S., Zeileis, A.: Collaborative software development using R-Forge. R J. **1**(1), 9–14 (2009)

Thomas, J.J., Cook, K.A.: Illuminating the Path: The Research and Development Agenda for Visual Analytics. IEEE Computer Society Press, Los Alamitos, CA (2005)

Thomas, J., Kielman, J.: Challenges for visual analytics. Inf. Vis. **8**(4), 309–314 (2009)

Trkman, P., McCormack, K., Valadares de Oliveira, M.P., Bronzo Ladeira, M.: The impact of business analytics on supply chain performance. Decis. Support. Syst. **49**(3), 318–327 (2010)

Turney, P.D., Pantel, P.: From frequency to meaning: vector space models of semantics. J. Artif. Intell. Res. **37**, 141–188 (2010)

Wandmacher, T.: How semantic is Latent Semantic Analysis? In: Proceedings of TALN/RECITAL, Dourdan, France (2005)

Wassermann, S., Faust, K.: Social Network Analysis: Methods and Applications. Cambridge University Press, Cambridge (1994)

Wertsch, J.: The primacy of mediated action in sociocultural studies. Mind Cult. Act. **1**(4), 202–208 (1994). Routledge

Whitelock, D.: New forms of assessment of learning for social TEL environments (ARV GCP 18). In: Fischer, F., Wild, F., Zirn, L., Sutherland, R. (eds.) Grand Challenge Problems in Technology Enhanced Learning: Conclusions from the STELLAR Alpine Rendez-Vous. Springer, New York, NY (2014a)

Whitelock, D.: Assessment and automated feedback (ARV GCP 7). In: Fischer, F., Wild, F., Zirn, L., Sutherland, R. (eds.) Grand Challenge Problems in Technology Enhanced Learning: Conclusions from the STELLAR Alpine Rendez-Vous. Springer, New York (2014b)

Whitelock, D.: Assessment and automated feedback (ARV GCP 7). http://www.teleurope.eu/pg/challenges/view/147638 (2012b). Accessed March 4, 2016

Whitelock, D.: New forms of assessment of learning in TEL environments (ARV GCP 18). http://www.teleurope.eu/pg/challenges/view/147649 (2012a). Accessed March 4, 2016

Wild, F., Sobernig, S.: Learning tools in higher education: products, characteristics, procurement. In: 2nd European Conference on Technology Enhanced Learning, CEUR Workshop Proceedings: RWTH Aachen, Vol. 280. ISSN 1613-0073 (2007)

Wild, F., Stahl, C., Stermsek, G., Neumann, G.: Parameters driving effectiveness of automated essay scoring with LSA. In: Proceedings of the 9th International Computer Assisted Assessment Conference (CAA), pp. 485–494. Loughborough, UK (2005)

Wild, F., Scott, P., Valentine, C., Gillet, D., Sutherland, R., Herder, E., Duval, E., Méndez, G., Heinze, N., Cress, U., Ochoa, X., Kieslinger, B.: Report on the State of the Art, Deliverable d7.1, STELLAR Consortium. http://www.stellarnet.eu/kmi/deliverables/20110307_d7.1___state-of-the-art___v1.3.pdf (2009). Accessed March 4, 2016

Wolpers, M., Grohmann, G.: PROLEARN: technology-enhanced learning and knowledge distribution for the corporate world. Int. J. Knowl. Learn. **1**(1–2), 44–61 (2005)

Chapter 2
Learning Theory and Algorithmic Quality Characteristics

A *theory* is a vehicle for understanding, explaining, and predicting particular subject matters. Theories are formed by their postulates and the derived sentences. The postulates are axioms that are assumed to hold for this theory, whereas the sentences and theorems are either logically derived or supported with evidence (Diekmann 2002, p. 141). Each theory has restricted validity and holds only for the subject matter studied.

Models are generally known to introduce order. A model is an abstract, conceptual representation of some phenomenon. Typically a model will refer to only some aspects of the phenomenon under investigation. Two models of the same phenomenon may be essentially different due to the modellers' decisions, different requirements of its users, or intended conceptual or aesthetic differences. Aesthetic differences may be, for example, the preference for a certain level of abstraction, preferences for probabilistic vis-à-vis deterministic models, or discrete versus continuous time. Therefore, users of a model have to understand the original purpose of the model and its underlying theory and the assumptions of its validity (cf. Diekmann 2002, p. 256ff, 169ff).

Theories allow deriving a class of possible models of reality from them (Balzer 1997, p. 50). Models are constructed in order to enable reasoning within an idealized logical framework of scientific theories. Idealised means that the model may make explicit assumptions that are known to be false in some detail. Those assumptions may be justified on the grounds that they simplify the model, while allowing productivity of acceptable accurate solutions. Models are used primarily as a reusable tool for discovering new facts, for providing systematic logical arguments as explicatory or pedagogical aids, for evaluating hypotheses theoretically, and for devising experimental procedures to test them.

This notion of models is not to be confused with the notion of *conceptual models*, which are theoretical constructs that represent physical, biological, or social processes with a set of variables and a set of logical or quantitative relationships between them. More precisely, for the computer and information sciences, a conceptual model can be defined as "a description of the proposed [software]

F. Wild, *Learning Analytics in R with SNA, LSA, and MPIA*,
DOI 10.1007/978-3-319-28791-1_2

system in terms of a set of integrated ideas and concepts about what it should do, behave and look like, that will be understandable by the users in the manner intended" (Preece et al. 2002, p. 40).

This chapter will not provide a conceptual model, but the theoretical foundations from which conceptual models for real learning analytics systems can be created. It will lead over to the one algorithmic model derived from this theory and further detailed in Chaps. 5 and 6, a novel algorithm called 'meaningful purposive interaction analysis'.

As already indicated above, the theory is neither restricted to this single algorithm implementation, nor to the conceptual model chosen in its implementation into a concrete software system presented and demonstrated in the Chaps. 8 and 9. It is not exclusive and does not rule out other classes of representational algorithms or different logical arrangements in other software systems.

Many different ways to formalise both theory and derived models are available that help in sharpening thinking. This is particularly useful, as natural language often contains ambiguities, is complex, and does not provide commonly accepted inference mechanisms (Balzer 1997, p. 60). Standardised language such as offered in logic and mathematics by the different types of propositional logic, predicate logic, and their reflection in set theory aim at supporting the development of methodically constructed, consistent, verifiable scientific theory.

Not all theories and models are formalised, but those that are provide structures as representations of the systems they are trying to explain. These structures consist of logical elements (variables, connectives, quantifiers, inference rules), mathematical elements (sets of numbers, mathematical spaces, relations, functions, terms), and empirical elements (generic terms, relations, functions, constants) (Balzer 1997, p. 61ff). The logical elements and the standard mathematical elements are stable across theories, so they do not need to be derived in the formulation of a theory (Balzer 1997, p. 64).

In this chapter, this theory of how learning happens is sketched out, with its postulates and logically derived foundations explicated. It is preparing the ground for its mathematical formalisation in one possible derived model, i.e. meaningful, purposive interaction analysis, as fleshed out in Chap. 5.

This theory is not new as such, as it is grounded in methodical culturalism (Janich 1998). Its novelty, however, is to extend this culturalist information theory with the constructs learning, competence, and performance. This way, novel theory is established: culturalist learning theory—as an extension of culturalist information theory.

It should be clearly noted, that the concept of 'purposiveness' (or being 'cooperatively successful', as in the words of Janich) is not derived from its seminal appearance in the context of behaviourist literature of the early twentieth century (Tolman 1920, 1925, 1928, 1948, 1955). In this arm of behaviourist learning theory, Tolman observes (1920, p. 222) that the "determining adjustment sets in readiness a particular group of subordinate acts" and "purpose [is defined] as interaction of determining adjustment and subordinate acts" (p. 233). The "satisfaction of purpose" then consists of "the removal of the stimulus to the determining adjustment as a result of one of the subordinate acts" (p. 233). Furthermore, so Tolman (1928, p. 524), "purpose is something which we objective teleologists have to infer from behavior", thereby clearly violating the primacy of methodical order, as introduced below in Sect. 2.3.[1]

It should also be denoted that the learning theory elaborated is not an instructional design theory (Reigeluth 2007; Reigeluth and Carr-Chellman 2009; Mandl et al. 2004). Though it can serve as a component in such, it, e.g., does not provide answers to questions about organisation form or questions of situational context and events.

Methodical culturalism is a form of constructivism. It states (see Sect. 2.1) that information is by nature socially constructed as it is grounded in human, communicative action. Competence is a potential that is built up through performance of such communicative information acts. This process of (co-)constructing[2] information is nothing else then 'learning'.

Within the structures of this theory, the subsequent Chap. 5 will derive a model for representing learning that combines the two algorithmic approaches latent semantic analysis and (social) network analysis. Chap. 8 will then substantiate the conceptual model, i.e. a model on how the representations possible with such algorithmic technology can actually be put into a practical software system for end-users.

Since every theory must be verifiable (or falsifiable!), this chapter concludes with a set of derived functional and non-functional requirements. These requirements are formulated in a way that they can be tested and the evaluation presented in Chap. 10 will refer back to it.

2.1 Learning from Purposive, Meaningful Interaction

Alternative, purist learning perspectives found today can be seen as extreme positions of different levels of analysis, then typically focussing solely on either 'associationist/empiricist' aspects (the modern variant of behaviourism), 'cognitive' aspects, or 'situative' aspects, with the latter two being further developed in the various strands of constructivist theory (Mayes and de Freitas 2004, pp. 7–10).

As already outlined in the introduction, this work defines technology-enhanced learning as the scientific study of the co-construction[3] of knowledge supported by tools, machines, techniques, and the like, with the aim of developing competence of an individual or groups of individuals. Consequently, this work bases its roots in

[1] This may seem less surprising considering the fact that Tolman's psychological research originates in animal studies (including his main work '*Purposive Behavior in Animals and Men*', 1932) and is then interpreted to the human context.

[2] 'Co-'construction further stresses the social element in construction of information.

[3] Aka 'social construction'.

constructivist (or more precisely: culturalist) epistemic theory. And, whereas 'technology' typically subsumes many different classes of artificial tools, machines, techniques, and the like, this work shall use it in a more restricted sense as referring to web-based information systems only. While the unit of analysis of the learning material could be expanded to cover multimedia elements (Mayer 2009) as well (following the deliberations about standardisation in Sect. 2.3 and introducing substitutes using annotation and meta-data), for this book they shall be restricted to text.

The foundational terms among the empirical entities considered within this theory are *persons*, *learning material* (texts or their substitutes), and *tools*. Within the co-construction of knowledge in web-based virtual learning environments, these entities act upon each other. From the viewpoint of an individual, the actions performed in this environment serve the purpose of positively influencing social, self, methodological, and professional *competence development*, i.e. they aim at building potential for 'performant' future action. When given opportunity, persons can put this potential into practice and *perform* to demonstrate their competence.

Following the principle of methodical order (see below, Sect. 2.3), human communication that is understood and fit for purpose forms the archetype of this interaction. From this perspective, 'information' is nothing else then *the abstraction from those communicative acts that serve the same purpose and that are equally meaningful* (see Sect. 2.3): information is the class of communication acts that are equally fit for purpose and equal in meaning—invariant of speaker, listener, and formulation. As will be shown in Sect. 2.6, this is the pivotal in the introduction of any machine-readable representation of 'learning' taking place in such web-based systems.

Competence is then, consequently, the outcome of such informing (see Sect. 2.5). The development of competence can be demonstrated in performance, i.e. through communicative acts that demonstrate understanding and fitness for purpose.

Learning tools help standardise and substitute information acts, which may be[4] more efficient than (re-)enacting 'informing' without the help of their technical substitutes (see also Sect. 2.3 on standardisation). This way of knowledge building in virtual learning environments has an advantage: the communication artefacts created and consumed are directly available in digital form and can thus be subjected to further analysis.

Analysing the communicative behaviour of people in web-based environments can be used to provide feedback on learning: feedback can be given on whether the performance demonstrated deviates from the norm or from expectation.

Since any model is typically a further abstraction derived from its underlying theory, further simplifications will be introduced in Sects. 2.10–2.12 that are correct

[4] Whether it actually *is* more efficient, is a different matter: as already mentioned in the introduction, very often learning tools lack thorough evaluation. Still and as outlined in Sect. 1.1, learning tools enjoy success in the market, even if the nature of their success is sometimes a mystery.

only to a certain degree. For example, they will propose to disregard word order, as it cannot be modelled in matrix algebra of vector spaces (at least not in a simple way). This paves the way for the MPIA model presented in Chap. 5, there serving the simplification of the model.

2.2 Introducing Information

In the following, culturalist information theory is introduced, based on which the concept pair of competence and performance are established. Learning then knits this idea of competence and performance together with the theory of communication acts being the constituting element for the underlying information construction.

Information is not a trivial concept, as the academic discourse shows (Hesse et al. 2008, p. 159). The concept can be attributed a certain duality in its usage traditions: works typically either side with a ‘naturalistic’ school of thought (heralded by Shannon and Weaver with their 1956 article on information theory), or they close up to a ‘culturalist’ school of thought (leading to Janich’s 1998 counter proposal), or they try to bridge between the two. For a recent extended overview on the academic discourse about ‘information’ see Hesse et al. (2008, p. 159), Capurro and Hjørland (2003) and for an earlier review by the author see Wild (2004, p. 69ff).

The naturalistic school of thought assumes that information exists in an external, real world and has something to do with the quality of a particular stimulus. Bridging positions assume that information comes into being only when processing this stimulus, thereby partly determined by the characteristics of its processing system. In such mediating position, information is defined as “the quality of a certain signal in relation to a certain release mechanism, the signal being a low energy phenomenon fulfilling some release specifications. The signal is thus the indirect cause, and the process of the release mechanism the direct cause of the resulting high-energy reaction” (Karpatschof 2000, p. 131).

In the culturalist school, more radically, information is seen as a logical abstractor (Janich 1998, 2003, 2006; Hesse et al. 2008; Hammwoehner 2005), i.e. a class of sentences of a communicatively and cooperatively successful exchange for which an equality relation can be established, thereby abandoning the idea of a naturalistic, material component of information.

Janich (1998, p. 179:§49–59) establishes this definition through invariance postulates for enquirer, respondent, and formulation, following the primacy of methodical order that postulates human communication to be the originating phenomenon of information (and therefore treats all other actions as derivate substitutes and other special cases). This will be detailed more precisely in the subsequent Sect. 2.3.

Following Janich, the primacy of a methodo-culturalistic view of information means that information is a social construction. We constitute meaning only through communicative interaction: information cannot exist without (successful) communication and cooperation.

Although the naturalist position has had a dominant position over the past decades in research and development, this more recently established culturalist position (and it's precursors leading up to it) allow for the development of novel, applied theories. Both of them are accepted schools of thought. Similar to Newtonian Mechanics versus Quantum Mechanics, these theories do not refute each other: their scope of validity and applicability, however, is restricted and not the same. In times of Social Software (Hippner and Wilde 2005), however, the culturalist position has much to offer.

2.3 Foundations of Culturalist Information Theory

Turning to the foundations of the theory in methodical culturalism (and its constructivist precursors), a precise definition of the concept can be elaborated.

Following the methodical order, communicative and cooperative successful action (i.e. 'performance' in its original sense) is a precondition for information to come into being (Janich 1998, p. 173: §27).

Meaning is actively constructed and information is an abstractor of those chains of communicative action, which are equivalent in the information conveyed (ibid, p. 179: §55; cf. Kamlah and Lorenzen 1967, p. 100ff).

Information is constructed when communication is successful and when it leads to the intended reaction, i.e. communication partners share understanding and do successfully cooperate in their exchange (Janich 1998, p. 178ff: §51–52; §54).

Janich (1998, p. 179:§54) explicitly roots information in both communicatively *and* cooperatively successful exchange.

Turning to the first part, *communicatively successful*, this means merely, that a question was understood. The simplest communicative exchange of a question and answer is an utterance or sentence pair that contains a single question and response. Developing this further, the response in this pair is a predicate of the question, if and only if the exchange is communicatively successful, i.e. the question was understood. The response, however, is not necessarily valid and in practice it is not always a satisfactory and relevant answer to the request uttered. In Janich's example (1998, p. 179: §54), a communicative successful exchange can be found in providing "I know very well, which train would be suited for you, but I won't tell you" as an answer to someone asking for travel information at a train station's information centre.

To inform, clearly, the second condition needs to be satisfied as well and *cooperation* of the respondent is required. Stating the next train that will help the inquirer to reach his destination in time would provide such a cooperative, valid answer.

Following Janich (1998, p. 179: §55), information is abstracted from these linguistical messages ("aus sprachlichen Mitteilungen") of communicatively and cooperatively successful exchange by rendering them invariant of speaker, listener, and formulation. In other words, the class of communicative exchanges that are equal in meaning and validity are '*an information*'.

They are—in Janich's words—"informationsgleich". Exchanges that are in such way information equivalent have to be both equivalent in meaning (= communicatively successful) and equivalent in purpose (= cooperatively successful or 'purposively' successful).

The according information equivalence relation defined for the equivalence class of 'a particular information' has to satisfy the criteria of reflexivity, symmetry, and transitivity[5] with respect to the information afforded.

As will be shown later-on in this Chapter, the equivalence relation can be defined with the help of a proximity measure in a vector space (such as the cosine measure) and the class of 'a particular information' can thus be substituted—for a given purpose (!)—as any set of vectors which are equal to each other with respect to this chosen proximity measure.[6]

For the bigger whole, i.e. the space of 'all information' equivalence classes, there exists a surjective function that maps any exchange to its according equivalence class.

Janich maps all other types of information, particularly those 'naturalistic', back to this basic equivalence equation, following the principle of methodical order (Janich 1998, p. 173: §27): when actions serve a purpose, their order cannot be reversed. The order of cooking an egg, peeling it, cutting it into halves, and decorating it may very well be reversed. When doing so, however, the chain of action no longer serves the purpose of preparing 'eggs a la russe' (Janich 1998, p. 173: §27).

Informing has a purpose that can be evaluated with respect to its success for humans, and thus any technical substitutes in an information process (such as when using computers or software systems) are means to an end—informing humans –, and can therefore be mapped back to the theoretical principles formulated above.

This also allows to refute that tying shoelaces (or work-shadowing a baker to learn how to make dough) would not be subjected to this primacy of methodical order: these cases are so-called standardisations (Janich 1998, p. 179: §53) of directives: they map distinct linguistic formulations to distinct compliant behaviour. The act of 'tying shoelaces' is only then successfully and cooperatively performed, if it results in a looped knot that can be opened easily again.

In consequence, this allows the following, pragmatic explanation of how non-communicative chains of action (that involve other modalities), that are often part of professional practice, relate back to such acts of communication. Practices

---

[5] It will be explained later in this Chapter, that transitivity is the only problematic one among these, when using geometric proximity relations as a substitute for such information equivalence relation.

[6] This may or may not involve a threshold level, above which two vectors are considered 'equal'. When working with thresholds, though, transitivity is only under certain circumstances ensured: for example, the equivalence class may be established through equivalence being set as the proximity to the centroid of a vector set.

are subject to this principle of methodical order and therefore they are mere standardisations as a result of successful and cooperative communication.

2.4 Introducing Competence

The concept of competence is of Latin origin (Jaeger 2001, p. 72; Erpenbeck and Rosenstiel 2007, p. XVIII), formed from 'cum' (with) and 'petere' (seek, strive for). It means to go after something with others at the same time—i.e. to 'compete' with others for something. The concept has an academic discourse with history: see Wild and Stahl (2007), and Stahl (2008, p. 16ff), for a more detailed discourse analysis.

Contemporary conceptualisations, such as found in Rychen and Salganik (2003, p. 43) characterise it as "the ability to successfully meet complex demands in a particular context through the mobilization of psychosocial prerequisites (including both cognitive and non-cognitive aspects)".

Competence is a human potential for action[7] and thus inaccessible for direct observation. It is demand-oriented, refers to abilities that can be learned, and involves cognitive and non-cognitive elements such as factual knowledge and procedural skills, but also internalised orientations, values, attitudes, and volitional aspects (Fiedler and Kieslinger 2006, p. 14; Stahl 2008, p. 16ff).

Empirical studies show that the construct of competence is rich in scope and can be further distinguished into four different classes: the class of professional, methodological, personal, and social competence (see, e.g., Bergmann 2007, p. 194ff.; Gay and Wittmann 2007, p. 635ff.; Kauffeld et al. 2007, p. 224ff; Schaper 2007, p. 160ff.; Kurz 2002, p. 601ff.; Jäger 2001, p. 78) (Fig. 2.1).

Thereby, professional competence (sometimes also called 'expertise') refers to both, elementary general knowledge plus (motoric, artistic, and technical) skills as

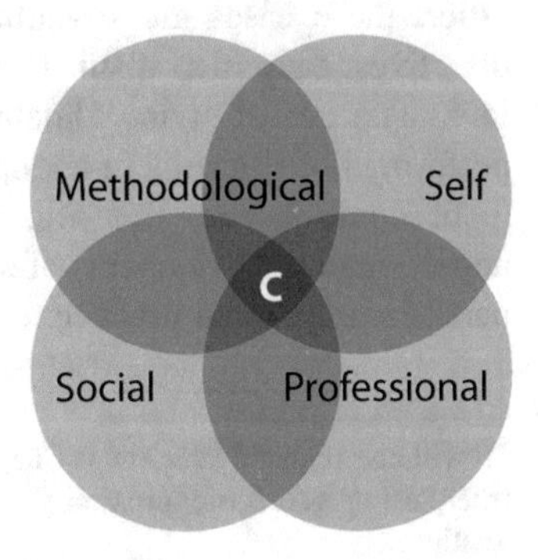

Fig. 2.1 The four basic competence classes

[7] Erpenbeck and Rosenstiel (2007, p. XIX) define it as "dispositions for self-organised action".

well as profession-dependant specialised knowledge plus skills (Jäger 2001, p. 131f).

Methodological competence is the ability to flexibly apply across situations those cognitive abilities needed for structuring problems, decision making, and the like (Kauffeld and Grote 2007, p. 314). It subsumes abilities required for self-directed analytical reasoning, structured, holistic, and systematic thinking (Jaeger 2001, p. 121). In essence, methodological competence is about creatively constructing, structuring, evaluating, and exploiting knowledge in a self-directed way (Kauffeld et al. 2007, p. 261ff).

Self-competence (also known as 'personal' competence) is "directed at a person's inner self" (Stahl 2008, p. 30) and is about those attitudes, values, and other character attributes that are needed to appraise one's own self, i.e. identity, individual fulfilment, independence, competence profile—and to reflect on and react upon this appraisal in order to pro-actively influence future further development (cf. Jaeger 2001, p. 104ff; Erpenbeck 2007, p. 489ff.).

Finally, social competence emphasises interaction between people. Social competence involves potential for action that "aim[s] at identifying, managing and mastering conflicts" (Erpenbeck 2003). Bearing social competence refers to exhibiting certain communicative abilities, interpersonal skills, a capacity for teamwork, and capabilities to manage conflicts (Jäger 2001, p. 83).

These competence classes are not independent of each other. For example, the latter three competence classes social, self, and methodological competence are strongly influencing the development of the first, professional competence—being both prerequisite and requirement to realise high levels of professional expertise. This work mainly deals with professional competence, though it is possible to apply the developed technique in the other areas as well (see, e.g., Wild and Stahl 2007).

Only in its reflection in performance, evidence for competence can be assessed. Performance, according to Stahl (2008, p. 24) is "the externalized demonstration of internalized competence": performance is competence in action. This does not imply, that performance is equivalent to competence. Competent persons do not always act according to their potential and—vice versa—high performance may happen incidentally rather than be driven by competence (Fig. 2.2).

The two constructs of competence and performance are strongly interlinked and assessment exercises aim at creating situations in which the observable performance reliably predicts future competent behaviour: the assessment exercise aims at gathering evidence of the presence and absence of the underlying competence by inspecting action in its observed behaviour and outcomes.

It is the verdict of the observer (Erpenbeck and Rosenstiel 2007, p. XIX) of performance that attributes the disposition for a particular competence to the

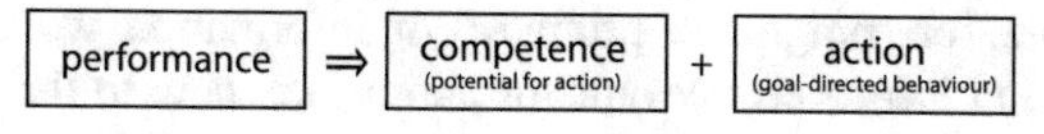

Fig. 2.2 Performance is competence in action

observed, independently acting individual. Performance is observable, purpose-wise successful action.

2.5 On Competence and Performance Demonstrations

Competence is a potential for action and only when turned into performance this potential becomes evident. Performance is demonstrated by giving a particular piece of information in a communicatively and cooperatively successful exchange.

According to this derivation, performance cannot occur by chance. An answer that—without intention—'informs', is by no means cooperative in serving its communicative purpose and thus excluded. This has to be taken into account, when constructing assessment exercises: it may look as if the person observed behaves in a competent way, but in fact it could just be coincidence—a problem that becomes particularly relevant, when turning to procedures of accreditation of prior learning.

Competence is the underlying disposition that causes performance. Without competence, there is no performance. The respondent is said to be competent to respond to this information need with purposive, communicative action. For all performances there is (or was) a person competent for this piece of information, who responded with an informative answer. An enquiry acted as trigger for a this person to respond competent with respect to a particular piece of information, therein demonstrating performance.

2.6 Competence Development as Information Purpose of Learning

Turning now to the exhibitor of the complementary utterance or sentence in the atomic tuple of a communicative exchange—the person inquiring –, the concept of 'learning' can be introduced.

'An information' is learnt, when the enquirer (= learner) engages in a successful exchange with a more knowledgeable other (or a substitute).

This knowing of 'an information' can subsequently be assessed by asking the learner for the correct answer (and validating it).

Quite obviously, a piece of information can be learnt only if the learner did not have this information already, which is constructed in the dialogue. Kuhlen calls this the postulate for 'pragmatic primacy' of information work (Kuhlen 1989, p. 16): information has to be relevant for action, i.e. new to the recipient, in the given context in which it is needed. Depending on appropriateness of instruction or instructional material, the learning curve of individuals can be steep or flat, thereby reflecting, how much information is actually created in offset of the learner's prior knowledge.

One could now argue that any exam situation does therefore not call for creation of information, as the inquiring person (the assessor) knows the answer already. This would, however, be then rooted in a misconception of what the information exchange in this case actually is about: the assessor is not interested in the assertions expressed in the answer, but interested in whether the person assessed is able to give a correct answer. The actual information exchanged in such assessment situation is thus a different one: the directive is the complicated case of 'I need information on whether you are able to give me an information-equal answer to my question, demonstrated by giving such answer"—and the actual information therefore the abstractor of such exchanges.

Moreover, when the enquirer is actually not enquiring and merely listening, and the response is 'dumped' on her without being understood and without fulfilling its information purpose, no learning is taking place. Nota bene, that such incidents of course can in fact be found in formal education not too rarely: the action of learning is in the end not successful, against all good intention.

When any such set of communicative actions (or their standardisation in substitutes) is successfully 'learnt', the person involved in them *develops the competence* to successfully understand and use their formulation in language, i.e. expressing their meaning while achieving the same purpose. Only in successful learning, the individual becomes competent both in acting and in reflecting about information: she develops the competence for that piece of information.

It is possible and not uncommon, that the learner immediately forgets the information, i.e. 'unlearns' the information. For now, this aspect of forgetting, however, shall be neglected, with the argument that memory is of no further relevance to the theory presented.

As already mentioned, competence is a potential for action, not the action itself—and thus not accessible for observation directly in any dialogue. Since it is not possible to analyse competence, the only way of validating, whether 'an information' was actually learnt (or instantly 'forgotten') is by testing the learner's performance: the learner has successfully learnt an information, when he is competent to perform, i.e. when he can demonstrate performance through meaningful, purposive communicative action.

The dialogue can provide evidence of the underlying disposition, though: success of learning can be assessed through textual re-enactment in an informing dialogue. In assessment, this is often done in a transfer situation, applying what has been learnt to a new context: for example, diagnostic knowledge learnt from a medicine lecture is applied to a real, medical case.

Memorising and repeating word by word an answer given by a different speaker in such learning dialogue—the special case of the plagiarism and cheating—is not sufficient to demonstrate performance: one cannot conclude that competence was developed, as competence requires comprehension and returning the response someone else has given in the past is possible without understanding. Though, formally, it does not exclude that the competence has been developed: it could be there, it is just not evident in such demonstration of performance. Quite precisely any memorised response would violate the formulation invariance condition of the basic information equation, as, technically, it is possible for any competent speaker to not only give a single competent answer, but—potentially—the full class of all possible answers.

Janich (2006, p. 155) postulates that permanent turn taking ("permanente[r] Rollenwechsel von Sprecher und Hoerer") is constitutive of human communication, as otherwise we would not be able to learn how to speak: only in this continuous interchange, it is possible to judge whether communicative acts have been communicatively and cooperatively successful.

Learning is change and this change is directed towards the (more or less) planful development of competence, which results in increasing the potential for future action.

It shall be noted that the competence definition developed here is bound to those specific information purposes, with which it is information equivalent, when put to action.

When learners construct knowledge from reading texts, understanding them, thereby fulfilling their information purpose, they develop the potential to write texts expressing this knowledge (i.e. that have the same meaning and that serve the same purpose). They become competent to 'perform' this information.

2.7 Performance Collections as Purposive Sets

Evidence of competence can be collected. In collecting the textual representations of such communicative performance demonstration, a graph can be formed between performances and competences, with vertices (also known as 'nodes') and edges (also called 'links') that connect the vertices with each other.

In the simplest case of such graph, the linking of competence to performance demonstration is a one to many mapping. Such case could be imagined, for example, with an exam question: The exam question triggers the production of responses of competent or incompetent learners. If such response demonstrates understanding of the question and is purposive in its message, then an ideal human rater would mark it as a 'correct' answer. Those correct answers now are known performance demonstrations of a certain competence. It is not required that this competence can be given a name, but typically and ideally within formal education such label can be found (in case of the purpose being the answer to one single exam question, this would admittedly be a rather fine granular competence specified) (Fig. 2.3).

In other cases, this mapping between performance demonstration and underlying disposition is not that simple. Different classes of information-equivalent performance demonstrations may map onto one or more competences: For example, the papers written by medical professionals are typically indexed in Medline against

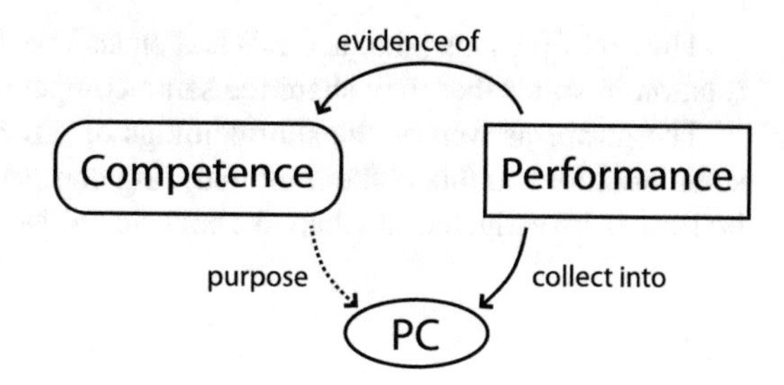

Fig. 2.3 Collecting evidence of competence: purposive sets

10–15 of the more than 25.000[8] medical subject headings (MeSH, see National Library of Medicine 2012).

This relation can be utilised to derive data about the linking of competences through inspecting the relation of their performance demonstrations. If two competences share a performance demonstration, then they are linked. Only in the case where two competences share exactly all performance demonstrations, such competences are identical.

2.8 Filtering by Purpose

With a graph given as defined above, extracting those performance demonstrations that are evidence of a particular competence can help extract performance classes (which then in turn can serve the classification of incoming new performance demonstrations).

Since this is more than just a generic query or filter, the more precise term 'purposive filtering' is used to describe it. Purposes have to be known to the system and usually cannot be derived from the response only.

2.9 Expertise Clusters

Competence is the human potential for action and performance is it's observable counterpart, when put into practice. If there is a group of people that is known to be competent for something and if it is possible to observe their thus purposive, communicative performance of this something, it is not required to further specify this underlying competence and assessment situations can be constructed, in which performance can be observed and compared against the given set of performances demonstrated.

[8] According to the MeSH fact sheet (http://www.nlm.nih.gov/pubs/factsheets/mesh.html), there are currently 27,149 descriptors (last updated: 9.12.2013).

The set of persons (aka 'actors') that enact a performance demonstration, enter a relation in so far that they share the same competence it demonstrates.

The power as well as the shortcomings of direct and indirect (qua association) social relations in this context of analysing competence and expertise clusters will be further investigated in Chap. 3 about Social Network Analysis.

2.10 Assessment by Equivalence

Since competence is postulated to be an abstractor and thus speaker and listener invariant, its performance demonstration has to be stable across actors and conversations. It is thus possible to collect sets of performance demonstrations of a certain competence with the set serving an equivalent purpose and sharing an equivalent meaning.

It also becomes possible to compare new performance demonstrations with an existing pool of collected performance demonstrations. If such new performance demonstration is equal in meaning to the existing set for which the underlying competence is already known and if and only if the purpose of the performance demonstration can be controlled, as it can be done in a learning situations in form of e.g. assessment exercises, presence of the underlying competence can be derived.

A communicative exchange meant to be a performance demonstration of a given competence, is a demonstration of that competence if and only if it is equal in meaning to the exchanges collected in the performance demonstrations class.

Since the purpose—demonstration of a particular competence—is given through the assessment situation, it is sufficient to test for meaning equality.

The other way round, those performances that are supposed to serve the same purpose, but are not equal in meaning to the demonstrations collected in such class, cannot be demonstrations of competence.

It is possible to collect those writings of competent persons that demonstrate a particular class of competence. If the amount of demonstrations collected is sufficiently big enough, the established set can be utilised to evaluate—in an automated way—yet unknown performance demonstrations.

In practice this means to govern the purpose of the performance demonstration (e.g. by posing an exam question) and to then check if the use of language is in line with the language model established by the collection. Any software system meant to automate the evaluation of performance demonstrations must be able to reliably and accurately differentiate demonstrations of competence from demonstrations of incompetence.

Such performance collections can be found, for example, in essay collections from exams. Other examples include publication databases: Medline (see above), for example, is a collection of millions of papers written by experts, which are indexed against their professional competences, i.e. subject areas (medical subject headings, MeSH) in the biomedical domain.

2.11 About Disambiguation

A professional language community (an expertise network) has a shared understanding. Linguistic relativity states that language mirrors cultural differences and this difference in language influences thinking (cf. Saeed 2009). Had linguistic relativity (Whorf 1956) in the earlier half of the twentieth century still been debated with respect to its applicability to language as such, it is now considered widely accepted (Saeed 2009). Consequently, this means that the differences in language and language use can be used to draw back conclusions on the underlying concepts. Studies show that this holds also for professional communities: terminology that is specific to certain communities can be automatically extracted (see also Heyer et al. 2002; Mielke and Wolff 2004; Quasthoff and Wolff 2002).

For example, a surgeon would know what an 'allograft' bone is, whereas most other people will have to look it up in a medical glossary. Homonyms such as 'service' in a restaurant and 'service' as in web-service often conceal a professional language culture existing in parallel to a more widely shared common language.

Obviously, humans manage to invent new words (and new concepts) with ease. Douglas Adam has put together an entertaining collection of new words[9] into his persiflage of a dictionary called 'The meaning of liff': for example, the word 'shoeburyness' stands for "the vague uncomfortable feeling you get when sitting on a seat, which is still warm from somebody else's bottom". Readers willing to include this new word (and maybe concept) in their own idiolect will for sure join a growing—admittedly less professional—community.

Professional communities use professional language: they create and make use of words with particular meanings (stored in documents or messages) and this use can be distinguished (disambiguated) from everyday language or from it's meaning in other language communities.

As Toulmin (1999) states, "language is the instrument that we use, during enculturation and socialization not merely to master practical procedures, but also to internalize the meanings and patterns of thought that are current in our culture or profession."

When professional texts have been disambiguated, their semantic representation can be utilised as a proxy to measure professional performance as evidence of the underlying competence(s).

There is a multitude of algorithms available to disambiguate and represent the meaning of texts. One particular such class of algorithms is formed by the ones rooted in matrix algebra, as further elaborated in the subsequent Chaps. 4 and 5.

It goes without saying that by grounding this work in linear algebra, limitations are introduced into the derived algorithmic model that are not part of the theory proposed in the sections above. For example, the restriction to bags of words and

[9] Following the imperative that place names are wasting sparse space in our mental dictionaries, the words are actually not 'new', but are repurposing existing place names—giving them new meaning to make them worth their while.

the complete negligence of syntactical structures is a limitation. Similarly, the pragmatic capabilities of the approaches elaborated are rather limited. Other natural language processing algorithms offer alternatives or could be used complementary to the matrix algebra algorithms chosen.

2.12 A Note on Texts, Sets, and Vector Spaces

The communicative exchange of text, i.e., words combined in the sentences of an utterance (Saeed 2009), allows for re-constructing meaning. Texts can be defined as sets of sentences, with an utterance being a unit of speech and its representation in language being the sentences out of which texts are composed. Same as its basic units, texts serve communication: they convey information, when understood and serving a purpose.

The set $\mathbb{U}$ of possible (written) sentences can also be defined within a dictionary $\mathbb{W}$. Sentences contain a distinct sequence of words (chosen from the set of all words $\mathbb{W}$), i.e. sentences are an (ordered) n-ary tuple. This introduces further simplifications: orthography and grammatical structure are ignored. Sentences u are treated as bags of words, disregarding word order—as postulated in vector space theory.

$$u = w_1, w_2, \ldots, w_n, \; w \in \mathbb{W} \tag{2.1}$$

Sentences can be converted to vectors, by counting the occurrence frequencies of their distinct words, and introducing a mapping function, that assigns a word label to the frequency in each vector position.

$$\begin{aligned} &\text{Vector } v: \; w_1, w_2, \ldots, w_n \mapsto f_1, f_2, \ldots, f_i, \; w_n \in \mathbb{D}, \; f_i \in \mathbb{R} \\ &\text{Label } l: \; n \in \mathbb{N} \mapsto w \in \mathbb{D} \\ &\text{Vector space } \mathbb{V}^n = \{f_1, f_2, \ldots, f_n \,|\, f_1, f_2, \ldots, f_n \in \mathbb{R}\} \end{aligned} \tag{2.2}$$

As introduced in Feinerer and Wild (2007), a multi set M can be defined, which holds a set of sentences in the vector space.

$$M = \{d | d \in \mathbb{V}^n\}$$

This multi-set has to satisfy the condition that all term vectors of represented utterances are in the same format, i.e. each frequency element maps to the same term label:

$$\forall v, w \in M : l(w) = l(v)$$

Thus, this multi-set qua format establishes a dimensional system of a vector space: the document-term matrix (aka 'text matrix') at the same time forms one possible implementation of the sentence space $\mathbb{U}$ introduced above in Eq. (2.1).

Each term constitutes a column vector in this matrix, each utterance a row vector. Other than the sentence space $\mathbb{U}$, however, the matrix M is in practice no longer infinite.

This lack of infinity poses two practical problems. First, only sentences that are contained in M can be represented, which is a significant restriction and can only be overcome by updating. Thereby the natural language sentences are processed into the vector format required, applying, for example, reduction to word stems or eliminating those words not contained in the controlled vocabulary coming with the original matrix M.

The second problem is even more critical: Phenomena such as polysemy, homography, and homonymy conceal the actual semantic structure of language. To overcome this problem, the vector space established with the multi-set M can utilise a lower dimensional, more semantic space from it, using a two-mode factor analysis such as singular value decomposition. This bears the advantage of reducing noise (introduced through choice of words), bringing the vector representations closer to their meaning.

Both LSA (presented in Chap. 4) and MPIA (fleshed out in Chap. 5) utilise matrix algebra and the foundational relations of words to vector spaces formally introduced in this section.

2.13 About Proximity as a Supplement for Equivalence

Building a collection of performance demonstrations for a given competence, a set can be defined within M, which satisfies the conditions of equivalence of meaning and purpose.

Such condition is the basic information equivalence relation introduced above in Sect. 2.3. Equality of purpose is given externally, as the performance demonstrations are taken from a situation that serves a given purpose.

In the vector space (foundational to Chaps. 4 and 5), meaning equivalence can be substituted with a proximity relation such as implemented by the Cosine measure.

In the social graph (foundational to Chap. 3), meaning equivalence cannot be assessed directly—but has to be provided explicitly. Once explicit, atomic relations are available, however, transitivity applies and proximity (and other types of inferences such as aggregation) can be calculated.

2.14 Feature Analysis as Introspection

Again applying measures of network analysis, it becomes possible to further investigate the graph structure given by a performance class and the edges established to the terms contained in the utterances it holds. For example, with the help of degree centrality, it is possible to identify those terms that play a more prominent role—and use them as descriptors of the underlying performance class. The instruments of network analysis to describe structural properties of graphs provide methods to further introspect the nature of the relations of performance demonstrations and their constituting features.

It is also possible to extract and further investigate those individuals and their performance demonstrations that deal with a certain competence. Thus, a method is established to identify experts and further investigate their expertise.

2.15 Algorithmic Quality Characteristics

A variety of non-functional, quality requirements and characteristics can be derived [cf. Jurafsky and Martin (2004, pp. 504–510)] to challenge validity of any implemented model of this *semantic appropriation theory* presented above.

Most notably, these include quality requirements for divergent and convergent validity as well as for its capabilities for introspection. Moreover, these initial three can be complemented with demands with regards to visual inspection and computational performance.

Thereby, *divergent validity* refers to the ability of a derived model to empirically differentiate demonstrations of performance from non-performance. *Convergent validity* then again refers to the element of representation equality. Any derived algorithm must be able to identify equal performance demonstrations and concept equality, both in the graph as well as graphically: proximity to a point, mapping to a point location on planar projection of a particular visualization format chosen. This can be done, for example, with an essay scoring or associative closeness experiment: to be considered valid, any algorithm developed must be able to map synonyms to identity and to evaluate essays on a level of near-human effectiveness.

Introspection refers to the ability to explain reasoning through feature analysis. For empirical testing, this, for example, means that any derived algorithmic model must be able to not only map texts to their competence representation, but also provide a description of this mapping (as well as instruments to explain relations to similar and unrelated evidence).

It is desirable that any model derived also provides means for *visual inspection* supporting both overview and detail in the graphical representation so as to help the analyst with intuitive displays for analysing learner positions and semantic (conceptual) appropriation of individuals and groups.

It is desirable that any chosen algorithmic model implementing this theory provides high computational *performance*, not only ensuring reasonable efficiency in set up and training, but also in use: it must be fast and easy to administer; requirements on collection availability should be kept small.

These requirements for quality characteristic of the algorithm presented in Chaps. 5 and 6 will be revisited in Chap. 10, thereby using the examples of Chap. 9 to illustrate the algorithm in action.

2.16 A Note on Educational Practice

What does this mean in practice? It means, that writing about knowledge and learning is key in demonstrating performance.

Loops of interchanging reading and writing activity have been postulated in the literature as being foundational to learning (Dessus and Lemaire 2002). They form an essential part of today's assessment regimes. This means, it is feasible and realistic to assume that automated support systems can be introduced to investigate the meaning structure and purposes afforded by texts born digital in the context of learning and its assessment.

By comparing with other texts known, it is possible to disambiguate meaningful structure particular to a certain competent group, separating it from generic language use. Any algorithm capable of implementing such disambiguation efficiently and effectively will be able to demonstrate meaning equality holds, while retaining precision and accuracy.

Since equal in meaning is an abstractor, this means that competent people, i.e. learners who developed competence with respect to a certain meaning, must be able to paraphrase information into differing texts. The special case, where wording of a learner written text is identical to the text learnt, is not sufficient to derive that a person is competent, since it can be memorised instead of comprehended.

2.17 Limitations

The theory does not cover *forgetting*: it does not offer a model that takes retention and forgetting over time into account. Though it deemed possible to extend the theory by modelling a decay function that is inline with concurrent memory models.

It is also clear, that this theory does not explain, *how difficult* it is to learn something. For example, studies have shown that the 'smarter' people are, the less open they are to develop new knowledge (Argyris 1991): learning often becomes an act of re-contextualising existing knowledge (single-loop learning), while acquiring truly new models (double-loop learning) becomes rarer when people know more already.

It is clear that *expertise networks* are much more complex and cannot be reduced to mere cultures of shared competence. Taste and style, for example, seem to be relevant properties that cannot be fully explained on a semantic level. Furthermore, pragmatic aspects such as discourse and argumentation play a crucial role in the constitution of professions. Still, disambiguated representations of professional texts can serve to represent professional performance thus allowing for the identification of where competence might be put into practice.

With respect to the acquisition of professional competence, it can be said that the work presented here will not answer the question about whether the *intelligent novice or the generalist expert* is the more favourable option in our information society of today. It merely allows making performance as evidence of the underlying competence more visible and working with representations of meanings expressed in professional language as a substrate for analysis.

The learning theory presented does not form an *instructional theory*, such as the ones discussed in length in Mandl et al. (2004, esp. pp. 7ff). Though it is deemed compatible with many of the approaches prevailing, to be used as a component.

2.18 Summary

Within this chapter, the theoretical foundations of learning from meaningful, purposive interaction were introduced. Branching off from methodo-culturalist information theory and following its principle of primacy of methodical order, information is defined as a logical abstractor, requiring both communicative and cooperative success in conversation to allow information to be constructed. In other words, only when meaning *and* purpose are established, information is conveyed. Competence is then introduced as the potential for informative action, observable only when turned into performance.

This has very practical implications, as derived in the subsequent Sections (2.7–2.10). First, it is possible to gather collections of performance demonstrations, if and only if their provenance context (their 'purpose') is known. Moreover, it allows constructing information systems that facilitate purposive filtering in the collections held. Furthermore, it is possible to extract expertise clusters using data analysis in such performance collections and utilising the principle of assessment by equivalence.

Generic deliberations for implementable models follow about disambiguation, vector spaces, and proximity as a supplement for equivalence. Moreover, thoughts on feature analysis as introspection are added.

To pave the way to evaluation, algorithmic quality characteristics are stated, implications for educational practice outlined, and limitations of the theory explicated.

While analysis of purposiveness and expertise clusters point in the direction of Social Network Analysis (see Sect. 1.4 on algorithmic roots) and while the analysis of equivalence of meaning points towards Latent Semantic Analysis (see Sect. 1.4), both dominant algorithmic strands fall short in investigating both together.

The subsequent two Chaps. 3 and 4 will now report on the state of the art in these algorithmic strands, while Chaps. 5 (and 6) will then bring them together in a novel fusion algorithm capable of resolving this shortcoming.

References

Argyris, C.: Teaching smart people how to learn. Harv. Bus. Rev. May–June (1991)
Balzer, W.: Die Wissenschaft und ihre Methoden: Grundsaetze der Wissenschaftstheorie. Alber, Muenchen (1997)
Bergmann, B.: Selbstkonzept beruflicher Kompetenz. In: Erpenbeck, J., von Rosenstiel, L. (eds.) Handbuch Kompetenzmessung. Schäffer-Poeschel, Stuttgart (2007)
Capurro, R., Hjørland, B.: The concept of information. Annu. Rev. Inf. Sci. Technol. **37**, 343–411 (2003). Chapter 8
Dessus, P., Lemaire, B.: Using production to assess learning. In: Cerri, S.A., Gouarderes, G., Paraguacu, F. (eds.) Proceedings of ITS 2002. LNCS 2363, pp. 772–781. Springer, Berlin (2002)
Diekmann, A.: Empirische Sozialforschung. Rowohlt, Hamburg (2002)
Erpenbeck, J.: Kompetenz-Diagnostik und -Entwicklung. In: Erpenbeck, J., von Rosenstiel, L. (eds.) Handbuch Kompetenzmessung. Schäffer-Poeschel, Stuttgart (2003)
Erpenbeck, J.: KODE®—Kompetenz-Diagnostik und -Entwicklung. In: Erpenbeck, J., von Rosenstiel, L. (eds.) Handbuch Kompetenzmessung. Schäffer-Poeschel, Stuttgart (2007)
Erpenbeck, J., Rosenstiel, L.: Einführung. In: Erpenbeck, J., von Rosenstiel, L. (eds.) Handbuch Kompetenzmessung. Schäffer-Poeschel, Stuttgart (2007)
Feinerer, I., Wild, F.: Automated coding of qualitative interviews with latent semantic analysis. In: Mayr, Karagiannis (eds.) Proceedings of the 6th International Conference on Information Systems Technology and its Applications (ISTA'07). Lecture Notes in Informatics, vol. 107, pp. 66–77. Gesellschaft fuer Informatik e.V., Bonn (2007)
Fiedler, S., Kieslinger, B.: iCamp Pedagogical Approach and Theoretical Background, Deliverable d1.1 of the iCamp Project, iCamp Consortium (2006)
Gay, F., Wittmann, R.: DISG-Persönlichkeitsprofil von persolog: Verhalten in konkreten Situationen. In: Erpenbeck, J., von Rosenstiel, L. (eds.) Handbuch Kompetenzmessung. Schäffer-Poeschel, Stuttgart (2007)
Hammwoehner, R.: Information als logischer Abstraktor? Überlegungen zum Informationsbegriff. In: Eibl, M., Wolff, C., Womser-Hacker, C. (eds.) Designing Information Systems, pp. 13–26. UVK Verlagsgesellschaft mbH, Konstanz (2005)
Hesse, W., Mueller, D., Ruß, A.: Information, information systems, information society: interpretations and implications. Poiesis Prax. **5**, 159–183 (2008)
Heyer, G., Quasthoff, U., Wolff, C.: Automatic analysis of large text corpora: a contribution to structuring web communities. In: Unger, H., Böhme, T., Mikler, A. (eds.) Innovative Internet Computing Systems. LNCS 2346, pp. 15–36. Springer, Berlin (2002)
Hippner, H., Wilde, T.: Social software. Wirtschaftsinformatik **47**(6), 441–444 (2005)
Jaeger, P.: Der Erwerb von Kompetenzen als Konkretisierung der Schlüs- selqualifikationen—eine Herausforderung an Schule und Unterricht. Dissertation, University of Passau (2001)
Janich, P.: Informationsbegriff und methodisch-kulturalistische Philosophie. Ethik und Sozialwissenschaften **9**(2), 212–215 (1998)
Janich, P.: Human nature and neurosciences: a methodical cultural criticism of naturalism in the neurosciences. Poiesis Prax. **2**(1), 29–40 (2003)
Janich, P.: Was ist Information? Kritik einer Legende. Suhrkamp, Frankfurt/Main (2006)
Jurafsky, D., Martin, J.H.: Speech and Language Processing. Pearson Education, Delhi (2004) (Third Indian Reprint)
Kamlah, W., Lorenzen, P.: Logische Propaedeutik. Bibliographisches Institut, Mannheim (1967)
Karpatschof, B.: Human Activity, Contributions to the Anthropological Sciences from a Perspective of Activity Theory. Dansk Psykologisk Forlag, Copenhagen (2000)
Kauffeld, S., Grote, S.: Alles Leben ist Problemloesen: Das Kasseler-Kompetenz-Raster. In: Schaefer, E., Buch, M., Pahls, I., Pfitzmann, J. (eds.) Arbeitsleben! Arbeitsanalyse—Arbeitsgestaltung—Kompetenzentwicklung, Kasseler Personalschriften, Band 6. Kassel University Press, Kassel (2007)

Kauffeld, S., Grote, S., Frieling, E.: Das Kasseler-Kompetenz-Raster (KKR). In: Erpenbeck, J., von Rosenstiel, L. (eds.) Handbuch Kompetenzmessung. Schäffer-Poeschel, Stuttgart (2007)

Kuhlen, R.: Pragmatischer Mehrwert von Information. Sprachspiele mit informationswissenschaftlichen Grundbegriffen, Konstanz, Oktober 1989, Bericht 1/89. http://www.kuhlen.name/documents.html (1989)

Kurz, C.: Innovation und Kompetenzen im Wandel industrieller Organisationsstrukturen. Mitteilungen aus der Arbeitsmarkt- und Berufsforschung **35**, 601–615 (2002)

Mandl, H., Kopp, B., Dvorak, S.: Aktuelle theoretische Ansätze und empirische Befunde im Bereich der Lehr-Lern-Forschung: Schwerpunkt Erwachsenenbildung. Deutsches Institut für Erwachsenenbildung, Bonn (2004)

Mayer, R.: Multimedia Learning, 2nd edn. Cambridge University Press, New York (2009)

Mayes, T., de Freitas, S.: JISC e-Learning Models Desk Study, Stage 2: Review of e-Learning Theories, Frameworks and Models. JISC (2004)

Mielke, B., Wolff, C.: Text Mining-Verfahren für die Erschliessung juristischer Fachtexte. In: Proceedings Internationales Rechtsinformatik Symposion, Verlag Österreich Salzburg (2004)

National Library of Medicine: What's the Difference Between Medline and Pubmed?. http://www.nlm.nih.gov/pubs/factsheets/dif_med_pub.html (2012). Accessed 07 Sept 2012

Preece, J., Rogers, Y., Sharp, H.: Interaction Design: Beyond Human-Computer Interaction. Wiley, New York (2002)

Quasthoff, U., Wolff, C.: The poisson collocation measure and its applications. In: Second International Workshop on Computational Approaches to Collocations, Wien (2002). 22./23 Jul 2002

Reigeluth, C.: Instructional-Design Theories and Models: Volume II. Routledge, New York (2007)

Reigeluth, C., Carr-Chellman, A.: Instructional-Design Theories and Models: Volume III: Building a Common Knowledge Base. Routledge, New York (2009)

Rychen, D.S., Salganik, L.H.: Key Competencies for a Successful Life and a Well-functioning Society. Hogrefe ☺ Huber, Göttingen (2003)

Saeed, J.: Semantics. Wiley-Blackwell, Chichester (2009)

Schaper, N.: Arbeitsproben und situative Fragen zur Messung arbeitsplatzbezogener Kompetenzen. In: Erpenbeck, J., von Rosenstiel, L. (eds.) Handbuch Kompetenzmessung. Schäffer-Poeschel, Stuttgart (2007)

Stahl, C.: Developing a Framework for Competence Assessment. Dissertation, Vienna University of Economics and Business, Austria (2008)

Tolman, E.: Instinct and purpose. Psychol. Rev. **27**(3), 217–233 (1920)

Tolman, E.: Purpose and cognition: the determiners of animal learning. Psychol. Rev. **32**(4), 285–297 (1925)

Tolman, E.: Purposive behavior. Psychol. Rev. **35**(6), 524–530 (1928)

Tolman, E.: Purposive Behavior in Animals and Men. Century, New York (1932)

Tolman, E.: Cognitive maps in rats and men. Psychol. Rev. **55**(4), 189–208 (1948)

Tolman, E.: Principles of performance. Psychol. Rev. **62**(5), 315–326 (1955)

Toulmin, S.: Knowledge as shared procedures. In: Engestroem, Y., Miettinen, R., Punamaeki, R.L. (eds.) Perspectives on Activity Theory, pp. 53–64. Cambridge University Press, Cambridge, MA (1999)

Whorf, B.: Language, thought, and reality. In: Carroll, J.B. (ed.) Selected Writings of Benjamin Lee Whorf. MIT Press, Cambridge, MA (1956)

Wild, F.: Visuelle Verfahren im Information Retrieval. Intelligentes Sehen als Grundlage der Informationserschließung. Universitaet Regensburg, Magisterarbeit (2004)

Wild, F.; Stahl, C.: Using latent semantic analysis to assess social competence. In: Mini-Proceedings of the 1st European Workshop on Latent Semantic Analysis in Technology-Enhanced Learning, Open University of the Netherlands: Herleen (2007)

Chapter 3
Representing and Analysing Purposiveness with SNA

Social Network Analysis is a powerful instrument for representing and analysing networks and the relational structure expressed in the incidences from which they are constructed. It is an excellent instrument to explore the 'power of purpose' of the concept introduced in the previous chapter and exemplified in the foundational and extended application cases included in this chapter.

Following a brief history and the definition of the main use cases for an analyst of learning applying social network analysis, the first foundational example illustrates a typical analysis embedded in a constructed usage scenario. The second, extended example then applies social network analysis in a real-life case of investigating a discussion board of a university learning management system.

While serving understanding of how learning purposes express the incidences that establish the required network structure on which the instruments and measures of social network analysis unfold their power, they also point towards its main shortcoming: the blindness to content and meaning. If source data do not already contain semantics, it cannot be further investigated. Moreover, even if such relations were added, maintenance often becomes a significant problem: with growing numbers of persons participating and with evolving topics, errors and inconsistencies are introduced at the cost of either quality or resources.

This shortcoming is not shared by Latent Semantic Analysis, the algorithm presented in the next Chap. 4—there, however, then falling short of facilities for complex analysis of resulting graph structures and there then falling short of analysing purposive, social context.

3.1 A Brief History and Standard Use Cases

Social network analysis can look back on a—for the information sciences—rather long-standing research tradition. Although the term 'social network analysis' emerged in the 1950s, with deliberate academic use flourishing only in the

F. Wild, *Learning Analytics in R with SNA, LSA, and MPIA*,
DOI 10.1007/978-3-319-28791-1_3

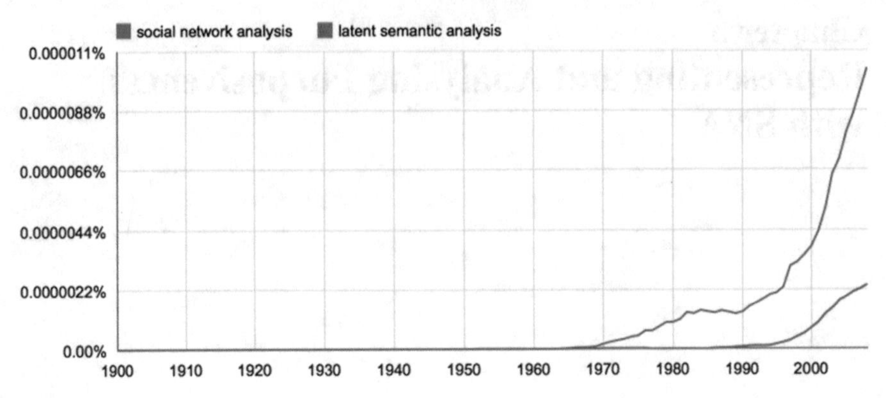

Fig. 3.1 N-gram statistics for SNA and LSA in the English book corpus from 1900 to 2011 (Google 2012)

subsequent decades, the idea of analysing social relationships in a structured way can be dated back at least to the 1920s and 1930s.

Figure 3.1 illustrates the uptake of the terminus technicus 'social network analysis' by depicting its rise in frequency in the English Book corpus available in Google's n-gram viewer: the phrase (i.e. n-gram) 'social network analysis' can be found in rising frequencies in the corpus beginning in the mid 1960s, after it first appears in the middle of the 1950s. The English book corpus used for the analysis indexes more than 8 million books (of which 4.5 million are written in English), which is about 6 % of all books ever published (Lin et al. 2012, p. 169, 170). While such n-gram analysis clearly is subject to many restrictions (see Michel et al. 2011, p. 181), it is indicative of "trends in human thought" (ibid, p. 176).

In the academic literature, the inception of social network analysis is often connected to Moreno's influential book 'Who shall survive?' (1934), which introduces the 'sociogram'. Less well-known predecessors in thought, however, can be found in the literature in the 1920s already (see Freeman 1996, p. 39; Carrington and Scott 2011, p. 1). Looking at the underlying mathematical theory, the roots of network analysis can even be dated back further, at least to Euler's (non) solution of the problem of the 'seven bridges of Koenigsberg', published at the beginning of the eighteenth century (see Carrington and Scott 2011, p. 4), which laid the foundations for mathematical graph theory.

While the idea dates back to the 1920s and 1930s, the term social network analysis itself is coined only in the 1950s, spreading quickly in the group of Manchester anthropologists where it emerged (Scott 2000, p. 5, 29) and marking off two decades of concerted efforts towards its formalization. Over these next decades, SNA is on the rise and by the 1980s social network analysis research reaches out into a wide variety of disciplines (see Freeman 2004, p. 148).

The basic idea of social network analysis is always identical. A social network consists of vertices and edges that connect them. Vertices are also known as 'nodes' or 'actors', whereas edges are alternatively called 'links' or 'ties'.

The nodes (and links) are not necessarily of the same type: only so-called one-mode data matrices are symmetrical, containing the same set of nodes in the columns and rows—such as found when analysing interpersonal relations. Two-mode networks involve different sets of nodes along the sides of the matrix, for example, when linking students with the courses they enrolled in (Carrington and Scott 2011, p. 4).

Such network spanning nodes and links can be further investigated to reveal structural properties. Measurement operations can be applied to evaluate and even assess certain structural characteristics of such networks. Furthermore, filtering operations can be used to zoom in on particular micro (e.g. circles), meso (groups and sub-networks), or macro (units, backbones) structures.

Over the years, a wide range of different filtering mechanisms and measures have been proposed and validated that serve in analysing social relationships in a wide variety of contexts, ranging from the friendship formation in a school class (Moreno 1934, p. 38) to political participation in campaigning for a presidential election on twitter (Barash and Golder 2011, p. 147), to name but a few.

Filtering operations help in selecting structures in the network that fulfil certain characteristics. For example, it is possible to select all students of a particular faculty from a social network in a university or to exclude the co-attendance relations of students for all sports courses. Such filters can be applied on nodes, edges, or connected structures (see Fig. 3.2). An example for the latter can be found in the analysis of so-called 'cliques' such as authorship circles or teams (Mutschke 2004, p. 14).

Measuring operations serve the inspection of characteristics of actors, their actions, and the therein hidden relationships. Again, measures can operate on different levels (see Fig. 3.2): a variety of node, network, and component measures have been proposed to describe and inspect characteristics on different aggregation levels. For example, the in-degree centrality of a node can be used to indicate the social status of a person in a social network. The in-degree is measured by the amount of directed edges pointing towards the person (see below for an example).

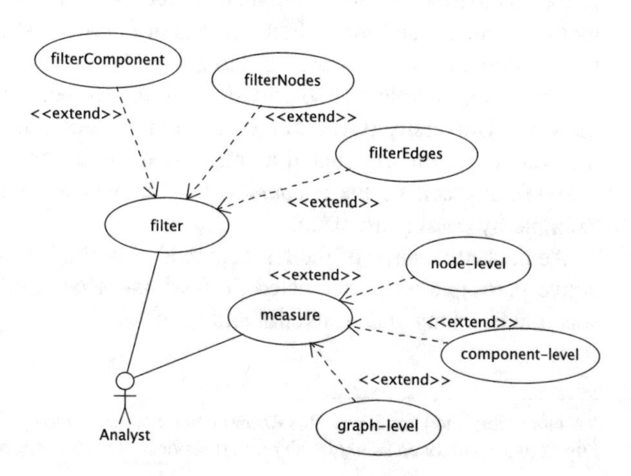

Fig. 3.2 SNA use cases of the analyst

Filtering and measuring operations can be formalised as a domain specific data manipulation language, such as proposed by Klamma et al. (2006, p. 466) and Klamma and Petrushyna (2008, p. 925) for identifying structural patterns in social networks. Within the R community and environment, such formal language is provided in the packages sna (Butts 2010), network (Butts et al. 2012), and igraph (Csardi and Nepusz 2006) and their extensions.

Social networks are—mathematically—modelled as graphs or as matrices. This is not to be mistaken for their visualization format: social network data that are mathematically captured in a matrix can be depicted with a graph-like social network diagram. Many other visualisation types, however, are possible. For example, the data represented in a social network graph can be visualized in tabular matrix format. Depending on the aim of analysis, a particular visualisation format can be more expressive than others (see Chap. 6 for more detail on visualisation formats).

With larger networks, the importance of this distinction between different representation formats for data, mathematical operations, and visualization becomes eminent: for example, converting between graph and matrix representations on a storage device or in working memory of the analyst's statistical processing environment can become a resource consuming operation.

3.2 A Foundational Example

In the following foundational example,[1] the human resource manager of a company has to look into the competence development profiles of nine employees in a particular unit of the company, who recently went through trainings, some of which offered by either of two universities from which the company procures part of their courses.

The usage scenario described is picked up again and extended in the next Chap. 4 in the foundation example provided in Sect. 4.3 in order to clearly work out the main advantages and main shortcomings of the two distinct methods. Chap. 9 will then describe how MPIA can help to get the best of both worlds (Sect. 9.2).

The two academic course providers are the University of Regensburg (UR) and the Open University (OU). The other trainings not commissioned exclusively for the company were conducted as massive open online courses (MOOCs, open to anyone). In each of these courses, the learners demonstrated performance, for example by passing an exam.

Additionally, some of the learners made informal learning experiences through active participation in dedicated professional groups the company set up on the social networking sites LinkedIn and FaceBook.

[1] A more simplified version of this example is made available by the author of this book online: http://crunch.kmi.open.ac.uk/people/~fwild/services/simple-sna.Rmw

The courses and informal learning groups are spread across the three subject areas Computer Science (CS, Informatics), Mathematics and Statistics (Math, Algebra, Statistics), and Pedagogy (Ped). An additional three are in the more interdisciplinary area of Technology-Enhanced Learning (TEL).

It is so prototypical for the company that besides course attendance, there is not much known about the actual content of the courses (and the skills and competences they motivate), as each training measure was signed off by a different line manager—and since there is no single shared system or catalogue in which such information would be captured across branches and locations.

The nine learners listed in Table 3.1 demonstrated their competence in twelve different contexts. For example, Joanna, Peter, and Christina demonstrated their disposition for pedagogy at the Open University, whereas Alba did so in massive open online courses.

Each of the nine persons shows a different set of dispositions: Thomas is regarded competent for 'Facebook-Algebra' and 'OU-Statistics', Simon for 'UR-Informatics', and so forth. Table 3.1 lists the relations as a so-called incidence matrix *im*: if a person has demonstrated performance in a certain competence, the cell value is '1', otherwise it is '0'.

The reason why the human resource manager is looking into the competence profiles of the employees of this unit is that one member, Christina, is off sick. Christina, however, was supposed to work on a job for a customer over the next month or two. The human resource manager now wants to know, who of the employees in this unit could be a worthy replacement with a similar competence profile to Christina. To find out, the following social network analysis can be helpful.

It is possible to calculate from this incidence matrix a so-called adjacency matrix that relates person with each other if they share a certain competence demonstration. The nodes connected that way will be 'adjacent' in the network calculated. The adjacency relation in this analysis is symmetrical, this similarity relation has no direction: Peter shares 'OU-PED' with Joanna and Joanna shares 'OU-PED' with Peter. Therefore, the adjacency can be calculated in R as shown in the following listing by multiplying the incidence matrix *im* with its transposed matrix im^T:

Listing 1 Calculating the adjacency matrix.

```
am = im %*% t(im)
```

Executing this line of code produces the adjacency data shown in Table 3.2: for example, Simon is connected to Paul, because both share the underlying competences 'CS' as demonstrated in the course at the University of Regensburg and on LinkedIn.

Although not relevant for the scope of this analysis, the matrix multiplication of Listing 1 creates values along the diagonal of the matrix as well: Thomas is related to Thomas with 2, since two competence demonstrations are shared.

Moreover, the adjacency matrix is symmetrical, as the underlying data is undirected: Peter is connected to Joanna (with 1) and Joanna is connected to Peter (with 1).

Table 3.1 Incidence matrix *im* of learners and their competence

	OU-CS	UR-Informatics	MOOC-PED	MOOC-TEL	MOOC-Math	OU-PED	MOOC-ocTEL	MOOC-LAK	OU-Statistics	Facebook-Statistics	Facebook-TEL	Linkedin-CS
Paul	*1*	*1*	*0*	*0*	*0*	*0*	*0*	*0*	*0*	*0*	*0*	*1*
Joanna	*0*	*0*	*1*	*0*	*0*	*1*	*0*	*0*	*0*	*0*	*1*	*0*
Maximilian	*0*	*0*	*0*	*0*	*1*	*0*	*0*	*0*	*0*	*1*	*0*	*0*
Peter	*0*	*0*	*0*	*1*	*0*	*1*	*1*	*1*	*0*	*0*	*0*	*0*
Christina	*0*	*0*	*1*	*1*	*0*	*1*	*1*	*1*	*0*	*0*	*0*	*0*
Simon	*0*	*1*	*0*	*0*	*0*	*0*	*0*	*0*	*0*	*0*	*0*	*1*
Ida	*0*	*0*	*0*	*0*	*0*	*0*	*0*	*0*	*1*	*1*	*0*	*0*
Thomas	*0*	*0*	*0*	*0*	*0*	*0*	*0*	*0*	*1*	*1*	*0*	*0*
Alba	*0*	*0*	*0*	*1*	*0*	*0*	*1*	*1*	*0*	*0*	*0*	*0*

Table 3.2 Adjacency matrix *am* of the learners

	Paul	Joanna	Maximilian	Peter	Christina	Simon	Ida	Thomas	Alba
Paul	3	0	0	0	0	2	0	0	0
Joanna	0	3	0	1	2	0	0	0	0
Maximilian	0	0	2	0	0	0	1	1	0
Peter	0	1	0	4	4	0	0	0	3
Christina	0	2	0	4	5	0	0	0	3
Simon	2	0	0	0	0	2	0	0	0
Ida	0	0	1	0	0	0	2	2	0
Thomas	0	0	1	0	0	0	2	2	0
Alba	0	0	0	3	3	0	0	0	3

This resulting adjacency matrix *am* can be converted in R with the help of the network package (Butts et al. 2012; Butts 2008) to an (undirected) graph *net* using the code in Listing 2.

Listing 2 Convert the adjacency matrix data to graph notation.

```
net=network(am, directed=FALSE)
```

In doing so, the matrix data is converted to node-node tuples that hold only the connections that are actually set, i.e., those links that are non-zero in the adjacency matrix, excluding the self-referential cases along the diagonal of *am*. Since it is an undirected graph, it will hold only one edge for each of the directed, but symmetrical relations listed in Table 3.2 above and below the diagonal. That way, the sparse matrix is stored more efficiently.

Listing 3 Accessing edge data of the graph object *net*.

```
# edge 1 goes from
net$mel[[1]]$outl
## [1] 6

# edge 1 goes to
net$mel[[1]]$inl
## [1] 1
```

Listing 3 shows example code of how to access the end-point data of the first edge: the edge connects node 1 with node 6 (and vice versa, since it is an undirected graph). Working with these data, it is possible to conduct measurements. It is, for example, possible to check whether particular nodes are connected: in the example: node 1 (i.e., 'Paul') and node 6 (i.e., 'Simon').

Listing 4 Plotting a network visualisation.

```
plot(net, displaylabels=TRUE)
```

While the data stored in *net* may still be difficult to read, their visualisation as a sociogram is—in such simple case—very intuitive to interpret (see also Chap. 6 for more background on the visualisations). The code in Listing 4 visualises the data of the social graph as so-called sociogram, by plotting a simple two-dimensional network diagram, resulting in the diagram of Fig. 3.3.

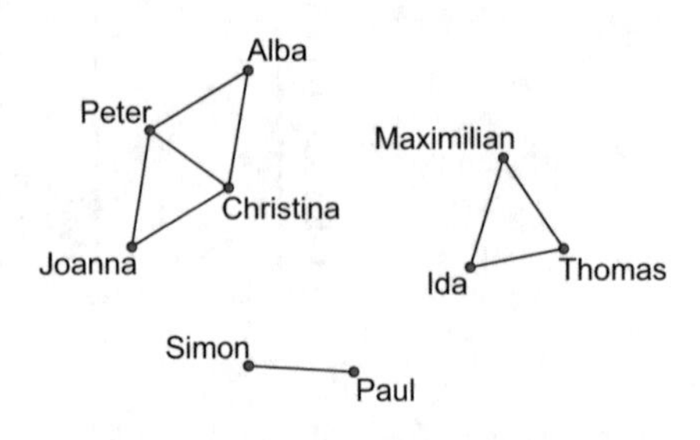

Fig. 3.3 Network visualization of *net*

From this visualisation, the inquiry of the human resource manager is very easy to answer. There are three persons who can fill in for Christina: Alba, Joanna, and Peter.

Listing 5 Manually conflating data of performance classes.

```
TEL=c("OU-CS", "UR-Informatics", "Linkedin-CS",
  "MOOC-PED","OU-PED", "MOOC-TEL","MOOC-ocTEL",
  "MOOC-LAK", "Facebook-TEL"
)
STATS=c("MOOC-Math", "OU-Statistics",
  "Facebook-Statistics"
)
im_new=cbind(
  rowSums( im[, TEL] ),
  rowSums( im[, STATS] )
)
colnames(im_new)=c("ALL-TEL", "ALL-STATS")
```

Let's assume now that 'TEL' is (as indicated in Chap. 1) an interdisciplinary area that connects amongst others 'CS' and 'PED'. This would then allow re-analysing the data presented in Table 3.1. For example, for the scope of looking at who can potentially deliver for the customer to 'TEL', both 'CS' and 'PED' are relevant and the data of all three can be conflated, as done with the code of Listing 5: the values of column 'TEL' are bound together with the sum of the other three columns into a new incidence matrix *im_new*.

The result of this operation is listed in Table 3.3: Paul previously had three out of the twelve competences, thus the aggregate ('ALL-TEL') results in a value of three. Maximilian had a value only for 'MOOC-Math' and 'Facebook-Statistics', thus now still listed with two.

For this *im_new* we can again calculate the adjacency matrix and plot the graph-converted data as a network visualisation analogously to Listing 2 and Listing 4.

Table 3.3 Conflated data

	ALL-TEL	ALL-STATS
Paul	*3*	*0*
Joanna	*3*	*0*
Maximilian	*0*	*2*
Peter	*4*	*0*
Christina	*5*	*0*
Simon	*2*	*0*
Ida	*0*	*2*
Thomas	*0*	*2*
Alba	*3*	*0*

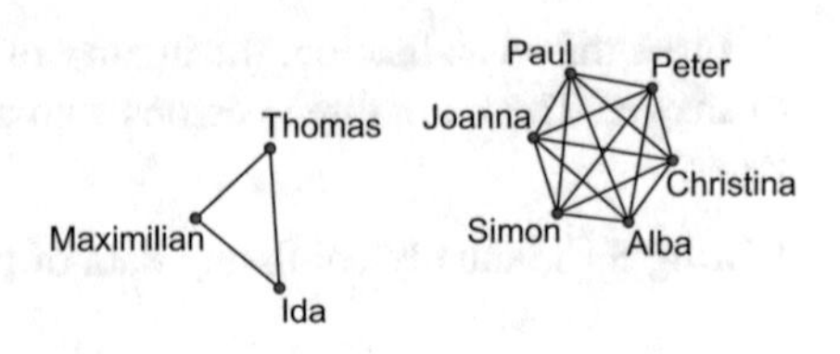

Fig. 3.4 Visualisation of the conflated adjacency data

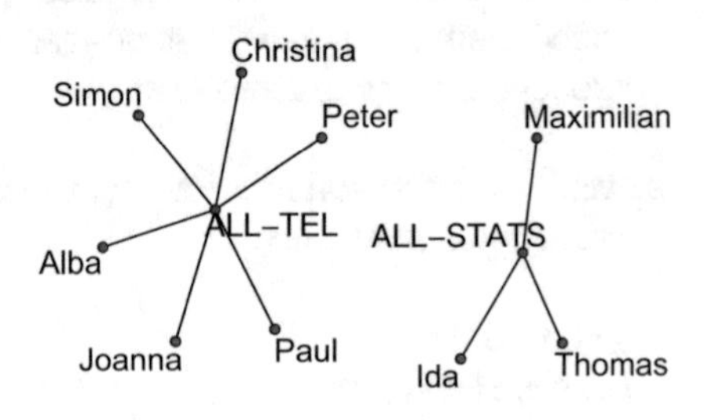

Fig. 3.5 Network plot for the bi-partite *im_new*

Now all the learners in the areas contributing to TEL share connections, as can be seen from the sociogram in Fig. 3.4.

This helps the human resource manager to discover additional potential replacements for the person off sick: Paul and Simon.

To visualise the bipartite network (as shown in Fig. 3.5) in order to display as nodes both competences as well as learners, the following code helps to force the conversion from the incidence matrix *im_new* into a bipartite graph retaining data of the two different modes (i.e., learners and competences).

Listing 6 Bi-partite graph from *im* and visualization.

```
net = network( im_new, directed = TRUE, bipartite = TRUE )
plot( net, displaylabels = TRUE )
```

Looking at Fig. 3.5, the nature of the relationships of learners to performance demonstrations of certain competences becomes more evident: seven learners are connected to 'ALL-TEL', whereas three show a disposition for 'ALL-STATS'.

The prototypical simple analysis process illustrated with this example is depicted in Fig. 3.6. The process starts off with filtering for those parts of the network, the analysis will focus on: in the case of the example this is done by filtering for nine persons and twelve competence demonstration contexts in the database.

The next step is to generate an incidence matrix, which is listed above in Table 3.1. Depending on whether focus is on the inspection of one-mode or two-mode data, this incidence matrix is either converted directly to the directed graph (Listing 6), or first an adjacency matrix is calculated (Listing 1) to then be converted to the undirected graph (Listing 2).

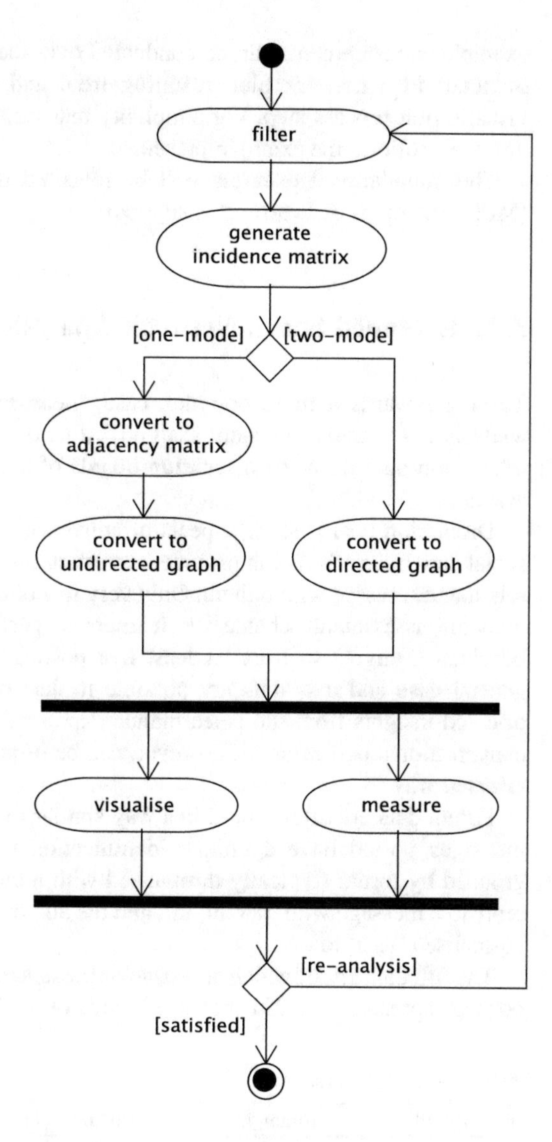

Fig. 3.6 Activity diagram of a simple social network analysis

Either way, measurements and visualisations can be conducted on the resulting graph data. The whole process is cyclic, and only when the analyst is satisfied and no further analysis is needed, the process concludes.

Depending on which software packages are used for extracting data from legacy systems,[2] for storing data internally, and for visualisation and measurement, some of these process steps may be hidden or their order may be even reversed: for

[2] Legacy systems are, e.g., the exam database of the learning management system of the Open University mentioned in the example above.

example, measurements can be conducted over the two-dimensional location data of nodes in a network plot, resulting from and not running in parallel to the visualisation process step. For simplicity reasons, the activity diagram presented, however, follows the example presented.

This foundational example will be revisited in Chap. 4 (LSA) and Chap. 9 (MPIA examples of Learning Analytics).

3.3 Extended Social Network Analysis Example

Turning towards a more complex case, measures and their interplay with the analysis and visualisation steps shall be further described with a realistic example[3]: interaction data from the discussion boards of a university learning management system.

Discussion boards show aspects of conversational performance in courses and social network analysis can provide interesting insights into the nature of the social relations expressed within them. Only very few of the courses investigated required with the assessment scheme the learners to perform in the forum (prescribing benchmark targets such as ‘at least five postings’). This, however, was not the general case and it is thus not possible to derive any summative, achievement-oriented insights from the performance expressed in the relational structure. The conversational performance, however, can be inspected in a formative, behaviour-oriented way.

Forum data are often stored in a way similar to what is shown in Table 3.4: the messages posted have a unique identification number ‘message_id’. They are grouped by forum (typically normalised with a unique ‘forum_id’), sometimes in reply to a message with ‘parent_id’, and the authoring user is typically stored by its normalised ‘user_id’.

The full data set, of which an extract is presented in Table 3.4, consists of 57.567 postings spread across 291 message boards of the learning management system of

Table 3.4 Sample forum data

message_id	forum_id	posting_date	parent_id	user_id
6355075	207398	Apr 24, 2006	6355031	4600
6355281	207398	Apr 24, 2006	6354978	3929
6355302	207398	Apr 24, 2006	6355281	4600
6355292	207398	Apr 24, 2006	6354978	3929
6355760	207398	Apr 24, 2006		4600
6361986	207398	Apr 25, 2006		4203

[3] The example presented has been made available by the author of this book on CRUNCH under the following URL: http://crunch.kmi.open.ac.uk/people/~fwild/services/forum-sna.Rmw

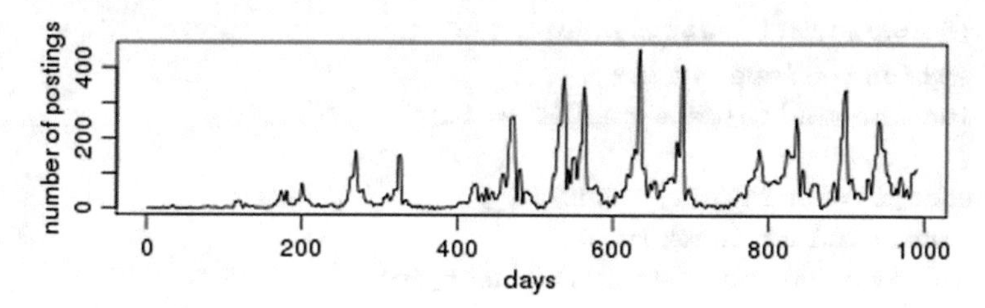

Fig. 3.7 Number of postings per day in the full data set. The frequencies depicted are slightly smoothened with Tukey's smoothers (Tukey 1977; implemented in the R base package 'stats') in order to emphasize the trend of the distribution more visible

the Vienna University of Economics and Business over the time frame from March 27, 2003, to April 25, 2006. All user names have been anonymised, giving them a unique alias that was randomly drawn from a generic first name database. For example, the user with the id 5780 was given the name 'Kerwinn'. No two users were given the same name and a mapping table against the original usernames was retained for analysis reasons.

In the 990 days on which messages were posted (of the total 1125 days in the given time frame), the number of postings per day varies, as can be seen from the frequency plot depicted in Fig. 3.7. Posting frequency is influenced by the lecture cycle of the academic year: for example, there are rather 'quiet' summer days visible in the chart from around day 362 to 388 (the month August in year 2004) and from 723 to 753 (the month August in year 2005).

Listing 7 Extraction of the incidence matrix from the raw data.

```
extractIncidences <- function(entries) {

  # all message ids
  message_ids = unique( c(
    entries$message_id,
    entries$parent_id
  ))

  # all users ids
  user_ids = unique( entries$user_id )

  # prepare and label an empty incidence matrix
  im = spMatrix(length(user_ids), length(message_ids))
  colnames(im) = message_ids
  rownames(im) = user_ids

  # populate the incidence matrix
  for (i in 1:nrow(entries)) {
```

```
      id=entries[i, "message_id"]
      # which row index in im?
      idcolnr=which(message_ids == id)

      user_id=entries[i, "user_id"]
      # which column index in im?
      uidrownr=which(user_ids == user_id)

      im[uidrownr,idcolnr]=im[uidrownr,idcolnr]+1
      parent_id=entries[i, "parent_id"]
      pidcolnr=which(message_ids == parent_id)
      if (!is.na(parent_id)) {
        im[uidrownr,pidcolnr] =
        im[uidrownr, pidcolnr]+1
      }

    } # end of for loop

    return(im)

} # end of function

im=extractIncidences(entries)
```

The first step of the analysis process is to evaluate the raw forum data, which therefore has to be parsed into an incidence matrix *im*: this matrix shall contain the message ids as incidences in the columns and the users as rows. For each message it is noted down which user posted it (i.e., its cell value is increased by +1). In case the message is a reply an additional incidence count is added (+1) for the id of the parent message. Listing 7 introduces a generalised function 'extractIncidences' to parse the forum data and assigns *im* to the incidence matrix extracted from the given forum data in its last line of code.

This extraction function works in the following way: First, an empty, sparse matrix is created with the right size. The right size in that case means that the matrix has as many columns, as there are distinct message ids (all distinct values of the union sets of message_id and parent_id, since messages can reply to messages that were posted earlier and which therefore are not included in the data set). Moreover, the matrix has as many rows, as there are distinct user ids.

The for loop in Listing 7 iterates through the forum raw data line by line, thereby adding +1 to each cell of the matrix, which (a) is at the row that stands for the user_id who authored the posting and at the column that stands for the message_id; and (b)—if the message is a reply and has a non-empty parent_id field—additionally adds +1 to the cell of the row reserved for the user_id and column reserved for the parent message id.

This sparse matrix *im* holds now all incidence data of the social network constituted by the message interaction: users in the rows, messages in the columns. Table 3.5 shows an extract of this incidence matrix: it lists which is involved in which message 'incident' either by originally posting the message or by replying to it.

The undirected adjacency matrix *am* can be calculated by multiplying the matrix with it's transposed self, see Listing 8. The new matrix *am* now holds data on how often each user interacted with each other user (i.e., is 'adjacent to'). To render the matrix more legible, instead of labelling rows and columns with the user ids, lines two and three of Listing 8 re-name the column and row labels to users' first names.

Listing 8 Calculating the adjacency matrix.

```
am = im %*% t(im)
colnames(am) = users[colnames(am)]
rownames(am) = users[rownames(am)]
```

This adjacency matrix *am* now looks like the extract presented in Table 3.6: the off-diagonal values are how often each user interacted with each other user. The values on the diagonal do not matter (and subsequent routines will ignore them): the matrix multiplication turns them into the sum of squares of the incidence values for this user. The user 'Meyer' listed in Table 3.6, for example, has—in that sample—been interacting only with Bendick.

Listing 9 shows how to convert the adjacency matrix *am* to a network object *net*: since *am* is a sparse matrix, this requires type casting to an object of class matrix.

Listing 9 Conversion to undirected network graph.

```
net = network(as.matrix(am), directed = FALSE)
```

Table 3.5 Sample of the incidence matrix

	3386743	5076632	3361197	5303512	6290712
2391	1	0	0	0	0
2917	0	1	0	0	0
3589	0	0	5	0	0
5792	0	0	0	2	0
3406	0	0	0	0	1

Table 3.6 Sample of the adjacency data

	Robbert	Lemmy	Bendick	Edsel	Meyer
Robbert	15	0	0	0	0
Lemmy	0	277	0	0	0
Bendick	0	0	2	0	1
Edsel	0	0	0	7	0
Meyer	0	0	1	0	212

As indicated above, the adjacency matrix still contains so-called loops (i.e., self references) along the diagonal. The network graph object *net* no longer stores these self references—and to remove them from the dataset, net can be simply casted back to an adjacency matrix (now without diagonal values), see Listing 10.[4]

Listing 10 Removing loops by type casting back to adjacency.

```
am=as.matrix.network(net, matrix.type="adjacency")
```

This social network is with more than 5.800 nodes still too big to process without considerate effort for a human analyst. Therefore, measuring operations help in gaining further insight. For example, basic measures show that although the graph is not very dense, it is very well connected: there are 79.993 edges in the graph, which is a low density of 0.005. Density of an undirected graph is calculated as the fraction of connected pairs (= unique edges) to the number of all 16,817,100 possible pairs (i.e. $\frac{n*(n-1)}{2}$). The Krackhardt connectedness score (Butts 2010; Krackhardt 1994) of the graph, however, is 88 %, see Listing 11, which indicates that the network is highly connected.

Listing 11 Examples of basic graph measures.

```
gden(net)
## [1] 0.004712664

connectedness(net)
## [1] 0.8792809
```

Though the overall graph is well connected, it contains a number of nodes that are not connected to any other node. This number of so-called isolates can be calculated as shown in Listing 12. It turns out, there are 316 among the 5827 nodes, which are not connected to other nodes. The function *which* returns the index values of those rows which have not a single non-zero value in their cells, thus summing up to a total of zero in their row sum.

Listing 12 Identifying isolates.

```
isolates=which(rowSums(am) == 0)
length(isolates)
## [1] 316
```

Since these isolates are not relevant for the visualisation of the network structure, they shall be excluded from the further analysis. This is done in Listing 13.

---

[4] More efficient (but for here also more complicated) would be to multiply the original adjacency matrix with an inverted diagonal matrix, thereby removing values from the diagonal.

Listing 13 Removing isolates from the adjacency matrix and graph.

```
am3=am[-isolates, -isolates]
net=network(am3, directed=FALSE)
```

Not all of the nodes are equally placed in the social network: some of them have a more exposed position. To investigate this, prestige scores are calculated for all nodes of the social graph, see Listing 14. Prestige is "the name collectively given to a range of centrality scores which focus on the extent to which one is nominated by others" (Butts 2010, function 'prestige'). High prestige scores indicate the "focal point[s] of communication" (Freeman 1979, p. 220), i.e. those nodes that are considered to be at the backbone of information flow. The default measure applied in the calculation of prestige in Listing 14 is the indegree centrality (Butts 2010, function 'degree'; Freeman 1979). The scores will be rescaled, i.e. they sum up to 1.

Listing 14 Calculating prestige scores.

```
prest=prestige(am3, rescale=TRUE)
```

From the rescaled values, it is possible to very quickly calculate an expansion factor for the node diameter in the visualisation (see Listing 15): each node's prestige score is multiplied with the total number of nodes (plus an offset of one, to avoid values below one); the logarithm of the resulting value (plus an offset of 0.5) is determined, thus giving the desired expansion factor.

The calculation of such expansion factor may vary from network to network and from analysis to analysis (see Chap. 5): in general it is recommended to avoid diversity in values (in order to allow the eye to distinguish different 'classes' of diameters) and to avoid scaling up node diameters to high (as then already a single node can clutter the whole display).

Listing 15 Calculating the node diameter for the visualization.

```
vcex=0.5+log( nrow(am3) * prest+1)
```

Working with rescaled prestige scores, introduces restrictions for the interpretation of the resulting network plot: the node diameter of two plots of two different networks cannot be measured quantitatively any more, as the diameter size now can only be interpreted as ordinals, relative to the size of the other nodes of the same plot. An example of such node diameters calculated is listed in Table 3.7.

Table 3.7 Sample of the node diameters calculated

Node	Diameter
Robbert	0.6
Lemmy	1
Bendick	0.7
Edsel	0.6
Meyer	1.1

This can then be visualised with the code of Listing 16, thereby using—per default—a Fruchterman-Reingold layout algorithm (Fruchterman and Reingold 1991) that places nodes in a spring-embedder system trying to optimise the mapping to the two-dimensional plane so that it reflects as closely as possible the underlying graph structure. More details about this planar projection will be described in Sect. 6.2, where also an extension of it (Kamada and Kawai 1989) will be discussed.

The parameter 'vertex.cex' sets the expansion factor for the node diameters.

Listing 16 Visualising the network.

```
plot( net, displaylabels=FALSE, vertex.cex=vcex )
```

The resulting visualization is depicted in Fig. 3.8, labels being suppressed to focus on the overall Gestalt of the network visualisation. As the figure shows, the resulting social graph is still a too densely connected one and does not allow for much of an insight into the underlying organisation structure.

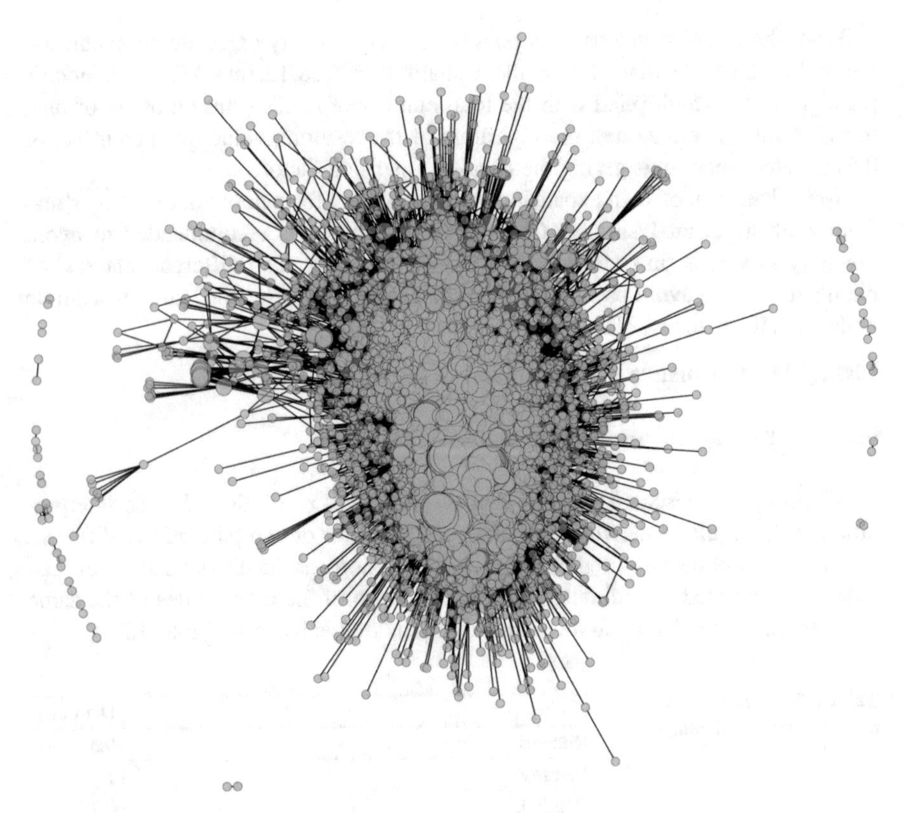

Fig. 3.8 Network plot depicting the overall Gestalt of the graph (full forum data with isolates removed)

A component analysis can help in zooming in on a particular component of the graph. For example, it is possible to focus on the social relations for a particular forum by extracting only those messages that have a particular forum_id.

To filter out an interesting forum for the further analysis, the number of messages per forum can be calculated as shown in Listing 17.

Listing 17 Calculation of the number of messages per forum.

```
msgDist=table(entries[, "forum_id"])
length( which(msgDist>1000) )
## [1] 13
```

As can be seen from the diagram depicted in Fig. 3.9 created with the code of Listing 18, there are only a few message boards that hold a larger number of postings: It turns out, there are 13 fora (of the total 291) that hold more than 1000 messages. The mean number of messages per forum is 198 with a standard deviation of 788 messages.

Listing 18 Visualising the number of messages per forum.

```
plot(msgDist, xlab="forum id",ylab="number of messages")
```

An example of a component analysis shall be demonstrated subsequently by zooming in on a single forum: the forum with the id '211611' is the forum of the course 'Financing I'. Listing 19 helps in extracting the entries from the raw forum data for this forum id. It re-uses the function 'extractIncidences' from above (Listing 7) to convert the raw data to a sparse incidence matrix.

Listing 19 Extract message data for a single forum.

```
fid=211611
eFin=entries[which(entries$forum_id==fid), ]
imFin=extractIncidences(eFin)
```

Listing 20 below provides the code needed to evaluate the incidence data into an adjacency matrix. The last two lines of the listing assure that the rows and columns are labelled with the users' first names (rather than their user ids).

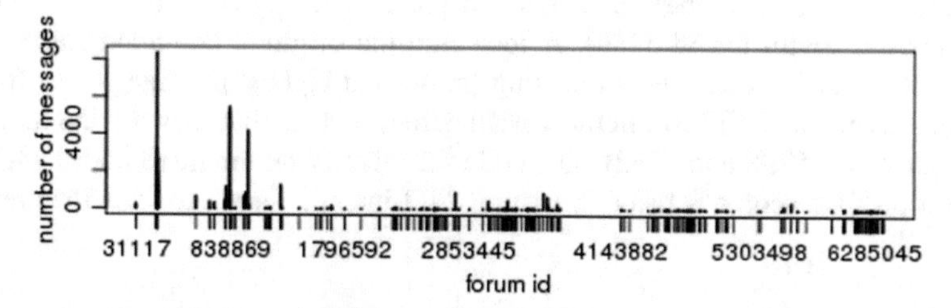

Fig. 3.9 Visualisation of the number of messages per forum

Listing 20 Calculating the adjacency matrix for this forum.

```
amFin=imFin %*% t(imFin)
colnames(amFin)=users[colnames(amFin)]
rownames(amFin)=users[rownames(amFin)]
```

The code in Listing 21 converts the adjacency matrix to a network graph (line one). Typecasting back to an adjacency matrix removes self-referential loops and helps to remove isolates, i.e. nodes that do not have any connection to any other node: since nodes are listed both in rows and columns, this identification can be done via the sum of the row values. If the node is not connected, its row sum will be zero.

Listing 21 Conversion to network graph and isolate removal.

```
netFin=network(as.matrix(amFin), directed=FALSE)
amFin=as.matrix.network(netFin,
  matrix.type="adjacency")
isolatesFin=which(rowSums(amFin) == 0)
am3Fin=amFin[-isolatesFin, -isolatesFin]
netFin=network(am3Fin, directed=FALSE)
```

Looking ahead at the visualisation performed in Listing 23, the degree centrality scores and a convenient expansion factor for the node diameters in the network visualisation is calculated in Listing 22 (in analogy to Listing 14 and Listing 15, see above).

Listing 22 Prestige scores and node diameter expansion factor.

```
prestFin=prestige(am3Fin, rescale=TRUE)
vcexFin=0.5+log( nrow(am3Fin) * prestFin+1 )
```

Listing 23 Visualisation of the social graph of the forum.

```
plot(netFin, vertex.cex=vcexFin)
```

In the visualisation presented in Fig. 3.10, the node with the highest prestige score is standing out in the centre of the display: the anonymised name of the node is ‘Kerwinn’ (with the id 5780). A look into the original, non-anonymised data reveals that this is a teacher. Inspecting the second highest prestige scores (nodes labelled Lynett and Yehudit in the visualization) reveals that they are the tutors of the course (ids 5798 and 5762). The code for identifying the three nodes with the highest prestige scores is listed in Listing 24. Line one therefore sorts the prestige

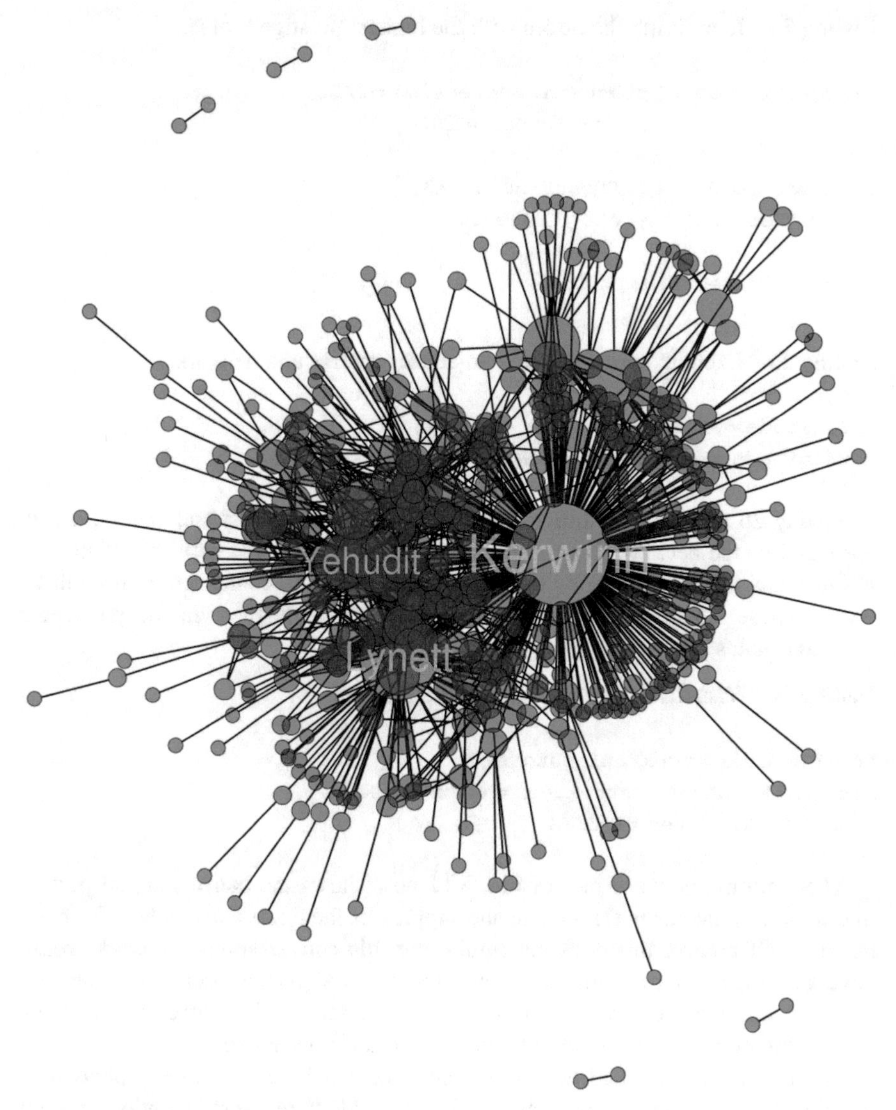

Fig. 3.10 Network plot of the sub-graph of the forum 'Financing I' (nodes scaled by prestige with the biggest three nodes labelled)

scores (while returning the sort index into the variable *ixPrestFin$ix*). Line two returns the names of these three nodes, while the last line of the listing stores their row and column index into *tutors*.

The rows and columns with the indexes of the three tutors can be dropped from the adjacency matrix as shown in Listing 25.

Listing 24 Identifying the nodes with the highest prestige scores.

```
ixPrestFin = sort( prestFin, decreasing = TRUE,
                   index.return = TRUE )

colnames(am3Fin)[ ixPrestFin$ix[1:3] ]
## [1] "Kerwinn" "Lynett" "Yehudit"

tutors = ixPrestFin$ix[1:3]
```

Listing 25 Exclude tutors from the adjacency matrix and network.

```
amFinStuds = am3Fin[-tutors, -tutors]
netFinStuds = network(amFinStuds, directed = FALSE)
```

Listing 26 removes the tutors' node diameters from the list and, subsequently, re-visualises the network. By cutting the three tutors from the network, additional isolates may have been created that were previously only connected to those three tutors. These isolate nodes can be suppressed from display via the parameter 'displayisolates'.

Listing 26 Visualising the network (without tutors).

```
vcexFinStuds = vcexFin[-tutors]
plot(netFinStuds, vertex.cex = vcexFinStuds,
  displayisolates = FALSE)
```

The resulting network plot of Fig. 3.11 now shows the conversational performance of the students in this course and suppresses the tutors and teachers from the network. Of course, this does not imply that this conversational network would have developed in the same way, had the facilitators not been there. On the contrary, through their instruction, the tutors and teacher influenced the students in showing certain (performant and non-performant) behaviour.

The examples presented here show, how the relational structure expressed in learning conversations can be analysed. It is possible to measure behaviour in such social network with structural coefficients that indicate, for example, how well connected the learners are or whether they positioned themselves in central positions in the social graph. Through filtering by the context of the courses the fora are associated with, it is possible to look into purposive relational data: the example presented shows, how learners interact in the context of the financing introductory course.

The analysis process of this example thereby is a variation of the simple process presented in Fig. 3.6 on page 14: raw data is manipulated (e.g. filtered by forum id), incidence matrices are generated, from where the adjacency data and network

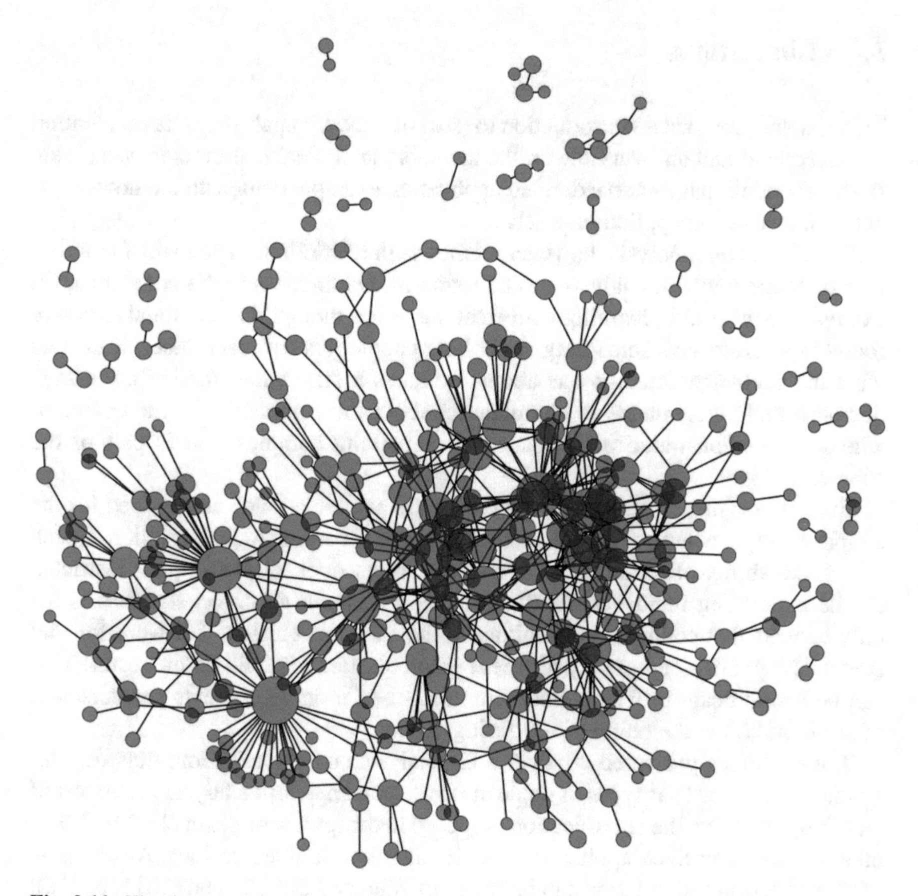

Fig. 3.11 The dense graph collapses, when removing the three tutors

graphs can be derived. Measuring and visualisation alternate, thereby informing the filtering of data in further analysis.

There are several specifications for exchange formats of graph data available that support the analyst in storing and sharing interim results of the analysis process. Regarding standards proposals, these are most notably Harwell-Boing (Duff et al. 1992, p. 7ff), the coordinate format for sparse and array format for dense matrices of the Matrixmarket exchange format (Boisvert et al. 1996, p. 2), and graphML (Brandes et al. 2004, p. 2; Brandes and Erlebach 2005). Several software packages provide their own, proprietary exchange and storage formats. The language and environment R additionally provides generic routines for storing native data objects via the 'save', 'load', and 'data' interfaces (R Core Team 2014).

3.4 Limitations

This chapter provided an introduction to (social) network analysis. Data preparation was described and an overview on the available analytical instruments was given. Both a foundational and an extended application example demonstrated how social network analysis is applied in practice.

Social network analysis, however, comes with restrictions. The main shortcomings of social network analysis can be found in its blindness to the content as, for example, expressed in learning conversations: even though the relational structure found in a forum says something about how people relate to each other, it matters what the exchange actually was about. It makes a difference whether a message demonstrated performance in complaining about the complexity of the course or whether it demonstrated performance of something complex that is part of the course.

Even where more data about the content are available (such as assumed for the simple example when course contexts are merged according to their relation to each other), the ability of social network analysis to discover relationships is restricted by the underlying incidence distribution: the Facebook group on statistics is the only context that connects Maximilian to the statistics cluster. If it were for that single data point, a completely different graph clustering would be the result! This can be a significant problem, since the number of training contexts per person is rather limited, but the number of offerings is not.

The examples presented conduct both analysis and visualisation, thus demonstrating two of the four types of application of social network analysis in support of learning, following the classification proposed in the review of Sie et al. (2012). The other two types of SNA application are simulations and interventions. An example of using SNA in a simulation can be found in Wild and Sigurdarson (2011, p. 412): the authors use a simulation model of a blog network among university students to examine the potential effects of certain pedagogical interventions.

The examples presented in this section show how purposive action can be analysed with the instruments provided by social network analysis, but they also clearly point out the main shortcoming found in the lack of instruments to analyse the semantics of the underlying 'conversations'.

If the source data does not already contain semantic relations between people and purposive contexts, it cannot be investigated. Even when available, maintenance of data often becomes a problem: Over time and with a growing number of people contributing data, errors and inconsistencies are introduced at the cost of either quality or resources.

The technique of choice, latent semantic analysis, for creating these missing data structures automatically will be described in the subsequent chapter. Since latent semantic analysis then again lacks facilities for complex analyses of the resulting graph structures and since such analyses are not trivial, the fusion of both instruments into meaningful, purposive interaction analysis (MPIA) will be described in Chap. 5.

References

Barash, V., Golder, S.: Twitter: conversation, entertainment, and information, all in one network! In: Hansen, D.L., Shneiderman, B., Smith, M. (eds.) Analyzing Social Media Networks with NodeXL, pp. 143–164. Morgan Kaufmann, Burlington, MA (2011)

Boisvert, R., Pozo, R., Remington, K.: The Matrix Market Exchange Formats: Initial Design, Internal Report, NISTIR 5935. National Institute of Standards and Technology (1996)

Brandes, U., Erlebach, T.: Network Analysis: Methodological Foundations. Springer, Berlin (2005)

Brandes, U., Eiglsperger, M., Lerner, J., Pich, C.: Chapter 18: graph Markup Language (GraphML). In: Tamassia, R. (ed.) Handbook of Graph Drawing and Visualization. Chapman and Hall/CRC, Boca Raton, FL (2004)

Butts, C.T.: Network: a package for managing relational data in R. J. Stat. Softw. **24**(2), 1–36 (2008)

Butts, C.T.: sna: Tools for Social Network Analysis, R Package Version 2.2-0. http://CRAN.R-project.org/package=sna (2010)

Butts, C.T., Hunter, D., Handcock, M.S.: Network: Classes for Relational Data, R Package Version 1.7-1, Irvine, CA. http://statnet.org/ (2012)

Carrington, P., Scott, J.: Introduction. In: Carrington, P.J., Scott, J. (eds.) The SAGE Handbook of Social Network Analysis. Sage, Los Angeles, CA (2011)

Csardi, G., Nepusz, T.: The igraph software package for complex network research. In: InterJournal: Complex Systems (CX.18), Manuscript Number 1695 (2006)

Duff, I., Grimes, R., Lewis, J.: Users' Guide for the Harwell-Boeing Sparse Matrix Collection (Release 1). ftp://math.nist.gov/pub/MatrixMarket2/Harwell-Boeing/hb-userguide.ps.gz (1992)

Freeman, L.: Centrality in social networks. Conceptual clarification. Soc. Networks **1**, 215–239 (1979)

Freeman, L.: Some antecedents of social network analysis. Connections **19**(1), 39–42 (1996)

Freeman, L.: The Development of Social Network Analysis: A Study in the Sociology of Science. Empirical Press, Vancouver (2004)

Fruchterman, T., Reingold, E.: Graph drawing by force-directed placement. Software. Pract. Exper. **21**(11), 1129–1164 (1991)

Google: Google N-Gram Viewer, Query 'social network analysis, latent semantic analysis', from 1900 to 2011, Corpus 'English' (20120701), Smoothing of 3. http://books.google.com/ngrams/graph?content=social+network+analysis%2Clatent+semantic+analysis&year_start=1900&year_end=2008&corpus=15&smoothing=3&share= (2012)

Kamada, T., Kawai, S.: An algorithm for drawing general undirected graphs. Inf. Process. Lett. **31**, 7–15 (1989)

Klamma, R., Spaniol, M., Denev, D.: PALADIN: a pattern based approach to knowledge discovery in digital social networks. In: Tochtermann, K., Maurer H. (eds.) Proceedings of 6th International Conference on Knowledge Management (IKNOW-06), pp. 457–464. Springer, Berlin (2006)

Klamma, R., Petrushyna, Z.: The troll under the bridge: data management for huge web science mediabases. In: Hegering, Lehmann, Ohlbach, Scheideler (eds.) Proceedings of the 38. Jahrestagung der Gesellschaft für Informatik e.V. (GI), INFORMATIK 2008, pp. 923–928. Köllen Druck + Verlag GmbH, Bonn (2008)

Krackhardt, D.: Graph theoretical dimensions of informal organizations. In: Carley, K., Prietula, M. (eds.) Computational Organization Theory, pp. 89–111. Lawrence Erlbaum and Associates, Hillsdale, NJ (1994)

Lin, Y., Michel, J., Lieberman Aiden, E., Orwant, J., Brockman, W., Petrov, S.: Syntactic annotations for the Google books Ngram corpus. In: Proceedings of the 50th Annual Meeting of the Association for Computational Linguistics, pp. 169–174. Jeju, Republic of Korea, 8–14 July 2012 (2012)

Michel, J., Shen, Y., Presser Aiden, A., Veres, A., Gray, M., The Google Books Team, Pickett, J., Hoiberg, D., Clancy, D., Norvig, P., Orwant, J., Pinker, S., Nowak, M., Lieberman Aiden, E.: Quantitative analysis of culture using millions of digitized books. Science **331**, 176 (2011)

Moreno, J.L.: Who Shall Survive? vol. xvi. Nervous and Mental Disease Publishing, Washington, DC (1934)

Mutschke, P.: Autorennetzwerke: Verfahren der Netzwerkanalyse als Mehrwertdienste für Informationssysteme, IZ-Arbeitsbericht Nr. 32, Informationszentrum Sozialwissenschaften der Arbeitsgemeinschaft Sozialwissenschaftlicher Institute e.V. (2004)

R Core Team: R: A Language and Environment for Statistical Computing, R Foundation for Statistical Computing, Vienna, Austria. ISBN 3-900051-07-0. http://www.R-project.org/ (2014)

Scott, J.: Social Network Analysis: A Handbook. Sage, London (2000)

Sie, R., Ullmann, T., Rajagopal, K., Cela, K., Bitter-Rijpkema, M., Sloep, P.: Social network analysis for technology-enhanced learning: review and future directions. Int. J. Technol. Enhanc. Learn. **4**(3/4), 172–190 (2012)

Tukey, J.W.: Exploratory Data Analysis. Addison-Wesley, Reading, MA (1977)

Wild, F., Sigurdarson, S.: Simulating learning networks in a higher education blogosphere—at scale. In: Delgado Kloos, C., et al. (eds.) EC-TEL 2011. LNCS 6964, pp. 412–423. Springer, Berlin (2011)

Chapter 4
Representing and Analysing Meaning with LSA

Semantics,—the study of meaning communicated through language (Saeed 2009)—, is usually defined to investigate the relation of signs to the objects they represent.

Such representation of objects and their relations, interactions, properties, as well as states can be created in various ways. For example, they emerge naturally in form of neural activity in the human brain. Expressing thoughts again in language and other formalisms materialises them intellectually. And they can be created automatically using data processing techniques and computation. One branch of such automated techniques for generating semantic representations uses the co-occurrence of the words in language to derive information on the semantic structure. This branch is often entitled 'heuristics-based approaches'. More recently, they are also called 'distributional semantics' (Sahlgren 2008). Latent Semantic Analysis (LSA) is one of the methods in this branch. LSA was introduced to facilitate the investigation of meaning in texts, originally in the context of indexing and information retrieval.

While Social Network Analysis, as presented in the previous chapter, is a powerful instrument to represent and analyse the purposive context of learning activity, Latent Semantic analysis is blind to such social and relational aspects. LSA lacks the elaborate instruments and measures provided by network analyses to further investigate the characteristics of structure found. Moreover, no clear guidance is provided on determining *before calculation* an optimal number of singular values to retain in the truncation of the dimensional system resolved.

Still, it is a time-tested algorithm for representing and analysing meaning from text, with its closeness in mathematical foundation being a natural candidate for further integration (see Chap. 5). These foundations as well as the analysis workflow with the *lsa* package developed and standard use cases are following in the Sects. 4.1 and 4.2.

Two demos are used within this chapter to foster understanding and allow derivation of the main restrictions applying to LSA. The foundational example presented in Sect. 4.3 picks up the usage scenario of the foundational SNA demo

F. Wild, *Learning Analytics in R with SNA, LSA, and MPIA*,
DOI 10.1007/978-3-319-28791-1_4

presented in Sect. 3.2. It will be revisited in the Chapter on application examples for MPIA (Sect. 9.2).

Following the summary of the state of the art in application of LSA to technology-enhanced learning in Sect. 4.4, a second, real-life application example in essay scoring will be added in Sect. 4.5. A summary outlining also the key limitations of LSA concludes this chapter in Sect. 4.6.

The academic discourse around latent semantic analysis started even more recently than that around social network analysis, as can be seen from the line plot depicted at the beginning of this the previous chapter in Fig. 3.1. The expression was coined and the method was developed by a group of researchers at the Bell Laboratories in the late 1980s (Deerwester et al. 1990) as an attempt to overcome synonymy and polysemy problems of—at that time—state-of-the-art information retrieval and navigation systems, following their interest in statistical semantic representation techniques expressed already in earlier articles (e.g. Furnas et al. 1983). The original patent for latent semantic analysis was granted in the US (Deerwester et al., patented 1989, filed 1988). The project leader of the group was Landauer (Landauer et al. 2008, p. ix).

The initial interest in LSA's application areas was on indexing, primarily in information retrieval (see, e.g., Dumais 1992). Though more complicated indexing applications follow soon: the Bellcore Advisor, for example, is an indexing system that uses technical memos to map human expertise by domains (Dumais et al. 1988, p. 285). From there focus broadens over the next years to encompass additional application areas such as information filtering (Foltz 1990), as a measure for textual comprehension (Foltz et al. 1998, p. 304), and technology-enhanced learning. Landauer et al. (1997), for example, investigated how well LSA can be used to evaluate essays. In two experiments with 94 and 273 participants, results of the LSA-based measures were found to have near human performance or to even outperform the human raters in their correlation to a "40 point short answer test" (p. 413 and p. 416). Foltz (1996, p. 200) finds that "grading done by LSA is about as reliable as that of the graders" in an experiment with four human graders and 24 essays.

In a special issue edited by Foltz (1998, p. 128, 129) about "quantitative approaches for semantic knowledge representation", a rich body of technology-enhanced learning applications is introduced: Landauer et al. (1998a) in particular describe several different methods for scoring essays with LSA (p. 279) and for assigning appropriate instructional material to learners (p. 280).

In the same issue, Rehder et al. (1998) investigate how the following set of influencing factors impact on scoring: focusing solely on the technical vocabulary (ignoring non-technical words) did not improve scoring, an essay length of 200 words was (under the conditions of the study) related to result in the best prediction of scores. Furthermore, the article finds both the cosine distance between the vectors representing model solution and essay, as well as the vector length of the essay itself to be important components in predicting scores.

Wolfe et al. (1998)—in the same special issue—investigate how text complexity influences learning, testing this with undergraduates as well as medical students and assessing high reliability and validity (p. 331). Different assessment tests were conducted: an assessment questionnaire was developed and scored, and the essays were evaluated with a score by professional graders from educational testing service. The cosine between a model solution and the essay was used as the LSA measure. The measure correlated with $r = 0.63$ to essay score and $r = 0.68$ to questionnaire score. This was quite similar to $r = 0.74$, with which the human grader score correlated to questionnaire score (and the $r = 0.77$ of the interrater correlation between the human evaluators), therefore leading Wolfe et al. to conclude that likely all three "were measuring largely the same thing" (p. 331).

4.1 Mathematical Foundations

The mathematical and statistical foundations are made accessible in a series of publications, most notably the seminal article Deerwester et al. (1990) and Berry et al. (1995).

The basic working principle of latent semantic analysis is to map texts to their bag-of-words vector space representation (of document vectors along term 'axes' and term vectors along document 'axes') and then rotate (and scale) the axes of the resulting vector space according to the insights gained from a two-mode factor analysis—using singular value decomposition. The rotation and scaling is done in such way that a new dimensional system factorizes those term and document vectors together along a new system of coordinate axes that appear frequently together. Typically, the resulting coordinate system is an approximation of the original vector space, deliberately neglecting the term and document loadings onto lower ranking factors.

The assumption behind this truncation of lower ranking factors is that it compensates for synonymy and other forms of word variation with same intended meaning, as this is suppressed through the approximation. At the same time it is assumed that the factorisation as such deals with polysemy and homonymy, splitting differing usage contexts across different factors or combinations thereof. This way, the resulting higher-order LSA vector space is assumed to better reflect the (latent) semantic structure previously obscured by the variability in word use.

In more detail, the following steps need to be conducted to produce a latent semantic analysis. First, a document-term matrix M is constructed from a given collection of n documents containing m terms (see Fig. 4.1). This matrix, also called 'text matrix', has m rows (representing the $t_1, t_2, \ldots, t_m$ terms) and n columns (representing the $d_1, d_2, \ldots, d_n$ documents of the corpus), thus denoting in its cells the frequency with which each term appears in each document. This text matrix M is typically sparse, i.e. few of the cells contain values greater than 0. The text matrix M holds the basic vector space of the corpus.

This text matrix M of size $m \times n$ is then resolved with *singular value decomposition* into three constituent matrices T, S, and D, such that their product is M (see Fig. 4.2). The constituent T thereby holds the left-singular vectors (term vector

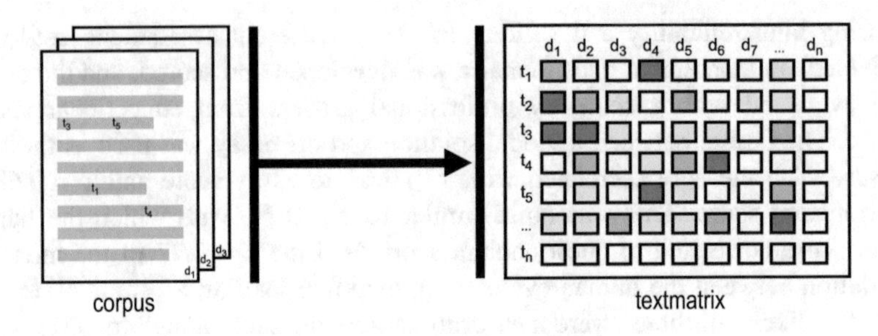

Fig. 4.1 Mapping of document collection to document-term matrix

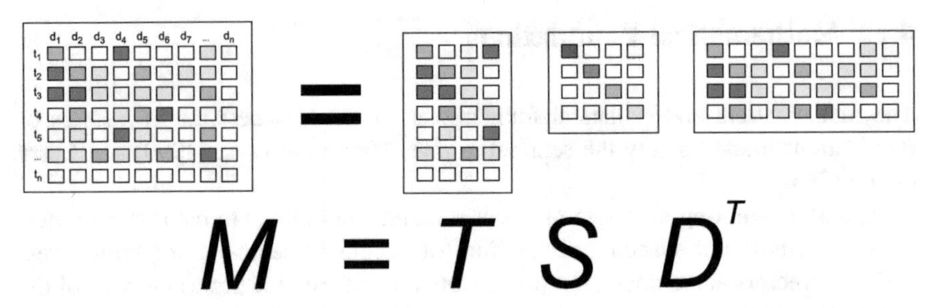

Fig. 4.2 Singular value decomposition (SVD, own graphic, modified from Wild et al. 2005b)

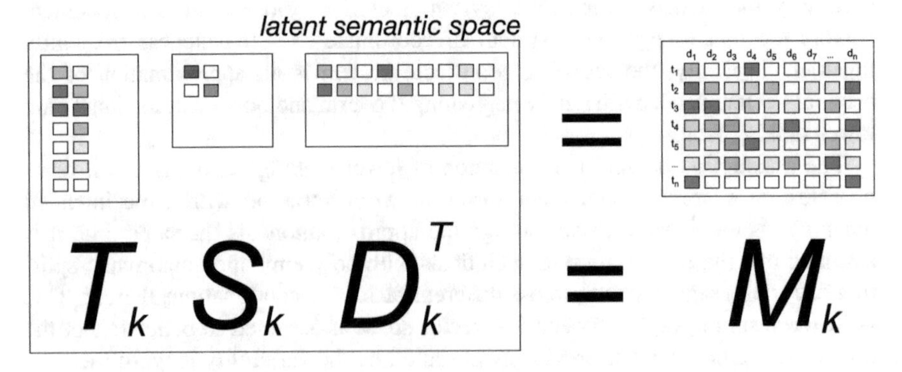

Fig. 4.3 Factor reduction (own graphic, modified from Wild et al. 2005b)

'loadings' onto the singular values in S) and the constituent D holds the right-singular vectors (the document vector 'loadings'). S is the orthonormal, diagonal matrix, listing the square roots of the eigenvalues of MM^T and M^TM (in descending order).

The diagonal matrix S is subsequently truncated to k diagonal values, effectively truncating T to T_k and D to D_k as well (see Fig. 4.3). This set of truncated matrices T_k, S_k, and D_k establishes the latent semantic space—the least-squares best-fit approximation of M with k singular values.

Multiplying T_k, S_k, and D_k^T produces the text matrix M_k, which is of the same format as M: rows representing the same terms and columns representing the same documents. The cells, however, now contain corrected frequencies that better reflect the (latent) semantic structure.

Since M_k is no longer sparse (other than M), in many cases it can be more memory efficient, to just hold the truncated space matrices to produce only those text matrix vectors in M_k that are actually needed.

The computationally costly part of the latent semantic analysis process is the step applying in the singular value decomposition. With rising corpus size through a larger number of documents and the typically resulting larger number of terms connected to it, the computation has a complexity of up to $O(mn \min\{m,n\})$, depending on the actual algorithm and basic linear algebra implementations used (see Menon and Elkan 2011, for a comparison of SVD algorithm complexity).

To avoid the bulk of calculations for the singular value decomposition, it is possible to project new documents into an existing latent semantic space—a process called 'folding in' (Berry et al. 1995, p. 577). This is particularly useful, where it is necessary to keep additional documents from changing the previously calculated factor distribution—for example, when evaluating low-scored student essays. Given that the reference corpus from which the latent semantic analysis is calculated were representative, such projection produces identical results.

To conduct a *fold-in*, the following equations need to be resolved [see Berry et al. 1995, p. 577, particularly Eq. (7)]. First, a document vector v needs to be constructed for the additional documents, listing their term occurrence frequencies along the controlled (and ordered!) vocabulary provided by T_k (and shared by M and M_k). This document vector v is effectively an additional column to the input text matrix M. The projections target document vector m' is then calculated by applying Eqs. (4.1) and (4.2), effectively mapping v to a new right-singular vector d' in D_k [Eq. (4.1)]—and then to a document-term vector m' in M_k [Eq. (4.2)].

$$d' = v^{\mathrm{T}} T_k S_k^{-1} \tag{4.1}$$

$$m' = T_k S_k d'^T \tag{4.2}$$

T_k and S_k thereby refer to the truncated space matrices from the existing latent semantic space.

Using fold in or not, the latent semantic space and its vectors allows for several ways to conduct *proximity measurement* of how close certain documents, terms, or documents and terms are.

Evaluations can be performed utilising the truncated partial matrices of the latent semantic space or—less memory efficient—in the re-multiplied text matrix M_k that reflects the underlying latent semantic structure. Same as in the 'pure' vector space model, various proximity measurement algorithms can be used: the cosine, for example, utilises the angle between vectors to provide a measure for their relatedness [see Fig. 4.4 and Eq. (4.3)].

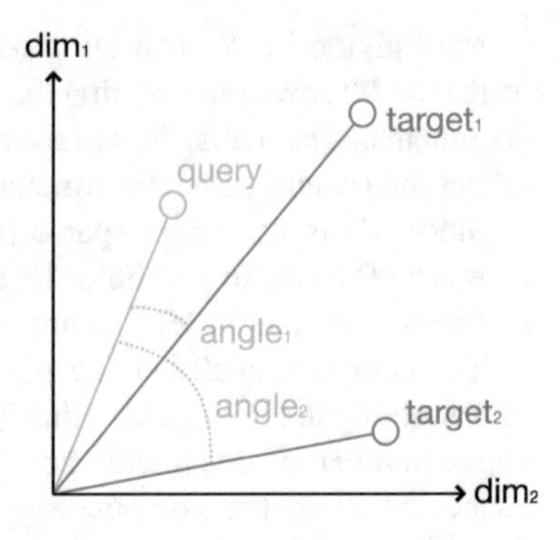

Fig. 4.4 Cosine proximity measurement of a query vector to two target vectors

When performed over the reconstituted text matrix M_k, the dimensions depicted in the figure relate to the terms (for term-to-term comparisons) and documents (for document-to-document comparisons). When performed in the latent semantic space, the dimensions refer to the factors of the singular value decomposition.

Other popular measures (see Leydesdorff 2005; Klavans and Boyack 2006; Tao and Zhai 2007) include, for example, Pearson's r, Euclidian distances, and the Jaccard coefficient. One of the advantages of the cosine measure is the reduced sensitivity for zeros (Leydesdorff 2005), a difference coming to effect particularly for large, sparse textmatrices.

$$\cos \propto = \frac{\sum_{i=1}^{m} a_i b_i}{\sqrt{\sum_{i=1}^{m} a_i^2}\sqrt{\sum_{i=1}^{m} b_i^2}} \tag{4.3}$$

The interpretation of value ranges provided by any of the measures depends largely on the underlying data. Given that the latent semantic space is valid in providing a representation of the meaning structures under investigation, high proximity values in the vector space between terms, documents, or both indicate associative closeness.

Only very high values indicate identity, whereas lower positive proximity values can be seen as to indicate whether certain features are associated, i.e., whether they are likely to appear in the same contexts.

For example, although the words 'wolf' and 'dog' are *semantically* very close, in a generic newspaper corpus, however, they cannot be expected to be *associatively* very close (not least to the widespread metaphoric use of 'wolf' and the rare use of wolfs as pets): it is much more likely that 'dog' will be found in closer proximity to any other popular pet (such as 'cat'), though not identical to them.

Landauer and Dumais (1997) demonstrated, that latent semantic spaces can be trained to perform a synonym test on a level required for the admission to U.S. universities in the Test Of English as a Foreign Language (TOEFL). Their findings, however, should not be overrated: it remains largely a question of selecting and sampling a representative corpus (and space) for the domain of interest.

4.2 Analysis Workflow with the R Package 'lsa'

In support of this book, the author has implemented the *lsa* package for R as Open Source (Wild 2014). This subsection provides an overview on the functionality covered by the package. More detailed documentation of the individual package routines can be found in Annex C.

The workflow of an analyst applying latent semantic analysis is depicted in Fig. 4.5. An 'analyst' thereby refers to any user of LSA, who applies the technique to investigate meaning structures of texts, for example being learner, tutor, teacher, faculty administrator, system designer, researcher, or the like.

Typically, analysis involves a set of filtering and other pre-processing operations (weighting, sanitising) in order to select and prepare the document collection to map to a text matrix, the calculation of the LSA space (singular value decomposition and factor truncation), and then—subsequently—the application of similarity measurement operations in interchange with further filtering of document and term vectors for the desired scope of analysis. This latter analysis may or may not involve folding in of additional documents.

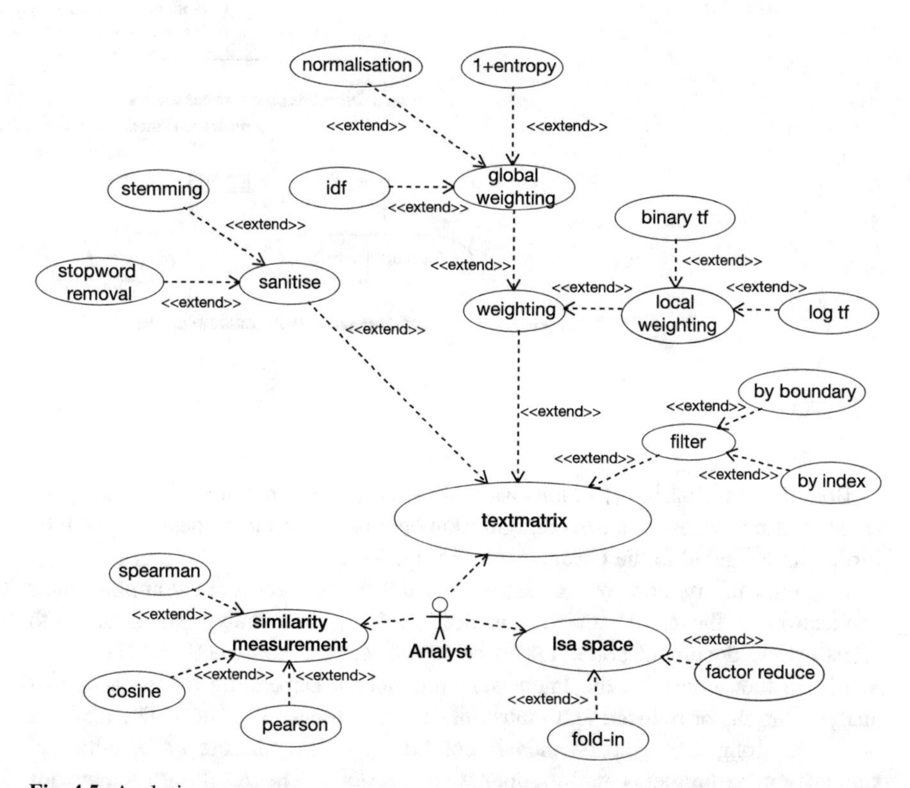

Fig. 4.5 Analysis use cases

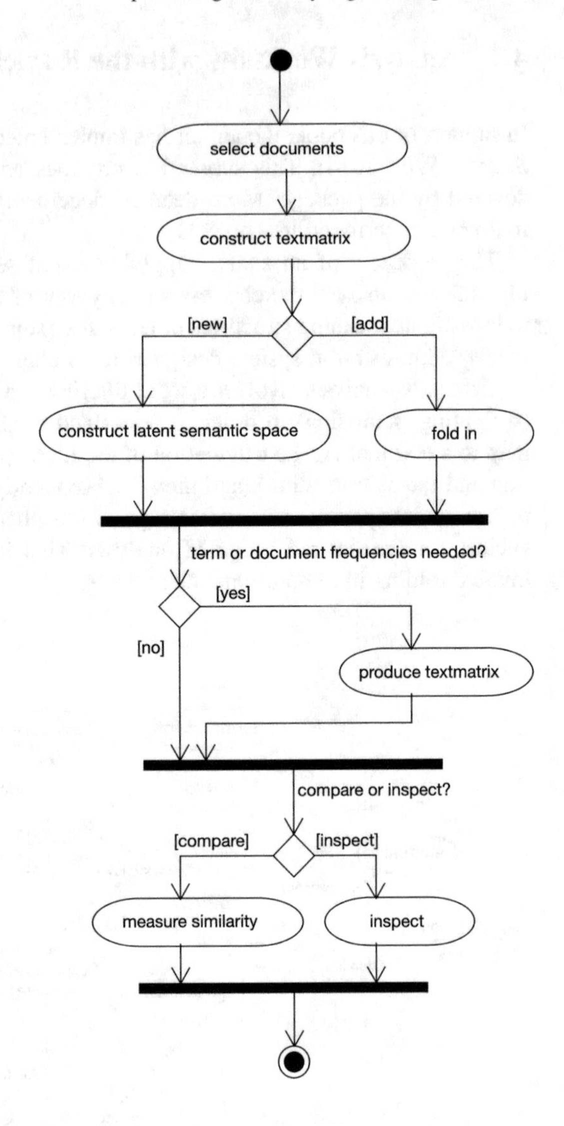

Fig. 4.6 Activity diagram of a simple latent semantic analysis

Both the text matrix operations as well as the factor reduction of the lsa space creation offer a wide choice in configuration options. Their interdependency will be further investigated in the Chaps. 5 and 7 (Fig. 4.6).

Although the number of use cases depicted in Fig. 4.5 looks complex, their application in the actual analysis is not. Typically, an analyst starts off with selecting the document collection to be analysed, then constructs the 'raw' text matrix to then construct the latent semantic space. Depending on the scope of analysis the factor reduced text matrix M_k can be produced (or not). The analysis of the resulting vector space then is conducted with a mixture of or either of similarity measurements and frequency inspections. The resulting frequencies

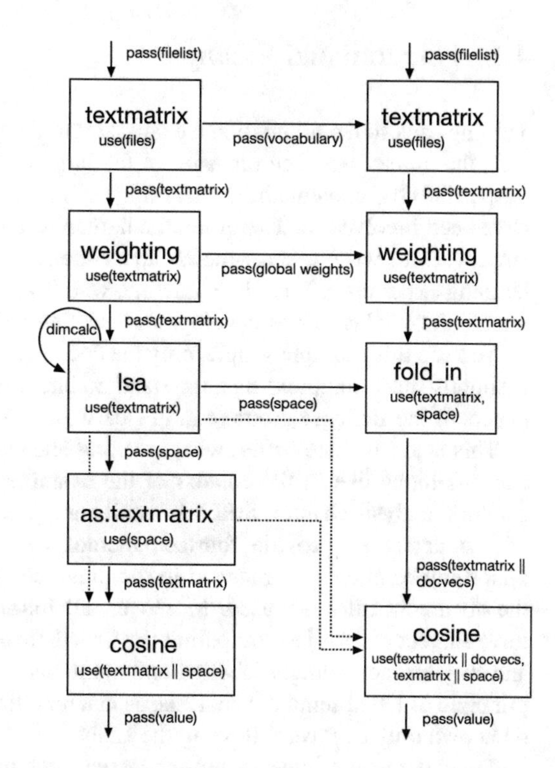

Fig. 4.7 Workflow data handling

thereby reflect the 'term activations' of each document vector in M_k and/or—in case of adding documents with fold-ins—m'.

The handling of the data required in the analysis steps involving fold-ins is thereby not trivial. Fig. 4.7 uses workflow data patterns (following the recommendation of Russell et al. 2004) to show what data are passed on in between the different functional activities.

The left hand side depicts the standard workflow, whereas on the right hand side, the fold in workflow is described. The standard workflow starts with parsing a list of files into a text matrix, weighting it, to then construct a latent semantic space. The fold-in process is dependent on this: the text matrix construction requires the controlled, ordered vocabulary of the original text matrix (otherwise projection is prevented). Moreover and noteworthy, if a global weighting schema was applied, then the resulting global weights have to be handed over to be used for the global weighting of the new text matrix to be projected into the existing latent semantic space. Bot these data hand-overs are depicted in the figure through a 'pass' statement.

When turning to the similarity measurement and text matrix production, it is evident, that either the space or the reconstructed text matrix are needed to allow for comparison and inspection.

More details on data handling are provided in Annex C, the 'lsa' package documentation and source code.

4.3 Foundational Example

Turning back to the foundational example introduced in the previous chapter (Sect. 3.2), the human resource manager of the large multinational decides now to look deeper into the contents of the learning activity the ten employees under scrutiny have been involved in. Therefore, the human resource manager has asked the nine employees to write a short memo about the fourteen trainings, summarising the learning experience. Luckily, Christina, who is still off sick, produced such memo already for her last career development meeting with the human resource manager.

To keep this example simple, only the document titles will be used. The example is slightly more complex[1] than the standard one used repeatedly in the introductory papers of the Bellcore group (e.g. in Deerwester et al. 1990, p. 396).

This is to illustrate better, what indexing and classification with latent semantic analysis looks like in the context of the example introduced above in the social network analysis chapter. Still it is small enough to follow.

The personnel provide fourteen memos for the fourteen different learning opportunities already introduced in the previous chapter. To keep it simple, only the document titles are used. Moreover, all fourteen documents visible fall into three subject areas—i.e. computing ('c'), mathematics ('m'), and pedagogy ('p')—and they are accordingly labelled and numbered. This is to illustrate the working principle of latent semantic analysis, as in a real life case, such classification would be known only *ex post*, following the analysis.

These document titles are pre-processed, such that only those terms are selected that appear in more than one document title (the underlined terms in the Table 4.1). The documents are filed in and converted to a document-term matrix using the `textmatrix` function of the *lsa* package, just as shown in Listing 1.

Listing 1 Reading text files into a document-term matrix.

```
dtm = textmatrix("lsa-example/", minWordLength = 1)
```

The result *dtm* is a sparsely populated text matrix such as the one depicted in Table 4.2. The order of terms and documents can vary slightly when running this example on different machines as it is basically constituted following the order of appearance in the documents (which again is driven by the file system ordering provided by the operating system used). Reshuffling rows and columns in this matrix is of course possible using the native R matrix manipulation routines. Since this has no impact whatsoever on the singular value decomposition, it will not be demonstrated here.

Listing 2 Singular-value decomposition.

```
space = lsa(dtm, dims = dimcalc_raw())
```

[1] 14 instead of 9 documents.

Table 4.1 Memos about the learning experiences

Titles
c1: A web interface for social media applications
c2: Review of access time restrictions on web system usage
c3: Content management system usage of the HTML 5 interface
c4: Error spotting in HTML: social system versus software system
c5: Barriers to access and time spent in social mobile apps
m1: The generation of random unordered trees
m2: A survey of divisive clustering along the intersection of partial trees
m3: Width and height of trees in using agglomerative clustering with Agnes
m4: Agglomerative clustering algorithms: a review
p1: The intersection of learning and organisational knowledge sharing
p2: A transactional perspective on teaching and learning
p3: Innovations in online learning: moving beyond no significant difference
p4: Tacit knowledge management in organisational learning
p5: Knowledge building: theory, pedagogy, and technology

Table 4.2 Document-term matrix

	c1	c2	c3	c4	c5	m1	m2	m3	m4	p1	p2	p3	p4	p5
interface	1	0	1	0	0	0	0	0	0	0	0	0	0	0
social	1	0	0	1	1	0	0	0	0	0	0	0	0	0
web	1	1	0	0	0	0	0	0	0	0	0	0	0	0
access	0	1	0	0	1	0	0	0	0	0	0	0	0	0
review	0	1	0	0	0	0	0	0	1	0	0	0	0	0
system	0	1	1	2	0	0	0	0	0	0	0	0	0	0
time	0	1	0	0	1	0	0	0	0	0	0	0	0	0
usage	0	1	1	0	0	0	0	0	0	0	0	0	0	0
html	0	0	1	1	0	0	0	0	0	0	0	0	0	0
management	0	0	1	0	0	0	0	0	0	0	0	0	1	0
trees	0	0	0	0	0	1	1	1	0	0	0	0	0	0
clustering	0	0	0	0	0	0	1	1	1	0	0	0	0	0
intersection	0	0	0	0	0	0	1	0	0	1	0	0	0	0
agglomerative	0	0	0	0	0	0	0	1	1	0	0	0	0	0
knowledge	0	0	0	0	0	0	0	0	0	1	0	0	1	1
learning	0	0	0	0	0	0	0	0	0	1	1	1	1	0
organisational	0	0	0	0	0	0	0	0	0	1	0	0	1	0

In the next step, this text matrix *dtm* is resolved using the singular-value decomposition, effectively resulting in the three partial matrices listed in Tables 4.3, 4.4, and 4.5. Typically, the resulting three 'space' matrices are immediately truncated to the desired number of factors. Together SVD and truncation from the core of the LSA process, the *lsa* package encapsulates them therefore in the `lsa` function.

Table 4.3 The term loadings *T* on the factors

	t_1	t_2	t_3	t_4	t_5	t_6	t_7	t_8	t_9	t_{10}	t_{11}	t_{12}	t_{13}	t_{14}
interface	−0.21	0.00	−0.06	0.14	−0.01	−0.69	0.05	−0.07	0.08	0.16	−0.22	0.14	−0.19	0.12
social	−0.31	0.07	−0.03	−0.06	0.69	−0.16	0.32	0.12	−0.09	0.09	0.08	−0.10	−0.16	0.21
web	−0.23	0.05	0.05	−0.31	0.03	−0.38	0.02	−0.25	0.38	−0.39	0.19	−0.05	0.36	−0.33
access	−0.23	0.06	0.08	−0.47	0.06	0.09	−0.16	0.14	−0.28	0.14	−0.08	0.06	0.06	0.14
review	−0.19	0.01	0.25	−0.28	−0.35	0.13	0.25	−0.07	0.26	0.01	0.15	−0.26	−0.68	0.00
system	−0.65	0.08	−0.08	0.31	0.01	0.44	−0.02	−0.09	0.15	−0.20	0.08	−0.01	0.18	0.10
time	−0.23	0.06	0.08	−0.47	0.06	0.09	−0.16	0.14	−0.28	0.14	−0.08	0.06	0.06	−0.35
usage	−0.31	0.02	0.01	−0.08	−0.43	−0.15	−0.29	−0.06	−0.02	0.07	−0.22	0.20	0.08	0.44
html	−0.32	0.02	−0.09	0.38	−0.01	0.06	0.01	0.04	−0.08	0.18	−0.17	0.09	−0.19	−0.40
management	−0.18	−0.23	−0.12	0.17	−0.31	−0.28	−0.04	0.27	−0.38	0.09	0.39	−0.36	0.07	−0.26
trees	−0.01	−0.12	0.5	0.19	0.21	−0.10	−0.47	−0.05	−0.27	−0.50	0.04	0.04	−0.30	0.00
clustering	−0.04	−0.12	0.62	0.14	−0.01	−0.01	0.16	0.03	0.03	0.26	0.01	−0.38	0.40	0.26
intersection	−0.02	−0.29	0.17	0.03	0.22	0.05	−0.41	−0.11	0.44	0.53	0.01	0.07	−0.04	−0.26
agglomerative	−0.03	−0.06	0.43	0.07	−0.14	0.01	0.49	0.11	−0.10	−0.03	−0.13	0.56	0.10	−0.26
knowledge	−0.05	−0.51	−0.11	−0.08	0.02	0.03	0.03	0.53	0.27	−0.30	−0.49	−0.18	0.01	0.00
learning	−0.05	−0.59	−0.13	−0.11	0.03	0.09	0.20	−0.66	−0.32	−0.01	−0.17	−0.07	0.00	0.00
organisational	−0.04	−0.45	−0.09	−0.07	0.01	0.02	0.02	0.19	0.09	0.01	0.59	0.46	−0.03	0.26

Table 4.4 The document 'loadings' D on the factors

	d_1	d_2	d_3	d_4	d_5	d_6	d_7	d_8	d_9	d_{10}	d_{11}	d_{12}	d_{13}	d_{14}
c1	−0.22	0.04	−0.01	−0.10	0.39	−0.75	0.27	−0.16	0.30	−0.15	0.06	−0.03	0.02	0.00
c2	−0.55	0.10	0.15	−0.59	−0.35	0.14	−0.24	−0.15	0.17	−0.23	0.06	0.00	0.13	0.00
c3	−0.50	−0.04	−0.13	0.42	−0.42	−0.38	−0.19	0.07	−0.20	0.31	−0.22	0.10	−0.10	0.00
c4	−0.58	0.09	−0.11	0.42	0.39	0.48	0.20	−0.02	0.11	−0.13	0.10	−0.05	0.02	0.00
c5	−0.23	0.06	0.05	−0.45	0.45	0.01	0.01	0.33	−0.51	0.37	−0.12	0.03	−0.11	0.00
m1	0.00	−0.04	0.19	0.09	0.12	−0.06	−0.32	−0.04	−0.22	−0.51	0.06	0.09	−0.72	0.00
m2	−0.02	−0.19	0.50	0.16	0.23	−0.04	−0.49	−0.10	0.16	0.29	0.09	−0.50	0.13	0.00
m3	−0.02	−0.10	0.60	0.18	0.03	−0.06	0.12	0.08	−0.27	−0.28	−0.13	0.43	0.46	0.00
m4	−0.07	−0.06	0.50	−0.04	−0.28	0.08	0.60	0.06	0.15	0.24	0.04	−0.14	−0.42	0.00
p1	−0.05	−0.65	−0.06	−0.11	0.16	0.12	−0.12	−0.03	0.39	0.22	−0.08	0.53	−0.14	0.00
p2	−0.02	−0.21	−0.05	−0.05	0.02	0.05	0.13	−0.53	−0.25	−0.01	−0.26	−0.14	0.01	0.71
p3	−0.02	−0.21	−0.05	−0.05	0.02	0.05	0.13	−0.53	−0.25	−0.01	−0.26	−0.14	0.01	−0.71
p4	−0.09	−0.62	−0.18	−0.04	−0.13	−0.08	0.14	0.27	−0.27	−0.21	0.48	−0.29	0.13	0.00
p5	−0.01	−0.18	−0.04	−0.04	0.01	0.02	0.02	0.43	0.21	−0.31	−0.72	−0.34	0.01	0.00

Table 4.5 The singular values S

	s_1	s_2	s_3	s_4	s_5	s_6	s_7	s_8	s_9	s_{10}	s_{11}	s_{12}	s_{13}	s_{14}
1	3.36	0	0	0	0	0	0	0	0	0	0	0	0	0
2	0	2.85	0	0	0	0	0	0	0	0	0	0	0	0
3	0	0	2.58	0	0	0	0	0	0	0	0	0	0	0
4	0	0	0	2.21	0	0	0	0	0	0	0	0	0	0
5	0	0	0	0	1.79	0	0	0	0	0	0	0	0	0
6	0	0	0	0	0	1.63	0	0	0	0	0	0	0	0
7	0	0	0	0	0	0	1.48	0	0	0	0	0	0	0
8	0	0	0	0	0	0	0	1.25	0	0	0	0	0	0
9	0	0	0	0	0	0	0	0	1.24	0	0	0	0	0
10	0	0	0	0	0	0	0	0	0	0.98	0	0	0	0
11	0	0	0	0	0	0	0	0	0	0	0.67	0	0	0
12	0	0	0	0	0	0	0	0	0	0	0	0.53	0	0
13	0	0	0	0	0	0	0	0	0	0	0	0	0.42	0
14	0	0	0	0	0	0	0	0	0	0	0	0	0	0

Listing 2 shows how this function is used to retain all available singular values, passing through the `dimcalc_raw` function to the dimensionality selection interface. The result *space* is a list with three slots containing the matrices decomposing *dtm* into the matrices T, S, and D as shown above in Fig. 4.2. The matrix T (accessible via *space$tk*) thereby holds the term 'loadings' onto the factors.

The partial matrix D (access via *space$dk*) contains the document 'loadings' onto the factors, with the columns holding the right singular vectors of the matrix decomposition. The according matrix from the example is presented in Table 4.4. The list S then contains the singular values of the decomposition, sorted descending. The values in S constitute a diagonal matrix depicted in Table 4.5.

As already indicated, this set of matrices (aka the 'latent semantic space') can be used to reconstruct the original document-term matrix, using the type casting operator `as.textmatrix` provided in the package, which effectively re-multiplies the three partial matrices again (as indicated in the second line in Listing 3).

Listing 3 Reconstruction of the document-term matrix from the space.

```
X = as.textmatrix(space)
X = space$tk %*% diag(space$sk) %*% t(space$dk)
```

To confirm, whether the re-multiplication in fact reconstructed the original document-term matrix, all elements in the reconstructed matrix X can be compared with all elements in *dtm* (values are rounded to three digits to avoid impact of minimal rounding errors resulting from the decomposition).

Listing 4 Confirming whether reconstruction succeeded.

```
X = round(X, 3)
all((dtm == X) == TRUE)
## [1] TRUE
```

The 'trick' of LSA is to factor-reduce this space, i.e. to eliminate those lower-ranking factors that obscure the semantic structure, while retaining those high-ranking factors that constitute the differences. In our example, a useful number of factors to retain is three.

Listing 5 Truncating the space.

```
space_red = lsa(dtm, dims = 3)
```

This reduced space now reflects better the semantic structure than the original document-term matrix, as can be seen from the following table which depicts the re-multiplied matrix. To be a bit more precise: this truncated space reflects better the 'latent semantic' structure contained in the documents—and of course construed by the boundaries of the semantics surfacing in this example (Table 4.6).

Table 4.6 Reconstructed document-term matrix of the factor-reduced space

	c1	c2	c3	c4	c5	m1	m2	m3	m4	p1	p2	p3	p4	p5
interface	0.2	0.4	0.4	0.4	0.2	0.0	−0.1	−0.1	0.0	0.0	0.0	0.0	0.1	0.0
social	0.2	0.6	0.5	0.6	0.2	0.0	−0.1	0.0	0.0	−0.1	0.0	0.0	0.0	0.0
web	0.2	0.5	0.4	0.4	0.2	0.0	0.1	0.1	0.1	−0.1	0.0	0.0	0.0	0.0
access	0.2	0.5	0.4	0.4	0.2	0.0	0.1	0.1	0.1	−0.1	0.0	0.0	−0.1	0.0
review	0.1	0.4	0.2	0.3	0.2	0.1	0.3	0.4	0.4	0.0	0.0	0.0	−0.1	0.0
system	0.5	1.2	1.1	1.3	0.5	0.0	−0.1	−0.1	0.0	0.0	0.0	0.0	0.1	0.0
time	0.2	0.5	0.4	0.4	0.2	0.0	0.1	0.1	0.1	−0.1	0.0	0.0	−0.1	0.0
usage	0.2	0.6	0.5	0.6	0.2	0.0	0.0	0.0	0.1	0.0	0.0	0.0	0.1	0.0
html	0.2	0.6	0.6	0.6	0.2	0.0	−0.1	−0.1	0.0	0.0	0.0	0.0	0.1	0.0
management	0.1	0.2	0.4	0.3	0.1	0.0	0.0	−0.1	−0.1	0.5	0.2	0.2	0.5	0.1
trees	0.0	0.2	−0.1	−0.1	0.1	0.3	0.7	0.8	0.7	0.1	0.0	0.0	0.0	0.0
clustering	0.0	0.3	−0.1	−0.1	0.1	0.3	0.9	1.0	0.8	0.1	0.0	0.0	−0.1	0.0
intersection	0.0	0.0	0.0	−0.1	0.0	0.1	0.4	0.4	0.3	0.5	0.2	0.2	0.4	0.1
agglomerative	0.0	0.2	−0.1	−0.1	0.1	0.2	0.6	0.7	0.6	0.0	0.0	0.0	−0.1	0.0
knowledge	0.0	−0.1	0.2	0.0	−0.1	0.0	0.1	0.0	0.0	1.0	0.3	0.3	1.0	0.3
learning	0.0	−0.1	0.2	0.0	−0.1	0.0	0.1	0.0	−0.1	1.1	0.4	0.4	1.1	0.3
organisational	0.0	−0.1	0.2	0.0	−0.1	0.0	0.1	0.0	0.0	0.8	0.3	0.3	0.8	0.2

Table 4.7 Proximity matrix for the original vector space (rounded)

	c1	c2	c3	c4	c5	m1	m2	m3	m4	p1	p2	p3	p4	p5
c1	1	0	0	0	0	0	0	0	0	0	0	0	0	0
c2	0	1	0	0	0	0	0	0	0	0	0	0	0	0
c3	0	0	1	1	0	0	0	0	0	0	0	0	0	0
c4	0	0	1	1	0	0	0	0	0	0	0	0	0	0
c5	0	0	0	0	1	0	0	0	0	0	0	0	0	0
m1	0	0	0	0	0	1	1	1	0	0	0	0	0	0
m2	0	0	0	0	0	1	1	1	0	0	0	0	0	0
m3	0	0	0	0	0	1	1	1	1	0	0	0	0	0
m4	0	0	0	0	0	0	0	1	1	0	0	0	0	0
p1	0	0	0	0	0	0	0	0	0	1	0	0	1	0
p2	0	0	0	0	0	0	0	0	0	0	1	1	0	0
p3	0	0	0	0	0	0	0	0	0	0	1	1	0	0
p4	0	0	0	0	0	0	0	0	0	1	0	0	1	0
p5	0	0	0	0	0	0	0	0	0	0	0	0	0	1

The document 'c3' contains now, for example, also a value of 0.4 for the term 'web', which previously was not there: the document is about the HTML interface of a content management system, but does not use the term 'web' in its title.

The effect of this truncation becomes even more evident, when looking at proximity relations between documents. The proximity of the documents is calculated as follows: in the first line for the original, non-truncated vector space (as already established in *dtm*); in the second line for the truncated space (Tables 4.7 and 4.8).

Listing 6 Calculating cosine proximities.

```
proximity = cosine(dtm)
proximitySpaceRed = cosine(as.textmatrix(space_red))
```

When looking at the proximity table of the documents in the original, unreduced vector space and compare them with the proximity of documents in the factor-reduced space, the difference becomes clearly visible: the 'computing' documents (starting with 'c') can be much better be differentiated for the latter space from the 'math' (starting with 'm') and 'pedagogy' documents (starting with 'p'). Moreover, the computing, math, and pedagogy documents respectively have become more similar within their own groups—and more dissimilar from each other.

Since this example uses only three factors, we can use the factor loadings of both documents and terms to draw a 3D perspective plot.

Table 4.8 Proximity table for the factor-reduced space (rounded)

	c1	c2	c3	c4	c5	m1	m2	m3	m4	p1	p2	p3	p4	p5
c1	1	1	1	1	1	0	0	0	0	0	0	0	0	0
c2	1	1	1	1	1	0	0	0	0	0	0	0	0	0
c3	1	1	1	1	1	0	0	0	0	0	0	0	0	0
c4	1	1	1	1	1	0	0	0	0	0	0	0	0	0
c5	1	1	1	1	1	0	0	0	0	0	0	0	0	0
m1	0	0	0	0	0	1	1	1	1	0	0	0	0	0
m2	0	0	0	0	0	1	1	1	1	0	0	0	0	0
m3	0	0	0	0	0	1	1	1	1	0	0	0	0	0
m4	0	0	0	0	0	1	1	1	1	0	0	0	0	0
p1	0	0	0	0	0	0	0	0	0	1	1	1	1	1
p2	0	0	0	0	0	0	0	0	0	1	1	1	1	1
p3	0	0	0	0	0	0	0	0	0	1	1	1	1	1
p4	0	0	0	0	0	0	0	0	0	1	1	1	1	1
p5	0	0	0	0	0	0	0	0	0	1	1	1	1	1

Listing 7 Generating the perspective plot of terms and documents.

```
p = persp(
   x = -1:1, y = -1:1,
   z = matrix(
      c(
        -1,0,1,
        -1,0,1,
        1,0,-1
      ),3,3),
   col = "transparent", border = "transparent",
   xlim = range(c(space$dk[,1], space$tk[,1])),
   ylim = range(c(space$dk[,2], space$tk[,2])),
   zlim = range(c(space$dk[,3], space$tk[,3])),
   theta = 35, phi = 20,
   xlab = "dim 1", ylab = "dim 2", zlab = "dim 3",
   expand = 0.5, scale = F,
   axes = TRUE, nticks = 10, ticktype = "simple"
)

points(trans3d(space$dk[,1], space$dk[,2],
   space$dk[,3], pmat = p), bg = "red", col = "red",
   pch = 22, cex = 1)

points(trans3d(space$tk[,1], space$tk[,2],
   space$tk[,3], pmat = p), bg = "blue", col = "blue",
   pch = 21, cex = 1)
```

```
text(trans3d(space$tk[,1], space$tk[,2],
  space$tk[,3], pmat = p), rownames(space$tk),
  col = "blue", cex = 0.8)

text(trans3d(space$dk[,1], space$dk[,2],
  space$dk[,3], pmat = p), rownames(space$dk),
  col = "red", cex = 0.8)
```

The code of Listing 7 thereby first creates an empty perspective plot using `persp` (to ensure that the limits of all axes are set up so that the projected points are all visible). Then the two commands `points` and `text` are used to plot the term and document positions (and according labels) into the perspective plot.

The resulting visualisation (Fig. 4.8) shows, how the factors separate the three clusters of documents and terms: to the top right, the math documents and math-related terms cluster together; in the bottom right corner, the pedagogy-related documents and terms; and in the top left corner, the computing ones. The axes thereby represent the new base of the Eigensystem resulting from the singular value decomposition.

One term is depicted in between the two clusters of pedagogy and computing, i.e. 'management'. This term is found in both a computing as well as a pedagogy document—bridging between the two clusters.

Calculation of spaces can be a resource intense endeavour, preventing recalculation in real-time: a space with a few million documents and terms can easily take one or more hours on a fast machine with big memory (not speaking of how long it can take on a slow machine with scarce memory).

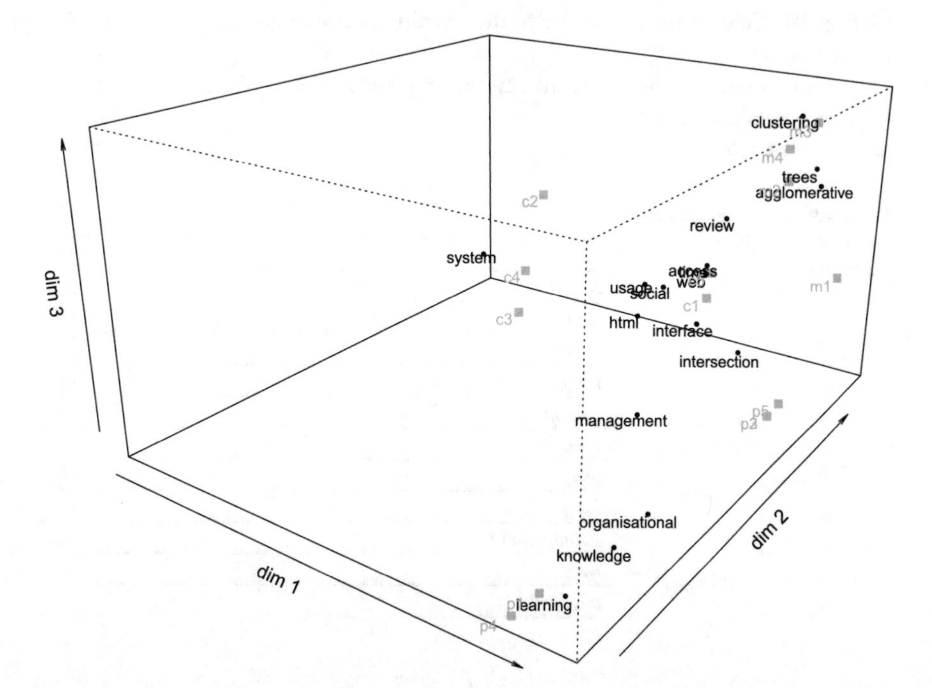

Fig. 4.8 Perspective plot of terms and documents (three factors)

There is, however, an efficient updating method that can be used to project new data into an existing space. Given that the space captures its semantic, this projection is almost lossless. Moreover, it prevents influencing the semantics of a given space, thus ensuring stability its semantics and validity.

To update, new data can be filed in (reusing the vocabulary of the existing document-term matrix, to ensure that the new data can be projected into the existing latent semantic space).

Listing 8 Reading additional data with a controlled vocabulary.

```
data = "Review of the html user interface of the system"
pdoc = query(data, rownames(dtm))
```

The title of the new document 'c6' to add is "Review of the html user interface of the system", which results in the following column vector (Table 4.9).

This new vector can be projected into the latent-semantic space using the `fold_in` procedure.

Listing 9 Folding in.

```
newY = fold_in(pdoc, space_red)
```

Once this is done, comparisons of the new document with the existing ones become possible. Therefore, the new column vector that was just folded in is bound to the reconstructed document-term matrix of the factor-reduced space—to then calculate all cosine proximities (see Listing 10).

Listing 10 Calculating proximity to the existing documents.

```
allY = cbind(newY, as.textmatrix(space_red))
allCos = cosine(allY)
allCos["c6",]
```

Table 4.9 'Raw' document vector of an additional document

	c6
interface	1
social	0
web	0
access	0
review	1
system	1
time	0
usage	0
html	1
management	0
trees	0
clustering	0

Table 4.10 Proximity of 'c6' to the existing documents (rounded to one digit)

	c6	c1	c2	c3	c4	c5	m1	m2	m3	m4	p1	p2	p3	p4	p5
c6	1	1	1	1	1	1	0	0	0	0.2	0	0	0	0.1	0

As visible from the last table, the new document is evaluated to be very close to the computing documents (and far from the math and pedagogy ones)—just as expected (Table 4.10).

4.4 State of the Art in the Application of LSA for TEL

While the application focus on essay scoring is taken up more widely over the next decade in the late 1990s and early 2000s with prototypes and original research contributed from research groups around the globe, a period of theoretical reflection follows in the research group in Colorado, ultimately leading up to the formulation of LSA as a representation theory.

In parallel, the mathematical/statistical foundation introduced above in Sect. 4.1 are further elaborated and generic implementations of the underlying linear algebra routines and its application in LSA/LSI are made available. First derivate algorithms such as probabilistic LSA and topic models arrive. Moreover, the parametric configuration of LSA such as applied in weighting schemes or dimensionality selection are starting to get investigated.

In this section, the state of the art of LSA in technology-enhanced learning and as a representation theory will be described. The mathematical and statistical foundations, as well as selected extensions were already described above in Sect. 4.1. The state of the art on parametric configuration of LSA and the derived MPIA is documented in Chap. 7: 'Calibrating for specific domains'.

To summarise the first area of growth in the body of literature, i.e. essay scoring, the following can be stated.

From the original research group at Bell Communications Research holding the patent, Dumais, Landauer, and Streeter/Lochbaum are the most active members, spinning off their own research teams.

Dumais focuses more, though not exclusively on the information retrieval aspects (as evident in Dumais et al. 1988, p. 282; Dumais 1991, 1992, p. 230). Dumais sometimes uses the notion 'LSI' for the application of LSA in indexing (Dumais 1995, p. 219). Dumais further investigates other information access problems such as categorisation and question answering using latent semantic analysis (see Dumais 2003, for a summary).

Landauer teams up with Foltz[2] and Laham in pursuing educational applications, reflected in multiple co-authorships as well as in the joint spin off company Knowledge Assessment Technologies (today: Pearson Knowledge Technologies):

[2] Landauer and Foltz first met at Bellcore: http://kt.pearsonassessments.com/whyChooseUs.php

as already introduced above, Landauer et al. (1997) provide experimental results on automated essay scoring; Landauer et al. 1998b, p. 48). Laham is based at the time—same as Landauer—at the University of Colorado at Boulder, whereas Foltz is based at New Mexico State University. Other prominent members of the LSA research group at the University of Colorado are Kintsch, Rehder, and Schreiner (see e.g. Landauer et al. 1997; Foltz et al. 1998; Wolfe et al. 1998; Rehder et al. 1998).

Streeter and Lochbaum both start publishing on further experiments and prototypes together with their army counter part Psotka in connection with their contract-research to develop tutoring systems for the army (Lochbaum et al. 2002; Lochbaum and Streeter 2002; Psotka et al. 2004; Streeter et al. 2002).

Of the remaining co-authors of the original patent, Furnas focused more on information visualisation research in his subsequent career (see his publication list[3]). Deerwester continued to Hongkong University of Science and Technology. He holds another patent on a related technology (indexing collections of items: Deerwester 1998) and filed another application (on optimal queries: Deerwester 2000), but stopped publishing about LSA-related topics and, ultimately, moved into a different career (see his Wikipedia entry[4]). Harshman focused before as after more on data analysis aspects (see publication list[5] and obituary Sidiropoulos and Bro 2009).

Regarding additional, new research groups, the following can be concluded. At the very end of the 1990ies, the Tutoring Research Group at the University of Memphis is formed around Graesser, visible often with co-authorship of P. -Wiemer-Hastings (aka Hastings in more recent years) and K. Wiemer-Hastings. They develop AutoTutor (and other prototypes and products): Wiemer-Hastings et al. (1998) and Graesser et al. (1999) introduce into the system architecture and way of functioning of AutoTutor, thereby putting more emphasis on tutorial dialogue moves as in the earlier works from the Bellcore-affiliated research groups. This is driven by the research interest of the group, as evident e.g. in Graesser et al. (1995), an extensive study of actual (human) tutorial dialogues and protocols.

With the beginning of the twenty-first century, the community using LSA in technology-enhanced learning quickly broadens: around 1999/2000, a French research group starts publishing at the Université Pierre-Mendès-France in Grenoble around Dessus and Lemaire follows, building the research prototype 'Assistant for Preparing Exams' (Apex, Dessus and Lemaire 1999; Dessus et al. 2000, p. 66ff; Lemaire and Dessus 2001, p. 310ff; Dessus and Lemaire 2002). Later, additional prototypes such as Apex-2 (Trausan-Matu et al. 2008, p. 46ff) and Pensum (Trausan-Matu et al. 2009, 2010, p. 9, 21ff) will follow.

[3] http://furnas.people.si.umich.edu/BioVita/publist.htm

[4] http://en.wikipedia.org/wiki/Scott_Deerwester

[5] http://psychology.uwo.ca/faculty/harshman/

In the UK, a research group at the Open University (Haley et al. 2003, 2005; Haley 2008) starts conducting LSA-based e-assessment research, building EMMA, in the context of the EC-funded eLeGi project (see Haley et al. 2007, p. 12).

In the Netherlands, a group at the Open University of the Netherlands around van Bruggen (2002) starts conducting essay scoring and question answering research. With Van Bruggen et al. (2004), they coin the term 'positioning' for assessing prior learning with the help of automated scoring techniques. Kalz et al. (2006) looks at how positioning can be done using LSA combined with a meta-data approach. van der Vegt et al. (2009) describe how to conduct placement experiments with the R implementation of the author of this book and with a PHP implementation made available.

With support from the Memphis group, van Lehn starts conducting essay-scoring research with LSA at the University of Pittsburgh, thereby creating Why2-Atlas (van Lehn et al. 2002) and CarmelTC (Rose et al. 2003).

At the University of Massachusetts, Larkey (1998) conducts essay-scoring experiments.

In Germany, Lenhard et al. (2007a, b, 2012), trained by the Kintschs and Landauer, builds up a research group at the University of Wuerzburg—focusing on a writing-trainer in German supported with automated feedback. The system is called conText.

In Austria, the author of this book Wild starts building up research around LSA and essay scoring (Wild et al. 2005a, b; Wild and Stahl 2007), thereby releasing the lsa package for R (Wild 2014), building an Essay Scoring Application (ESA) module for .LRN/OpenACS (Wild et al. 2007b; Koblischke 2007). Later this will lead to the joint development of LEApos and Conspect (LTfLL d4.3), when Wild moves on to the Open University. Conspect focuses on monitoring conceptual development from student writings, whereas LEApos sets focus on short answer scoring, but with a stronger focus on knowledge-rich technology with LSA playing a limited role.

In Australia, Williams and Dreher at the Curtin University of Technology in Perth start development of MarkIT (Williams 2001; Palmer et al. 2002; Williams and Dreher 2004, 2005; Dreher 2006).

In Germany, at the Goethe University of Frankfurt, Holten et al. (2010) start using LSA to assess professional expertise and application domain knowledge.

In Spain, a cross-university group forms staffed by members from mainly the Autonomous University of Madrid and from the National Distance Education University (UNED). From 2006 onwards, mainly Olmos, Jorges-Botano, and Leon publish about their findings (Leon et al. 2006; Olmos et al. 2009; Jorge-Botana et al. 2010a, b; Olmos et al. 2011). The system developed is named Gallito.

Several special events support community building over time. Besides the 1998 special issue introduced above, a second dedicated set of articles is published in 2000. It features seven articles (spread out across two issues). Landauer and Psotka (2000) introduce into the seven articles, thereby also providing an short introduction into LSA. Foltz et al. (2000) document different essay scoring methods implemented in a real system and its evaluation with undergraduates. Graesser et al. (2000) describe how AutoTutor uses LSA for the dialogue interaction with the learners. Wiemer-Hastings and Graesser (2000) show how LSA is used in Select-a-Kibitzer to provide feedback on student writings. Kintsch et al. (2000) describe the systems and evaluation results of summary writing tools called 'Summary Street' and 'State the Essence'. Laham et al. (2000) describe a system for matching job and competence profiles with the help of LSA. Freeman et al. (2000) describe how to model expert domain knowledge from written short texts.

The Handbook on Latent Semantic Analysis (Landauer et al. 2008) brings together contributions from the original and extended circle.

In 2007, the 1st European Workshop on Latent Semantic Analysis takes place (Wild et al. 2007a), from which a European research group around Wild, Kalz, Koper, van Rosmalen, and van Bruggen forms, teaming up with Dessus for the EC-funded LTfLL project (funded from 2008 to 2011).

Several review articles and debates support spreading of research over time. Whittington and Hunt (1999) compare LSA with shallow surface feature detection mechanisms offered by PEG (Page 1966) and hybrid systems that take syntactical or even discourse-oriented features into account. They acknowledge LSA "impressive results" (p. 211), but at the same time criticise its need for large amounts of data and computational resources.

Hearst (2000) moderates a debate for the column 'Trends and Controversies' of IEEE Intelligent Systems, with position statements from Kukich (2000) (the director of the natural language processing group at Educational Testing Service), Landauer et al. (2000), and (MITRE, Hirschmann et al. 2000), rounded up with a comment by Calfee (Dean of the School of Education at UC Riverside). The message of the debate is clear: it is surprising how well the automated techniques work, but there is still "room for improvement" (Calfee 2000, p. 38). Yang et al. (2001) review validation frameworks for essay scoring research. Miller (2003) provides a review of the use of LSA for essay scoring and compares the technique against earlier, computer-based techniques. Valenti et al. (2003) review findings for ten different essay-scoring systems. They particularly criticize the lack of test and training collections, a problem common to all techniques surveyed. Dumais (2005) provides an extensive review over a wide range of LSA applications, including educational ones. Landauer and Foltz (2007) review the applications available at the University of Colorado and at Pearson Knowledge Technologies.

In parallel to the application-oriented research, basic research leads to establishing LSA as a representation theory.

Following the seminal articles listed above, research is conducted to reflect on the capabilities of LSA to be a semantic representation theory, starting the first attempt in Landauer and Dumais (1997): using Plato's poverty of stimulus problem,[6] they investigate the validity of LSA as a theory of "acquired similarity and knowledge representation" (p. 211). They come to the conclusion that LSA ability

[6] The poverty of stimulus problem is paraphrased as: "How do people know as much as they do with as little information as they get?" (Landauer and Dumais 1997, p. 211)

to use local co-occurrence stimuli for the induction of previously learnt global knowledge provides evidence of its applicability.

Kintsch (1998) discusses how LSA can be used as a surrogate for the construction of a propositional network of "predicate-argument structures with time and location slots" (p. 411), thereby grounding LSA in this representation theory.

Landauer (1998) proposes that word meaning is acquired from experience and therefore it is valid to represent them through their positions in a semantic vector space. When the space is calculated from a text body large enough, similar performance to humans can be found. He reports results from an experiment where an LSA space was trained from a psychology text book, then used to evaluate a multiple-choice test used in real assessments: LSA passed.

In Landauer (1999), in a reply to Perfetti's critique (1998), defends LSA as a theory of meaning and mind. He particularly defends the criticised grounding in co-occurrences (though not mere 1st order co-occurrences) as an fundamental principle in establishing meaning from passages. He also clearly states that LSA is not a complete theory of discourse processing, as this would include turning perceived discourse into meaning and turning ideas into discourse. In other words, one might say, it would have to provide means to model purpose.

Kintsch (2000, 2001) and Kintsch and Bowles (2002) discuss the interpretation of the mathematical and statistical foundations of LSA with respect to "metaphor interpretation, causal inferences, similarity judgments, and homonym disambiguation" (Kintsch 2001, p. 1).

Landauer (2002) further elaborates on LSA as a theory of learning and cognition, again basing its fundamental principle in the co-occurrence of words within passage contexts and in the analysis of these. They characterise the shortcomings of LSA as a theory of language and verbal semantics (p. 28) to lie in the lack of a model of production and the dynamic processes of comprehension, discourse, and conversation conventions (p. 28). The chapter also discusses the relation of LSA to visual perception and physical experience.

Quesada et al. (2001, 2002a, b, 2005) and Quesada (2003) elaborate on complex problem solving with LSA, implemented—quite similar to case-based reasoning—as contextual similarity.

4.5 Extended Application Example: Automated Essay Scoring

Automated essay scoring is one of the popular application areas of latent semantic analysis, see the overview presented below in Table 4.11. Over time, a wide variety of scoring methods were proposed and evaluated. While many roads lead to Rome, already naïve scoring approaches work astoundingly well—such as comparing a student-written essay with a model solution (a.k.a. gold standard). This subsection

Table 4.11 The state of the art of TEL applications of LSA: Summary

Application	System(s)	Type[a]	Group
Expertise mapping, people recommender	Bellcore advisor	P	Dumais
Essay scoring	Intelligent Essay Assessor, Summary Street, WriteToLearn, Open-Cloze, Meaningful Sentences, Team Communications, Knowledge Post	PEC	Landauer
Learning object search, summaries	SuperManual	P	Landauer
Identifying learning standards	Standard seeker	P	Landauer
Matching learning experience and training programmes	Career map	P	Landauer
Tagging learning objects	Metadata tagger	P	Landauer
Summary writing	Apex, Apex-II, Pensum	PE	Dessus
Essay scoring	EMMA	PE	Haley
Essay scoring	conText	PE	Lenhard
Locating tutors	ASA-ATK	PEC	Van Rosmalen
Dialogue tutoring	AUTOTUTOR, Select-a-Kibitzer, State the Essence	PE	Graesser
Assessing conversations	PolyCAFe	PE	Trausan-Matu
Essay scoring	ESA, R	PEC	Wild
Monitoring conceptual development	CONSPECT	PE	Wild
Essay scoring	MarkIT	P	Dreher
Essay scoring	Gallito	PE	Jorge-Botano, Olmos, Leon

P TEL Prototype, *E* TEL Evaluation study, *C* Configuration

provides an exemplification of such naïve scoring method.[7] The subsequent Chap. 9 on MPIA's application examples will revisit this example to unveil the here unspoken assumptions underlying this approach.

Emphasis thereby will lie on the algorithmic details and therefore the following limitations apply: From a didactical, instructional perspective, delivering a 'naked' single score to learners is more than just a bit questionable. Assessment for learning requires much more than that and even assessment of learning should—if not for

[7] The author made this example available online at: http://crunch.kmi.open.ac.uk/people/~fwild/services/lsa-essay-scoring.Rmw

accuracy reasons, then for acceptance reasons—better rely on the advice and guidance of skilled human evaluators.

Still, to improve quality in assessment, similar scoring systems are in use around the globe (Steedle and Elliot 2012, p. 3), mostly where assessment situations can be standardised to scale, so the investment into training an LSA-based system with high precision pays off. In such scenario, the machine scores typically serve as an extra rater in addition to the human ones (and mediating reviews are called where human and machine differ).

Figure 4.9 gives an overview on the process of analysis. First, a training corpus is created, holding generic, domain specific documents, or both. In case of this example, only domain specific documents were used. They had been gathered from the first hits of a major search engine, thereby splitting larger documents in about paragraph sized units (yielding 71 documents). Additionally, copies of the three model solutions were added to the training corpus. In step 1, a text matrix is constructed from this document collection, to then—step 2—calculate the latent semantic space from it. Step 3 allows to inspect the LSA-revised resulting text matrix.

The student essays (and—for convenience—the three model solutions) are subsequently filed in (step A) and then folded into this space (step B), not least to avoid bad essays from distorting the associative closeness proximity relations in the space. The resulting appendable document vectors are then used to calculate the proximity of each student essay with the three model solutions (step C). The average Spearman Rho rank correlation is thereby used as the score for each essay.

The 74 documents used for training the latent semantic space are converted to their document-term matrix representation with the help of the *lsa* package's `textmatrix` routine.

Listing 11 Filing in the training corpus.

```
corpus_training = textmatrix("corpus/corpus.6.base",
   stemming = FALSE, minWordLength = 3, minDocFreq = 1)
```

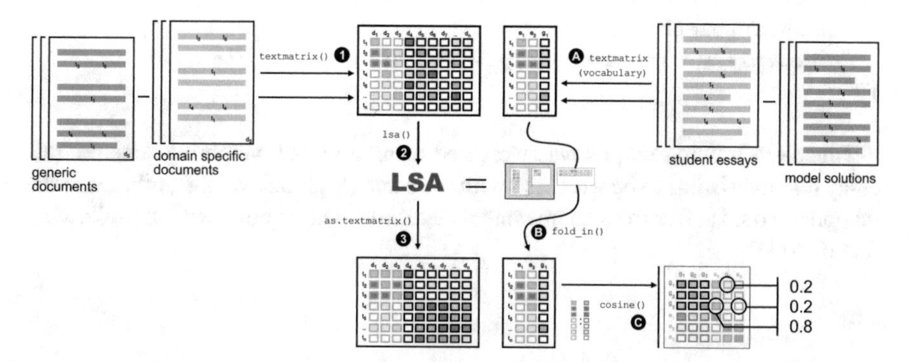

Fig. 4.9 Naïve essay scoring process (revised and extended from Wild and Stahl 2007, p. 388)

Table 4.12 The text matrix of the training corpus

Term	data6_18.txt	data6_19.txt	data6_20.txt	data6_21.txt
zerlegt	0	1	0	1
zweite	0	1	1	0
art	3	1	1	1
auftretens	0	0	0	0
ausgerichtet	0	0	0	0

The resulting text matrix has 1056 terms in the rows and 74 documents in the columns. A subset of these is shown in Table 4.12: four documents and their term frequencies for the subset of five terms.

Following the construction of the text matrix, weighting measures are applied and the space is calculated (see Listing 12). The chosen weighting measure is 1 + entropy (see package documentation in the annex for more detail).

Listing 12 Applying a weighting measure and calculating the latent semantic space.

```
weighted_training = corpus_training *
    gw_entropy(corpus_training)
space = lsa(weighted_training,
    dims = dimcalc_share(share = 0.5))
```

The next step is to map the essays into this existing latent semantic space, in order to enable their analysis in the structure of the space. This prevents 'bad' essays from distorting the structure of the space. Therefore, a text matrix representation of the essays is constructed using the controlled and ordered vocabulary of the training corpus is done with the line of code shown in Listing 13.

Listing 13 Text matrix construction with a controlled, ordered vocabulary.

```
corpus_essays = textmatrix("essays/",
    stemming = FALSE,
    minWordLength = 3,
    vocabulary = rownames(corpus_training)
)
```

Since the training corpus was weighted using a global weighting scheme, this essay text matrix has to be weighted with the same (!) global weights, otherwise the mapping would suffer from an unwanted distortion. This is done with the following line of code.

Listing 14 Weighting of essay corpus with existing global weights.

```
weighted_essays = corpus_essays *
   gw_entropy(corpus_training)
```

Subsequently, the weighted essay matrix can be mapped into the existing latent semantic space with the following command.

Listing 15 Fold in of the essay text matrix into the space.

```
lsaEssays = fold_in(weighted_essays, space)
```

As a naïve scoring method, the average Spearman rank correlation of each student essay to the three model solutions (the three 'gold standards') is used:

Listing 16 Assigning a score to each essay.

```
essay2essay = cor(lsaEssays, method = "spearman")

goldstandard = c("data6_golden_01.txt",
   "data6_golden_02.txt", "data6_golden_03.txt")

machinescores = colSums(essay2essay[goldstandard,])/3
```

To evaluate, how well the machine-assigned scores perform in comparison to the human raters, first the human-assigned scores for each of the essays in the collection are loaded.

Listing 17 Filing in the scores of the human raters.

```
corpus_scores = read.table("corpus/corpus.6.scores",
   row.names = "V1")
```

Then the correlation of machine-assigned to human-assigned scores is calculated. The human scores range from 0 to 4 points in half-point steps and the machine scores with a real number between 0 and 1. Consequently, their correlation is best measured with Spearman's rank correlation coefficient Rho, as shown in Listing 18.

Listing 18 Measuring the correlation between human and machine scores.

```
cor.test(
    humanscores[names(machinescores),],
    machinescores,
    exact = FALSE,
    method = "spearman",
    alternative = "two.sided"
)
```

```
##
## Spearman's rank correlation rho
##
## data: humanscores[names(machinescores),] and
   machinescores
## S = 914.6, p-value = 0.0001049
## alternative hypothesis: true rho is not equal to 0
## sample estimates:
## rho
## 0.6873
```

Listing 18 also lists the output of the correlation test: a measured Spearman's Rho of 0.69 for this case of scoring essays in the latent semantic space.

The interesting question is, would a comparable result have been reached without the base change and singular value decomposition: how well is the 'pure' vector space model performing? The following Listing 19 provides an answer to this question. In fact, the pure vector space performs with a measured Spearman's Rho of 0.45 visibly lower.

Listing 19 Correlation of human and machine scores in the 'pure' vector space.

```
essay2essay = cor(corpus_essays, method = "spearman")

machinescores = colSums(essay2essay[goldstandard,])/3

cor.test(
   humanscores[names(machinescores),],
   machinescores,
   exact = FALSE,
   method = "spearman",
   alternative = "two.sided"
)

##
## Spearman's rank correlation rho
##
## data: humanscores[names(machinescores),]
   and machinescores
## S = 1616, p-value = 0.02188
## alternative hypothesis: true rho is not equal to 0
## sample estimates:
## rho
## 0.4475
```

4.6 Limitations of Latent Semantic Analysis

In this chapter, an introduction to LSA was delivered, covering the foundations and application techniques. Comprehensive examples illustrate the description. The open source implementation of LSA provided by the author was described in further detail. Two application demos illustrate the use of the *lsa* package for R.

Further information on how to use the 'lsa' package with sparse matrices (using the *tm* package, see Feinerer et al. 2008) and partial SVDs for Linux and Mac OsX (interfacing with the svdlibc of Rhode 2014) is available from the author. The same applies for the binding of the lsa R routines to a REST-ful (Fielding 2000) web service using, for example, the Apache web server (Wild et al. 2008, p. 19ff).

The core restriction of the means for content analysis provided by LSA are its blindness to purpose and social relations and the instruments for interaction analysis that SNA is so popular for.

Moreover, there is no clear rule available, which number of factors to retain and to which to truncate, a shortcoming to which a lot of the criticism of the method can be attributed, often leading to unsuccessful attempts of utilising LSA.

Both these shortcomings will be resolved in the subsequent chapter, which introduces meaningful, purposive interaction analysis as implementation model of the theoretical foundations presented before in Chap. 2.

References

Berry, M., Dumais, S., O'Brien, G.: Using linear algebra for intelligent information retrieval. SIAM Rev. **37**(4), 573–595 (1995)

Calfee, R.: To grade or not to grade. In: Hearst, M. (ed.) The Debate on Automated Essay Grading. IEEE Intelligent Systems, Sep/Oct 2000, pp. 35–37 (2000)

Deerwester, S.: Method and system for revealing information structures in collections of data items, US patent number 5,778,362, dated July 7, 1998, filed June 21, 1996 (1998)

Deerwester, S.: Apparatus and method for generating optimal search queries, Application (deemed withdrawn), Number EP0978058, European Patent Office (2000)

Deerwester, S., Dumais, S., Furnas, G., Harshman, R., Landauer, T., Lochbaum, K., Streeter, L.: Computer information retrieval using latent semantic structure, United States Patent, No. 4,839,853, Appl. No.: 07/244,349, filed: September 15, 1988, Date of Patent: June 13 (1989)

Deerwester, S., Dumais, S., Furnas, G., Landauer, T., Harshman, R.: Indexing by latent semantic analysis. J. Am. Soc. Inf. Sci. **41**(6), 391–407 (1990)

Dessus, P., Lemaire, B.: Apex, un système d'aide à la préparation d'examens (Apex, a system that helps to prepare exams). Sciences et Techniques Éducatives **6–2**, 409–415 (1999)

Dessus, P., Lemaire, B.: Using production to assess learning: an ILE that fosters self-regulated learning. In: Cerri, S.A., Gouarderes, G., Paraguacu, F. (eds.) ITS 2002, LNCS 2363, pp. 772–781. Springer, Berlin (2002)

Dessus, P., Lemaire, B., Vernier, A.: Free-text assessment in a Virtual Campus. In: Zreik, K. (ed.) Proceedings of the Third International Conference on Human System Learning (CAPS'3), Europia, Paris, pp. 61–76 (2000)

Dreher, H.: Interactive on-line formative evaluation of student assignments. Issues Inform. Sci. Inform. Technol. (IISIT) **3**(2006), 189–197 (2006)

Dumais, S.: Improving the retrieval of information from external sources. Behav. Res. Methods Instrum. Comput. **23**(2), 229–236 (1991)

Dumais, S.: Enhancing performance in Latent Semantic Indexing (LSI) retrieval. Bellcore Technical memo (sometimes this is dated 1989, not 1992). Online at: http://citeseerx.ist.psu.edu/viewdoc/summary?doi=10.1.1.50.8278 (1992)

Dumais, S.: Latent Semantic Indexing (LSI): TREC-3 Report. In: Harman, M. (ed.) The Third Text REtrieval Conference (TREC3), NIST Special Publication 500–226, pp. 219–230 (1995)

Dumais, S.: Data-driven approaches to information access. Cogn. Sci. **27**(3), 491–524 (2003)

Dumais, S.: Latent semantic analysis. Annu. Rev. Inform. Sci. Technol. **38**(1), 188–230 (2005)

Dumais, S., Furnas, G., Landauer, T., Deerwester, S., Harshman, R.: Using latent semantic analysis to improve access to textual information, In: CHI '88 Proceedings of the SIGCHI Conference on Human Factors in Computing Systems, pp. 281–285, ACM, New York, NY (1988)

Feinerer, I., Hornik, K., Meyer, D.: Text mining infrastructure in R. J. Stat. Softw. **25**(5), 1–54 (2008)

Fielding, R.T.: Architectural styles and the design of network-based software architectures. Doctoral dissertation, University of California, Irvine (2002)

Foltz, P.: Latent semantic analysis for text-based research. Behav. Res. Methods Instrum. Comput. **28**(2), 197–202 (1996)

Foltz, P.: Quantitative approaches to semantic knowledge representation. Discource Process. **25** (2–3), 127–130 (1998)

Foltz, P.: Using latent semantic indexing for information filtering. In: COCS'90: Proceedings of the ACM SIGOIS and IEEE CS TC-OA conference on Office information systems, pp. 40–47, ACM, New York, NY (1990)

Foltz, P., Kintsch, W., Landauer, T.: The measurement of textual coherence with latent semantic analysis. Discource Process. **25**(2-3), 285–307 (1998)

Foltz, P., Gilliam, S., Kendall, S.: Supporting content-based feedback in on-line writing evaluation with LSA. Interact. Learn. Environ. **8**(2), 111–127 (2000)

Freeman, J., Thompson, B., Cohen, M.: Modeling and diagnosing domain knowledge using latent semantic indexing. Interact. Learn. Environ. **8**(3), 187–209 (2000)

Furnas, G., Landauer, T., Dumais, S., Gomez, L.: Statistical semantics: analysis of the potential performance of key-word information systems. Bell Syst. Tech. J. **62**(6), 1753–1806 (1983)

Graesser, A., Person, N., Magliano, J.: Collaborative dialogue patterns in naturalistic one-to-one tutoring. Appl. Cogn. Psychol. **9**, 295–522 (1995)

Graesser, A., Wiemer-Hastings, K., Wiemer-Hastings, P., Kreuz, R., Tutoring Research Group: AutoTutor: a simulation of a human tutor. J. Cogn. Syst. Res. **1**, 35–51 (1999)

Graesser, A., Wiemer-Hastings, P., Wiemer-Hastings, K., Harter, D., Tutoring Research Group, Person, N.: Using latent semantic analysis to evaluate the contributions of students in AutoTutor. Interact. Learn. Environ. **8**(2), 129–147 (2000)

Haley, D.: Applying latent semantic analysis to computer assisted assessment in the computer science domain: a framework, a tool, and an evaluation. Dissertation, The Open University, Milton Keynes (2008)

Haley, D., Thomas, P., Nuseibeh, B., Taylor, J., Lefrere, P.: E-Assessment using Latent Semantic Analysis. In: Proceedings of the 3rd International LeGE-WG Workshop: Towards a European Learning Grid Infrastructure, Berlin, Germany (2003)

Haley, D., Thomas, P., De Roeck, A., Petre, M.: A research taxonomy for latent semantic analysis-based educational applications. Technical Report 2005/09, The Open University, Milton Keynes (2005)

Haley, D., Thomas, P., Petre, M., De Roeck, A.: EMMA—a computer assisted assessment system based on latent semantic analysis. In: ELeGI Final Evaluation, Technical Report 2008/14, The Open University, Milton Keynes (2007)

Hearst, M.: The Debate on Automated Essay Grading. IEEE Intelligent Systems, Sep/Oct 2000, pp. 22–37 (2000)
Hirschmann, L., Breck, E., Light, M., Burger, J., Ferro, L.: Automated grading of short-answer tests. In: Hearst, M. (ed.) The Debate on Automated Essay Grading. IEEE Intelligent Systems, Sep/Oct 2000, pp. 31–35 (2000)
Holten, R.; Rosenkranz, C.; Kolbe, H.: Measuring application domain knowledge: results from a preliminary experiment. In: Proceedings of ICIS 2010, Association for Information Systems (2010)
Jorge-Botana, G., Leon, J., Olmos, R., Escudero, I.: Latent semantic analysis parameters for essay evaluation using small-scale corpora. J. Quant. Linguist. **17**(1), 1–29 (2010a)
Jorge-Botana, G., Leon, J., Olmos, R., Hassan-Montero, Y.: Visualizing polysemy using LSA and the predication algorithm. J. Am. Soc. Inf. Sci. Technol. **61**(8), 1706–1724 (2010b)
Kalz, M., Van Bruggen, J., Rusman, E., Giesbers, B., Koper, R.: Positioning of learners in learning networks with content analysis, metadata and ontologies. In: Koper, R., Stefanov, K. (eds.) Proceedings of International Workshop "Learning Networks for Lifelong Competence Development" pp. 77–81, Mar 30–31, 2006, TENCompetence Conference, Sofia, Bulgaria (2006)
Kintsch, W.: The representation of knowledge in minds and machines. Int. J. Psychol. **33**(6), 411–420 (1998)
Kintsch, W.: Metaphor comprehension: a computational theory. Psychon. Bull. Rev. **7**(2), 257–266 (2000)
Kintsch, W.: Predication. Cogn. Sci. **25**(2001), 173–202 (2001)
Kintsch, W., Bowles, A.: Metaphor comprehension: what makes a metaphor difficult to understand? Metaphor. Symb. **17**(2002), 249–262 (2002)
Kintsch, E., Steinhart, D., Stahl, G., LSA Research Group, Matthews, C., Lamb, R.: Developing summarization skills through the use of LSA-based feedback. Interact. Learn. Environ. **8**(2), 87–109 (2000)
Klavans, R., Boyack, K.: Identifying a better measure of relatedness for mapping science. J. Am. Soc. Inf. Sci. **57**(2), 251–263 (2006)
Koblischke, R.: Essay Scoring Application @ DotLRN—Implementierung eines Prototyps, Diploma Thesis, Vienna University of Economics and Business (2007)
Kukich, K.: Beyond automated essay scoring. In: Hearst M (ed.) The Debate On Automated Essay Grading. IEEE Intelligent Systems, Sep/Oct 2000, pp. 22–27 (2000)
Laham, D., Bennett, W., Landauer Jr., T.: An LSA-based software tool for matching jobs, people, and instruction. Interact. Learn. Environ. **8**(3), 171–185 (2000)
Landauer, T.: Learning and representing verbal meaning: the latent semantic analysis theory. Curr. Dir. Psychol. Sci. **7**(5), 161–164 (1998)
Landauer, T.: Latent semantic analysis: a theory of the psychology of language and mind. Discource Process. **27**(3), 303–310 (1999)
Landauer, T.: On the computational basis of learning and cognition: arguments from LSA. Psychol. Learn. Motiv. **41**(2002), 43–84 (2002)
Landauer, T., Dumais, S.: A solution to Plato's problem: the latent semantic analysis theory of acquisition, induction, and representation of knowledge. Psychol. Rev. **1**(2), 211–240 (1997)
Landauer, T., Psotka, J.: Simulating text understanding for educational applications with latent semantic analysis: introduction to LSA. Interact. Learn. Environ. **8**(2), 73–86 (2000)
Landauer, T., Laham, D., Rehder, B., Schreiner, M.: How well can passage meaning be derived without using word order? A comparison of latent semantic analysis and humans. In: Proceedings of the 19th annual meeting of the Cognitive Science Society, pp. 412–417, Erlbaum, Mahwah, NJ (1997)
Landauer, T., Foltz, P., Laham, D.: An introduction to latent semantic analysis. In: Discource Process. 25(2–3), 259–284 (1998a)
Landauer, T., Laham, D., Foltz, P.: Learning human-like knowledge by singular value decomposition. In: Jordan, M.I., Kearns, M.J., Solla, S.A. (eds.) Advances in Neural Information

Processing Systems 10. Proceedings of the 1997 Conference, pp. 45–51, The MIT Press (1998b)

Landauer, T., Laham, D., Foltz, P.: The intelligent essay assessor. In: Hearst M. (ed.) The Debate on Automated Essay Grading. IEEE Intelligent Systems, Sep/Oct 2000, pp. 27–31 (2000)

Landauer, T., McNamara, D., Dennis, S., Kintsch, W.: Handbook of latent semantic analysis. Lawrence Erlbaum Associates, Mahwah (2008)

Landauer, T., McNamara, D., Dennis, S., Kintsch, W.: Preface. In: LAndauer, T., McNamara, D., Denns, S., Kintsch, W. (eds.) Handbook of Latent Semantic Analysis. Lawrence Erlbaum Associates, Mahwah (2008)

Larkey, L.: Automated essay grading using text categorization techniques. Proc SIGIR **98**, 90–95 (1998)

Lemaire, B., Dessus, P.: A system to assess the semantic content of student essays. J. Educ. Comput. Res. **24**(3), 305–320 (2001). SAGE Publications

Lenhard, W., Baier, H., Hoffmann, J., Schneider, W.: Automatische Bewertung offener Antworten mittels Latenter Semantischer Analyse. Diagnostica **53**(3), 155–165 (2007a)

Lenhard, W., Baier, H., Hoffmann, J., Schneider, W., Lenhard, A.: Training of Summarisation skills via the use of content-based feedback. In: Wild, F., Kalz, M., van Bruggen, J., Koper, R. (eds.) Mini-Proceedings of the 1st European Workshop on Latent Semantic Analysis in Technology-Enhanced Learning, pp. 26–27, Open University of the Netherlands, Heerlen (2007b)

Lenhard, W., Baier, H., Endlich, D., Lenhard, A., Schneider, W., Hoffmann, J.: Computerunterstützte Leseverständnisförderung: Die Effekte automatisch generierter Rückmeldungen. Zeitschrift für Pädagogische Psychologie **26**(2), 135–148 (2012)

Leon, J., Olmos, R., Escudero, I., Canas, J., Salmeron, L.: Assessing short summaries with human judgments procedure and latent semantic analysis in narrative and expository texts. J. Behav. Res. Methods **38**(4), 616–627 (2006)

Leydesdorff, L.: Similarity measures, author cocitation analysis, and information theory. J. Am. Soc. Inf. Sci. **56**(7), 69–772 (2005)

Lochbaum, K., Psotka, J., Streeter, L.: Harnessing the power of peers. In: Interservice/Industry, Simulation and Education Conference (I/ITSEC), Orlando, FL (2002)

Lochbaum, K., Streeter, L.: Carnegie Hall: an intelligent tutor for command-reasoning practice based on latent semantic analysis. United States Army Research Institute for Behavioral and Social Sciences, ARI Research Note 2002-18 (2002)

Menon, A.K., Elkan, C.: Fast algorithms for approximating the singular value decomposition. ACM Trans. Knowl. Discov. Data **5**(2), 136 (2011)

Miller, T.: Essay assessment with latent semantic analysis. Technical Report (2003)

Olmos, R., Leon, J., Jorge-Botana, G., Escudero, I.: New algorithms assessing short summaries in expository texts using latent semantic analysis. J. Behav. Res. Methods **41**(3), 944–950 (2009)

Olmos, R., Leon, J., Escudero, I., Jorge-Botana, G.: Using latent semantic analysis to grade brief summaries: some proposals, In: Int. J. Cont. Eng. Educ. Life-Long Learn. **21**(2/3), 192–209 (2011)

Page, E.: The imminence of grading essays by computer. Phi Delta Kappan **47**(5), 238–243 (1966)

Palmer, J., Williams, R., Dreher H.: Automated essay grading system applied to a first year university subject—how can we do it better. In: Proceedings of the Informing Science and IT Education (InSITE) Conference, pp. 1221–1229, Cork, Ireland (2002)

Psotka, J., Robinson, K., Streeter, L., Landauer, T., Lochbaum, K.: Augmenting electronic environments for leadership. In: Advanced Technologies for Military Training, RTO meeting proceedings, MP-HFM-101, pp. 307–322 (2004)

Quesada, J.: Introduction to latent semantic analysis and latent problem solving analysis: chapter 2. In: Latent Problem Solving Analysis (LPSA): a computational theory of representation in complex, dynamic problem solving tasks, pp. 22–35, Dissertation, Granada, Spain (2003)

Quesada, J., Kintsch, W., Gomez, E.: A computational theory of complex problem solving using the vector space model (part I): latent semantic analysis, through the path of thousands of ants. Technical Report (2001)

Quesada, J., Kintsch, W., Gomez, E.: A computational theory of complex problem solving using latent semantic analysis. In: Gray, W.D., Schunn, C.D. (eds.) 24th Annual Conference of the Cognitive Science Society, pp. 750–755, Lawrence Erlbaum Associates, Mahwah, NJ (2002a)

Quesada, J., Kintsch, W., Gomez, E.: A computational theory of complex problem solving using the vector space model (part II): latent semantic analysis applied to empirical results from adaptation experiments. Online at: http://lsa.colorado.edu/papers/EMPIRICALfinal.PDF (2002b)

Quesada, J., Kintsch, W., Gomez, E.: Complex problem-solving: a field in search of a definition? Theor. Issues Ergon. Sci. **6**(1), 5–33 (2005)

Rehder, B., Schreiner, M., Wolfe, M., Laham, D., Landauer, T.K., Kintsch, W.: Using latent semantic analysis to assess knowledge: Some technical considerations. Discourse Process. **25** (2–3), 337–354 (1998)

Rhode, D.: SVDLIBC: A C library for computing singular value decompositions, version 1.4. Online at: http://tedlab.mit.edu/~dr/SVDLIBC/ (2014). Last access 31 Jan 2014

Rose, C., Roque, A., Bhembe, D., VanLehn, K.: A hybrid text classification approach for analysis of student essays. In: Proceedings of the HLT-NAACL'03 workshop on Building educational applications using natural language processing, Vol. 2, pp. 68–75, ACM (2003)

Russell, N., ter Hofstede, A., Edmond, D., van der Aalst, W.: Workflow data patterns. QUT Technical report, FIT-TR-2004-01, Queensland University of Technology, Brisbane (2004)

Saeed, J.: Semantics. Wiley-Blackwell, Chichester (2009)

Sahlgren, M.: The distributional hypothesis. Rivista di Linguistica **20**(1), 33–53 (2008)

Sidiropoulos, N., Bro, R.: In memory of Richard Harshman. J. Chemom. **23**(7–8), 315 (2009)

Steedle, J., Elliot, S.: The efficacy of automated essay scoring for evaluating student responses to complex critical thinking performance tasks. Whitepaper, Council for Aid to Education (CAE), New York (2012)

Streeter, L., Psotka, J., Laham, D., MacCuish, D.: The credible grading machine: automated essay scoring in the DOD. In: Interservice/Industry, Simulation and Education Conference (I/ITSEC), Orlando, FL (2002)

Tao, T., Zhai, C.: An exploration of proximity measures in information retrieval. In: SIGIR'07, July 23–27, 2007, Amsterdam, The Netherlands (2007)

Trausan-Matu, S., Dessus, P., Lemaire, B., Mandin, S., Villiot-Leclercq, E., Rebedea, T., Chiru, C., Mihaila, D., Gartner, A., Zampa, V.: Writing support and feedback design, Deliverable d5.1 of the LTfLL project, LTfLL consortium (2008)

Trausan-Matu, S., Dessus, P., Rebedea, T., Mandin, S., Villiot-Leclercq, E., Dascalu, M., Gartner, A., Chiru, C., Banica, D., Mihaila, D., Lemaire, B., Zampa, V., Graziani, E.: Learning support and feedback, Deliverable d5.2 of the LTfLL project, LTfLL consortium (2009)

Trausan-Matu, S., Dessus, P., Rebedea, T., Loiseau, M., Dascalu, M., Mihaila, D., Braidman, I., Armitt, G., Smithies, A., Regan, M., Lemaire, B., Stahl, J., Villiot-Leclercq, E., Zampa, V., Chiru, C., Pasov, I., Dulceanu, A.: Support and feedback services (version 1.5), Deliverable d5.3 of the LTfLL project, LTfLL consortium (2010)

Valenti, S., Neri, F., Cucchiarelli, A.: An overview of current research on automated essay grading. J. Inf. Technol. Educ. **2**(2003), 319–330 (2003)

Van Bruggen, J.: Computerondersteund beoordelen van essays. Technical Report, OTEC 2002/1, Open Universiteit Nederland, Heerlen (2002)

Van Bruggen, J., Sloep, P., van Rosmalen, P., Brouns, F., Vogten, H., Koper, R., Tattersall, C.: Latent semantic analysis as a tool for learner positioning in learning networks for lifelong learning. Br. J. Educ. Technol. **35**(6), 729–738 (2004)

van der Vegt, W., Kalz, M., Giesbers, B., Wild, F., van Bruggen, J.: Tools and techniques for placement experiments. In: Koper, R. (ed.) Learning Network Services for Professional Development, pp. 209–223. Springer, London (2009)

van Lehn, K., Jordan, P., Rosé, C., Bhembe, D., Böttner, M., Gaydos, A., Makatchev, M., Pappuswamy, U., Ringenberg, M., Roque, A., Siler, S., Srivastava, R.: The architecture of why2-atlas: a coach for qualitative physics essay writing. In: Cerri, S.A., Gouardères, G., Paraguaçu, F. (eds.) ITS 2002, LNCS 2363, pp. 158–167, Springer, Berlin (2002)

Whittington, D., Hunt, H.: Approaches to the computerized assessment of free text responses. In: Proceedings of the Third Annual Computer Assisted Assessment Conference (CAA'99), pp. 207–219, Loughborough University (1999)

Wiemer-Hastings, P., Graesser, A.: Select-a-Kibitzer: a computer tool that gives meaningful feedback on student compositions. Interact. Learn. Environ. **8**(2), 149–169 (2000)

Wiemer-Hastings, P., Graesser, A., Harter, D., Tutoring Research Group: The foundations and architecture of AutoTutor. In: Proceedings of the 4th International Conference on Intelligent Tutoring Systems, pp. 334–343, San Antonio, TX, Springer, Berlin (1998)

Wild, F.: lsa: Latent Semantic Analysis: R package version 0.73 (2014). http://CRAN.R-project.org/package=lsa

Wild, F., Stahl, C.: Investigating unstructured texts with latent semantic analysis. In: Lenz, H.J., Decker, R. (eds.) Advances in Data Analysis, pp. 383–390. Springer, Berlin (2007)

Wild, F., Stahl, C., Stermsek, G., Neumann, G.: Parameters driving effectiveness of automated essay scoring with LSA. In: Proceedings of the 9th International Computer Assisted Assessment Conference (CAA), pp. 485–494, Loughborough (2005a)

Wild, F., Stahl, C., Stermsek, G., Penya, Y., Neumann, G.: Factors influencing effectiveness in automated essay scoring with LSA. In: Proceedings of the 12th International Conference on Artificial Intelligence in Education (AIED), Amsterdam, The Netherlands (2005b)

Wild, F., Kalz, M., van Bruggen, J., Koper, R.: Latent semantic analysis in technology-enhanced learning. In: Mini-Proceedings of the 1st European Workshop, Mar 29–30, 2007, Heerlen, NL (2007a)

Wild, F., Koblischke, R., Neumann, G.: A Research prototype for an automated essay scoring application in .LRN. In: OpenACS and .LRN Spring Conference, Vienna (2007b)

Wild, F., Dietl, R., Hoisl, B., Richter, B., Essl, M., Doppler, G.: Services approach & overview general tools and resources, deliverable d2.1, LTfLL consortium (2008)

Williams, R.: Automated essay grading: an evaluation of four conceptual models. In: Kulski, M., Herrmann, A. (eds.) New Horizons in University Teaching and Learning: Responding to Change. Curtin University of Technology, Perth (2001)

Williams, R., Dreher, H.: Automatically grading essays with Markit©. Issues Inform. Sci. Inform. Technol. (IISIT) **1**(2004), 693–700 (2004)

Williams, R., Dreher, H.: Formative assessment visual feedback in computer graded essays. Issues Inform. Sci. Inform. Technol. (IISIT) **2**(2005), 23–32 (2005)

Wolfe, M., Schreiner, M., Rehder, B., Laham, D., Foltz, P., Kintsch, W., Landauer, T.: Learning from text: matching readers and texts by latent semantic analysis. Discourse Process. **25**(2–3), 309–336 (1998)

Yang, Y., Buckendahl, C., Juszkiewicz, P.: A review of strategies for validating computer automated scoring. In: Proceedings of the Annual Meeting of Midwestern Educational Research Association, Chicago, IL (2001)

Chapter 5
Meaningful, Purposive Interaction Analysis

The governance of conversations through meaning and purpose is at the heart of the learning theory laid out in the Chap. 2. For example, when constructing knowledge as an individual, 'knowing' is the ability to perform whenever challenged and invariant of actual conversation partner or formulation. In the challenge to perform we thus find purpose and in conversational understanding we find meaning. Chapter 2 above already provided motivation, theoretical foundation, and considerations for an algorithmic implementation of this theory, which shall be further substantiated in this chapter.

This chapter tends now to the algorithm required for achieving the first two objectives of this work, namely '*to automatically represent conceptual development evident from interaction of learners with more knowledgeable others and resourceful content artefacts; to provide the instruments required for further analysis*' (Sect. 1.2). The third and final objective on re-representation including visualisation will be dealt with in the subsequent Chap. 6, with comprehensive application examples in learning analytics following in Chap. 9.

The proposed algorithm—meaningful, purposive interaction analysis (MPIA)—builds on latent semantic analysis and (social) network analysis, exploiting their advantages, while at the same time overcoming some of their shortcomings.

The conceptual indexing provided by latent semantic analysis as well as the facilities to index social context deployed in social network analysis share the same matrix theory foundations with MPIA, enhanced by—amongst others—visualization, identity measurement, and introspection capabilities that help understand and explain competent verbal action.

Moreover, this chapter derives a mathematically sound way of determining an optimal number of Eigendimensions that explains the desired amount of variability and which is calculable *ex ante* from the trace of the involved matrix, thereby providing significant efficiency gains (see Sect. 10.3.5 for an investigation of performance gain). While providing better efficiency in calculation, this overcomes another significant shortcoming of its predecessor LSA, which is often responsible for failing to produce effective spaces.

F. Wild, *Learning Analytics in R with SNA, LSA, and MPIA*,
DOI 10.1007/978-3-319-28791-1_5

Within this branch of algorithms, matrix theory offers the advantage of providing an efficient and effective instrument to investigate the network-like structures spanned by graphs through their vertices and the edges connecting them.

In learning, focus of analysis can zoom in on a variety of vertex types (such as learners, lecturers, books, essays, or even concepts). The incidence types, from which relationships and other forms of interaction emerge, are similarly unbounded: reading activity, writing activity, in a forum, an essay, etc. As introduced above in Chap. 2, the number of model elements, however, is not endless and many different foci of analysis can be mapped to the elements of the theoretical model elaborated.

This chapter introduces to the mathematical foundation of MPIA in matrix theory, setting relevant links to the theory outlined in Chap. 2 where required. The shared mathematical foundation with (social) network analysis and latent semantic analysis will be explored in order to make visible where MPIA goes beyond the sum of these predating methods.

The chapter thereby is organised as follows. First, essential theorems about matrix algebra are revisited, as a side effect introducing also the basic mathematic vocabulary used in this section. Then, Eigensystem calculation and singular value decomposition are deduced, adding detail that will serve understanding of the MPIA analysis, visualisation and inspection methods that make use of it.

The proximity and identity relations are at the core of this and consequently their working principles and foundations are discussed in the following. Where possible, working examples will illustrate how the matrix theory foundations (and the higher level transformations making use of it) relate to the incidence matrices of the social network analysis and latent semantic analysis constituents.

5.1 Fundamental Matrix Theorem on Orthogonality

Figure 5.1 introduces the basic concepts of Kernel, Domain, Codomain, linear Mapping, Image, and Cokernel (cf. Strang 2009). The Kernel of a linear mapping *Kerf* (aka the 'Nullspace') is where all x are mapped by the rule f to 0, i.e. $f(x) = 0$ with $f(x)$: $x \in X$. In other words, 0 is embedded ('↪') in the *Ker f*, which again is embedded in the domain X.

The Image *Imf* of X, itself a surjective map ('⇒') of X, is the set of elements created in the codomain when ignoring the Kernel *Ker f*. In other words, $X/(Ker f) \cong Im f$.

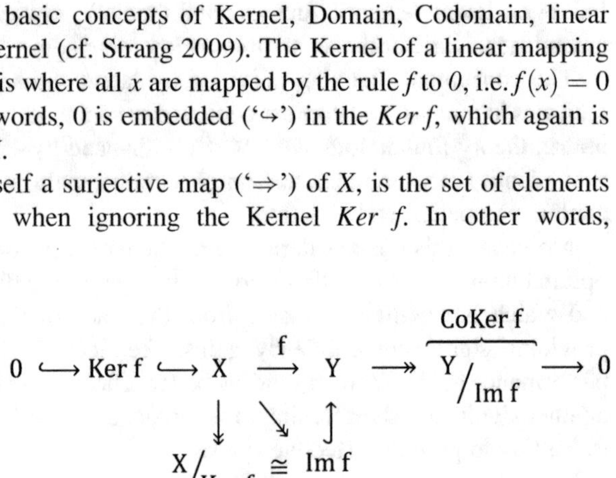

Fig. 5.1 Relation of kernel, mapping, image, and cokernel

$$0 \hookrightarrow \mathrm{Ker\,f} \hookrightarrow \mathrm{X} \xrightarrow{\mathrm{f}} \mathrm{Y} \twoheadrightarrow \overbrace{\mathrm{Y}/\mathrm{Im\,f}}^{\mathrm{CoKer\,f}} \longrightarrow 0$$

$$\mathrm{X}/\mathrm{Ker\,f} \cong \mathrm{Im\,f}$$

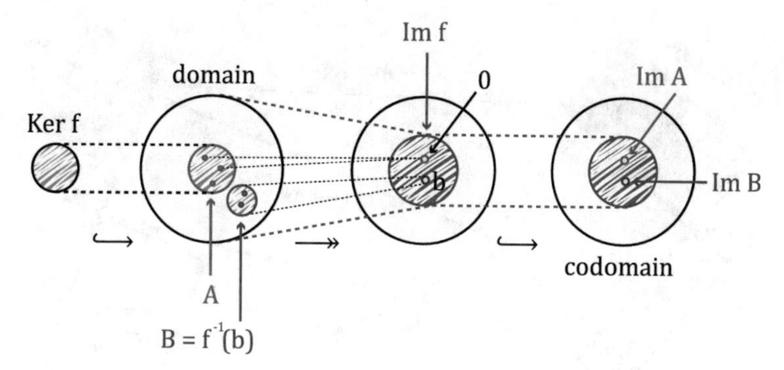

Fig. 5.2 *Ker f* is embedded in the domain, which is mapped by *Im f* to the Image *Im A* in the codomain

The Image *Im f* is embedded ('↪') in the Codomain *Y*, which—when ignoring the Cokernel$CoKer f = Y/(Im f)$—is mapped ('→') to *0*.

Figure 5.2 depicts these relationships. The subset *A*, for example, is mapped to *0* by *Im f*, since it is a subset in the Nullspace*Ker f*. Subset *B*, however, is mapped to *b*, indicating that this is not an injective mapping, as two distinct elements in *B* are mapped to a single element *b*. If the mapping were preserving such difference, it would be called injective. If the Image *Im A* covers the full Codomain *Y*, it would be called surjective. A mapping that is both injective and surjective is called bijective, as only then it is reversible (otherwise only allowing for pseudoinversion).

In the context of *MPIA*, the domain is the set of all uttered cooperatively and communicatively successful exchanges *C*, with any subset such as *A* or *B* holding such exchanges $c \in C$ (see Sect. 2.3). Moreover, in its bag-of-words binding to the vector space model (see Sect. 2.12), the image relation *Im f* becomes possible, which transforms the matrix representation of a document-term matrix to their images (see Sect. 2.3), which are embedded in the codomain of 'all meanings expressed', for which, as we will see subsequently, the Eigensystem calculation and singular value decomposition provide a mathematically sound transformation (see also Sect. 2.3).

For all matrices, the *Rank* of the Kernel *Ker f* and the *Rank* of the Image *Im f* together add up to the *Rank* of the domain. The rank is a measure of the largest number of linear independent columns (or rows for that matter) of a matrix (Barth et al. 1998, p. 23/C). The Rank of the Kernel (aka Nullspace) is defined as the number of independent directions (the number of linearly independent 'dimensions'). The Rank of a subspace is defined as the dimensionality of the linear span of its vectors.

Every matrix can be split into row and column space, as indicated in Fig. 5.3. Thereby, the rank of the row space plus the rank of the Nullspace give the number of columns (Strang 1988, p. 138; Strang 2009).

Moreover, the vectors *x* themselves can be split into their 'row space component' and 'Nullspace component', so that $x = x_r + x_n$. When multiplied with A, the

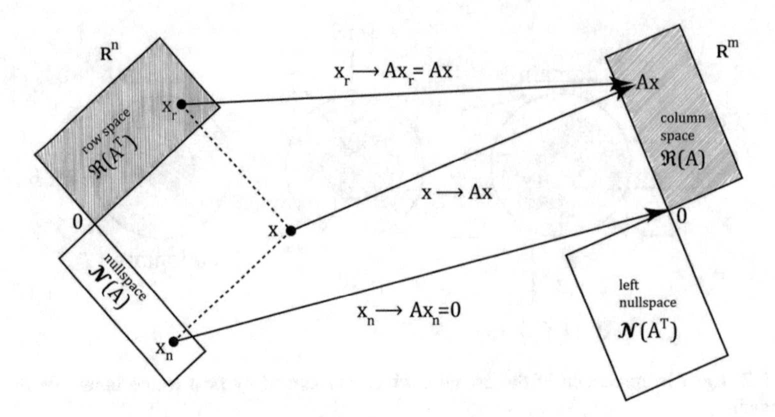

Fig. 5.3 The fundamental theorem of linear algebra: "every matrix transforms its row space to its column space" (redrawn from Strang 1988, p. 140)

Nullspace component carries to 0 and the row space component to the column space (Strang 1988, p. 140).

Effectively, this means that every row space can be mapped to its column space (and vice versa). This is the key fundamental theorem of linear algebra, which is applied subsequently for the Eigensystem calculation—an operation that helps in rotating, shifting, and scaling a matrix, such that the basis of the actual matrix can be determined and made use of in the mapping.

This theorem is further illustrated in Fig. 5.4: each row vector x in a matrix A can be split into the component on x_n on *Ker A* and the row space component x_r.

Since in the transformation the Nullspace components on *Ker A* are neglected, the mapping illustrated in Fig. 5.4 effectively reduces all points on the x_r line to their counterpoint in the column space *Col A*. The pseudoinverse can invert this operation, when ignoring the *Ker A* components: all col space vectors are mapped by the pseudoinverse onto their row space components again.

The application of this transformation will be shown in Sect. 5.1.2. This transformation from row spaces to column spaces is the fundamental theorem behind the mapping operations that turn incidence matrices into affiliation matrices (in SNA) and that map terms (rows) and documents (columns) into the same basis in LSA and MPIA.

To calculate the Kernel *Ker A* of a matrix, the following Eq. (5.1) has to be satisfied.

$$Ax = 0 \tag{5.1}$$

For a matrix A with m rows and n columns, this means that (Strang 1988, p. 136, Eq. 6):

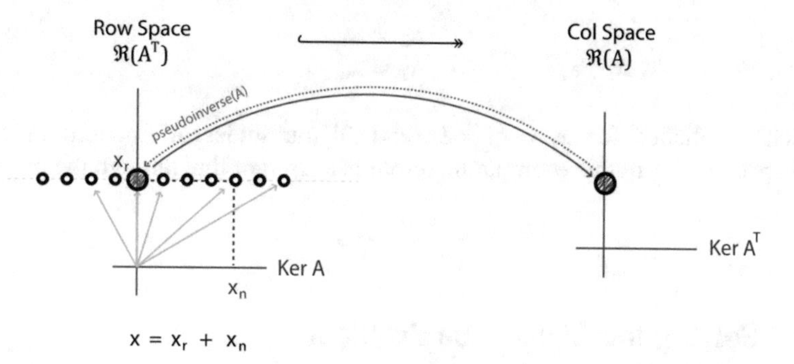

Fig. 5.4 Illustration of the transformation from row space to column space: the *Ker A* components are dropped

$$\begin{pmatrix} - & r_1 & - \\ & \vdots & \\ - & r_m & - \end{pmatrix} \begin{pmatrix} x_1 \\ \vdots \\ x_n \end{pmatrix} = \begin{pmatrix} 0 \\ 0 \\ 0 \\ 0 \end{pmatrix} \tag{5.2}$$

This can be rewritten as a system of linear equations [see Eq. (5.3)], for which a set of solutions can be found (a set since scaling the solution with a scalar does not change the system of equations). As Strang (1988, p. 136) mentions, "the main point is already in the first equation: row 1 is orthogonal to x. Their inner product is zero; that is the equation". The same applies for the second row, and so forth. Each x in the Nullspace is orthogonal to each vector in the row space.

$$\begin{aligned} r_{11}x_1 + r_{12}x_2 + \cdots + r_{1n}x_n &= 0 \\ &\vdots \\ r_{m1}x_1 + r_{m2}x_2 + \cdots + r_{mn}x_n &= 0 \end{aligned} \tag{5.3}$$

Consider, for example, the matrix

$$A = \begin{pmatrix} -2 & 2 \\ 2 & -2 \end{pmatrix} \tag{5.4}$$

To satisfy Eq. (5.1), rewriting Eq. (5.2) leads to the following system of linear equations [as indicated in Eq. (5.3)]:

$$\begin{aligned} -2x_1 + 2x_2 &= 0 \\ 2x_1 + (-2)x_2 &= 0 \end{aligned} \tag{5.5}$$

Reformulated, this leads to

$$\begin{aligned} x_2 &= x_1 \\ x_1 &= x_2 \end{aligned} \tag{5.6}$$

which is satisfied for $x_1 = x_2 = 1$ (and all multitudes or fractions of 1). The Nullspace *Ker A* in this example therefore is a straight line through the origin of R^2.

5.2 Solving the Eigenvalue Problem

For symmetric, square matrices, Eigenvectors and Eigenvalues can be calculated such that any multiplication of the matrix B with an Eigenvector x yields a constant multiple of the Eigenvector, scaled by the Eigenvalue λ (Barth et al. 1998, p. 90/E): $Bx = \lambda x$.

From the system of Eigenvectors, a so-called Eigenbasis can be constructed that consists of the linearly independent Eigenvectors. The Eigenvalue calculation moves the background (the basis of a matrix), thereby valuing symmetry.

A document-term matrix A can be multiplied with its transpose to establish a symmetric matrix $B = A^T A$:

$$\left(A^T A\right)^T = A^T A \tag{5.7}$$

Since it is symmetric, it has Real number eigenvalues and a base of orthogonal eigenvectors, which can be normalized, such that there is an orthonormal base:

$$\exists \text{ON base } \{e_1, \ldots e_n\} \text{ for } A^T A \text{ such that } A^T A e_i = \lambda e_i \tag{5.8}$$

From Eq. (5.8) follows:

$$\begin{aligned} A^T A e_i &= \lambda e_i && |\text{move } \lambda e_i \text{ to the left} \\ A^T A e_i - \lambda e_i &= 0 && |\text{pull out } e_i \text{ using identity matrix}^1 I \\ \left(A^T A - \lambda I\right) e_i &= 0 \end{aligned}$$

The <u>eigenvectors</u> for the symmetric mapping $B = A^T A$ are the non-zero solutions x to the equation $Bx = \lambda x$. The corresponding scalar λ is called an <u>eigenvalue</u>. This means for each x_i:

$$\begin{aligned} A^T A\, x_i &= \lambda_i x_i && |\text{multiply with } x^T \text{from the left} \\ x_i^T A^T A x_i &= x_i^T \lambda\, x_i && |\text{pull out } \lambda, \text{ since it is a number} \end{aligned} \tag{5.9}$$

$$x_i{}^T A^T A x_i = \lambda\, x_i{}^T x_i \tag{5.10}$$

If x is coordinatized by using a basis E:

$$[x_i]_E = \begin{pmatrix} x_{i,1} \\ \cdots \\ xi,_n \end{pmatrix}_E$$

and since this basis E is orthonormal, then the length of x is:

$$||x_i|| = \sqrt{(x_{i,1^2} + x_{i,1^2} + \cdots x_{i,n}{}^2)} \tag{5.11}$$

because $e_i e_i = 1$ and $e_i e_k = 0$

$$x_i^T x_i = (x_{i,1} x_{i,2} \cdots x_{i,n}) \begin{pmatrix} x_{i,1} \\ x_{i,2} \\ \vdots \\ x_{i,n} \end{pmatrix} = x_{i,1}^2 + x_{i,2}^2 + \cdots + x_{i,n}^2 = ||x_i||^2 \tag{5.12}$$

From Eq. (5.10):

$$\begin{aligned} x_i^T A^T A x_i &= \lambda\, x_i^T x_i && | \text{ from Eq. 12 : } x^T x \text{ is equal} \\ & && \text{to square of length of x} : ||x||^2 \\ x_i^T A^T A x_i &= \lambda ||x_i||^2 && | \text{ transposition reversal : } (AB)^T = B^T A \\ (Ax_i)^T\, A x_i &= \lambda ||x_i||^2 && | \text{ using Eq. 12 on the left now} \\ ||Ax_i||^2 &= \lambda ||x_i||^2 && \\ ||Ax_i|| = \sqrt{\lambda}\, ||x_i|| &= \sigma\, ||x_i|| && | \text{ note : repeat this for all } x_i, \\ & && \text{to get to } \Sigma = \text{diag}(\sigma_1, \sigma_2, \ldots \sigma_k) \end{aligned}$$

From Eq. (5.9) follows:

$$\begin{aligned} A^T A\, x_i &= \lambda_i\, x_i && | - \lambda_i\, x_i \\ A^T A\, x_i \lambda_i\, x_i &= 0 && | \text{factor out } x_i \\ (A^T A - \lambda_i I) x_i &= 0 && \end{aligned} \tag{5.13}$$

This is the case if and only if λ_i is a root of A, satisfying 'the characteristic equation' (Barth et al. 1998, p. 91/E1.):

$$\det(A^T A - \lambda_i I) = 0 \tag{5.14}$$

The determinant theorems state that AB and BA have the same eigenvalues:

$$\det(\mathrm{AB} - \lambda \mathrm{E}) = 0$$
$$\det(\mathrm{BA} - \lambda \mathrm{E}) = 0$$
$$\det(\mathrm{AB}) = \det(\mathrm{A})\det(\mathrm{B})$$

Therefore:

$$\begin{aligned}\det\left(\mathrm{AA}^{\mathrm{T}} - \lambda I_m\right) &= 0 \\ \det\left(\mathrm{A}^{\mathrm{T}}\mathrm{A} - \lambda I_{\mathrm{n}}\right) &= 0\end{aligned} \tag{5.15}$$

5.3 Example with Incidence Matrices

A document-term matrix A such as the one presented in the previous chapter in Table 4.3 (Sect. 4.4) results in the following matrix AA^T, when multiplied with its transpose. Thereby, the document-term matrix is an undirected incidence matrix similar to the incidence matrix of persons and their course attendance presented in Table 3.1 of Chap. 3 (Sect. 3.2).

Multiplying with its transpose results in the number of shared 'incidences' per row entry—in case of the SNA example shared incidences per pair of persons, in case of the LSA example (see Table 5.1 below), the number of shared document occurrence per pair of terms.

For example, 'social' co-occurs with 'system' two times (in a single document), whereas 'time' co-occurs with 'access' two times as well (but in two different documents).

Multiplying the other way round—the transpose of the document-term matrix with the document-term matrix—the shared terms for all pairs of documents are calculated, resulting in the symmetric matrix A^TA depicted in Table 5.2.

To keep the illustrating calculation simple, we will look just at a subspace in the symmetric matrix AA^T: the co-occurrence between the words 'web' and 'access', which is (see Table 5.1):

$$AA^T[3:4,3:4] = B = \begin{pmatrix} 1 & 2 \\ 2 & 1 \end{pmatrix}$$

Let x be an eigenvector of the matrix B. This means that x is mapped by B to a multiple of itself, and such multiple is the corresponding eigenvalue λ [see Eq. (5.9)].

$\mathrm{Bx} = \lambda \mathrm{x}$

This equation is equivalent to [see Eq. (5.13)]:

$(\mathrm{B} - \lambda \mathrm{I})\mathrm{x} = 0$

Table 5.1 AA^T: number of shared document incidences per term pair

	interface	social	web	access	review	system	time	usage	html	management	trees	clustering	intersection	agglomerative	knowledge	learning	organisational
interface	2	1	1	0	0	1	0	1	1	1	0	0	0	0	0	0	0
social	1	3	1	1	0	2	1	0	1	0	0	0	0	0	0	0	0
web	1	1	2	1	1	1	1	1	0	0	0	0	0	0	0	0	0
access	0	1	1	2	1	1	2	1	0	0	0	0	0	0	0	0	0
review	0	0	1	1	2	1	1	1	0	0	0	1	0	1	0	0	0
system	1	2	1	1	1	6	1	2	3	1	0	0	0	0	0	0	0
time	0	1	1	2	1	1	2	1	0	0	0	0	0	0	0	0	0
usage	1	0	1	1	1	2	1	2	1	1	0	0	0	0	0	0	0
html	1	1	0	0	0	3	0	1	2	1	0	0	0	0	0	0	0
management	1	0	0	0	0	1	0	1	1	2	0	0	0	0	1	1	1
trees	0	0	0	0	0	0	0	0	0	0	3	2	1	1	0	0	0
clustering	0	0	0	0	1	0	0	0	0	0	2	3	1	2	0	0	0
intersection	0	0	0	0	0	0	0	0	0	0	1	1	2	0	1	1	1
agglomerative	0	0	0	0	1	0	0	0	0	0	1	2	0	2	0	0	0
knowledge	0	0	0	0	0	0	0	0	0	1	0	0	1	0	3	2	2
learning	0	0	0	0	0	0	0	0	0	1	0	0	1	0	2	4	2
organisational	0	0	0	0	0	0	0	0	0	1	0	0	1	0	2	2	2

Table 5.2 A^TA: Shared term incidences for document pairs

	c1	c2	c3	c4	c5	m1	m2	m3	m4	p1	p2	p3	p4	p5
c1	3	1	1	1	1	0	0	0	0	0	0	0	0	0
c2	1	6	2	2	2	0	0	0	1	0	0	0	0	0
c3	1	2	5	3	0	0	0	0	0	0	0	0	1	0
c4	1	2	3	6	1	0	0	0	0	0	0	0	0	0
c5	1	2	0	1	3	0	0	0	0	0	0	0	0	0
m1	0	0	0	0	0	1	1	1	0	0	0	0	0	0
m2	0	0	0	0	0	1	3	2	1	1	0	0	0	0
m3	0	0	0	0	0	1	2	3	2	0	0	0	0	0
m4	0	1	0	0	0	0	1	2	3	0	0	0	0	0
p1	0	0	0	0	0	0	1	0	0	4	1	1	3	1
p2	0	0	0	0	0	0	0	0	0	1	1	1	1	0
p3	0	0	0	0	0	0	0	0	0	1	1	1	1	0
p4	0	0	1	0	0	0	0	0	0	3	1	1	4	1
p5	0	0	0	0	0	0	0	0	0	1	0	0	1	1

This equation has non-zero solutions x iff

$$\begin{aligned} \det(\mathrm{B}-\lambda\mathrm{I}) &= 0 = \\ &= \begin{vmatrix} 1-\lambda & 2 \\ 2 & 1-\lambda \end{vmatrix} = (1-\lambda)\cdot(1-\lambda) - 2\cdot 2 = \\ &= 1+\lambda^2 - 2\lambda - 4 = \lambda^2 - 2\lambda - 3 = \\ &= (\lambda+1)(\lambda-3) = (\lambda-(-1))\cdot(\lambda-(+3)) = 0 \end{aligned}$$

This condition is only fulfilled for the eigenvalues $\lambda = -1$ and $\lambda = 3$. Note that for this equation, the eigenvectors x are not required for the calculation of its solution. The eigenvalues can be determined through the determinant and the eigenvectors can then be found in the subsequent step. Any eigenvector x that is *0* is in the Nullspace.

To calculate the corresponding eigenvector for the first eigenvalue $\lambda = -1$, Eq. (5.13) can be set in as follows:

$$(B - \lambda I)x = \begin{pmatrix} 1-(-1) & 2 \\ 2 & 1-(-1) \end{pmatrix} x = \begin{pmatrix} 2 & 2 \\ 2 & 2 \end{pmatrix} x = 0$$

Rewriting this into a system of linear equations gives:

$$\begin{aligned} 2x_1 + 2x_2 &= 0 \quad &&|\text{(first row)} \\ 2x_1 + 2x_2 &= 0 \quad &&|\text{(second row); divide by 2} \\ x_1 + x_2 &= 0 \quad &&|\text{move to left with} - x_2 \\ x_1 &= -x_2 && \end{aligned}$$

This equation system is satisfied, for example, by $x_1 = 1$ and $x_2 = -1$, which then gives

$$(B - \lambda I)x = \begin{pmatrix} 2 & 2 \\ 2 & 2 \end{pmatrix} x = 0 = \begin{pmatrix} 2 & 2 \\ 2 & 2 \end{pmatrix} \begin{pmatrix} 1 \\ -1 \end{pmatrix} = \begin{pmatrix} 2 \cdot 1 + 2 \cdot -1 \\ 2 \cdot 1 + 2 \cdot -1 \end{pmatrix} = \begin{pmatrix} 0 \\ 0 \end{pmatrix}$$

To calculate the second eigenvector for the second eigenvalue $\lambda = 3$, the second eigenvalue is put in again into Eq. (5.13):

$$\begin{aligned} (B - \lambda I)\mathrm{x} &= \begin{pmatrix} 1-3 & 2 \\ 2 & 1-3 \end{pmatrix} x = \begin{pmatrix} -2 & 2 \\ 2 & -2 \end{pmatrix} x \\ &= \begin{pmatrix} -2x_1 & 2x_2 \\ 2x_1 & -2x_2 \end{pmatrix} = 0 \end{aligned}$$

Solving the system gives the following constraint on the second eigenvector subspace (in this case a line through the origin):

$$x_1 = x_2$$

This is satisfied, for example, by $x_1 = x_2 = 1$, or better—(normalized by intersecting with the unit circle!)—by $x_1 = x_2 = 1/\sqrt{2}$. It should be noted that when binding the eigenvectors together into an eigenvector matrix, the eigenproblem in Eq. (5.9) of course has to be formulated to fit the column vector binding. All eigenvectors x_i found via

$$\begin{aligned} Ax_1 &= \lambda_1 x_1 \\ Ax_2 &= \lambda_2 x_2 \\ &\dots \\ Ax_n &= \lambda_n x_n \end{aligned}$$

are appended to an eigenvector matrix Q such that the basic eigenproblem equation states

$$AQ = Q\Lambda \tag{5.16}$$

5.4 Singular Value Decomposition

To extend this to non-symmetrical matrices (such as any m by n matrix), singular value decomposition comes into play: Any m by n matrix can be factored into $A = U\Sigma V^T$ (with $A \in^{m \times n}$) see Golub and van Loan (1983, p. 285) and Strang (1976, p. 142/3R). Thereby, U holds the left-singular eigenvectors and V the right-singular eigenvectors: they are created from symmetric matrices using the trick of multiplying the original (document-term) matrix A with its transpose: U are the eigenvectors of A^TA and V those of AA^T.

$$A = U\Sigma V^T \tag{5.17}$$

Setting in Eq. (5.9) with A^TA and AA^T gives the following linear equations for the eigenvectors and eigenvalues:

$$A^TAU = \lambda U \tag{5.18}$$

$$AA^TV = \mu V \tag{5.19}$$

Following the instructions in Sect. 5.1.2, the Eigenvalue problem can be solved by first computing the determinant of $A^TA - \lambda I$ and $AA^T - \mu I$, then finding the eigenvalues for this polynomial by solving eigenvectors by $det(A^TA - \lambda I) = 0$ and $det(AA^T - \mu I) = 0$, to then finally calculate the solving $(A^TA - \lambda I)x = 0$ and $(AA^T - \mu I)x = 0$.

The resulting eigenvector matrix literals of such operation are listed below in Eqs. (5.20) and (5.21).

$$U = Eigenvectors(A^TA) \tag{5.20}$$

$$V = Eigenvectors(AA^T) \tag{5.21}$$

Thereby, λI and the same valued though holding additional *0* values μI (A^TA and AA^T are rotations of each other!) denotes the diagonal matrix of eigenvalues for AA^T and A^TA, the square root of which gives the Σ needed for A, see subsequent Eq. (5.24).

Turning to the other possible transformations, the following Eqs. (5.22), (5.23), (5.28), and (5.29) can be derived.

$$\begin{array}{ll} A = U\Sigma V^T & |\text{Eq. (5.17), multiply from left with } U^T \\ U^TA = U^TU\Sigma V^T & |\text{multiply with } V \text{ from the right} \\ U^TAV = U^TU\Sigma V^TV & |U^TU = I_u \text{and } V^TV = I_V, \text{ thus drop out} \\ U^TAV = \Sigma & |\text{rewrite} \\ \Sigma = U^TAV & \end{array} \tag{5.22}$$

$$\begin{array}{ll} A = U\Sigma V^T & |\text{Eq. (5.17), left multiply with } A^T \\ A^TA = (U\Sigma V^T)^T(U\Sigma V^T) = & |\text{transpose reverses order}: (AB)^T = B^TA^T \\ = (V^T)^T\Sigma^TU^TU\Sigma V^T = & |U^TU = I, \text{ therefore dropped} \\ = V\Sigma^T\Sigma V^T = A^TA & \end{array} \tag{5.23}$$

$$\begin{aligned} & && \text{right multiply with } V \\ &= V\Sigma^T\Sigma V^T V = A^T A\, V && V^T V = I, \text{ therefore drops out} \\ &= V\Sigma^T\Sigma = A^T A\, V && \end{aligned} \tag{5.24}$$

Equation (5.24) is nothing else than the eigenvalue matrix equation $AQ = Q\Lambda$. Thereby, Λ is the diagonal matrix holding the eigenvalues.

For all diagonal matrices, Eq. (5.23) applies.

$$\Sigma^T\Sigma = \Sigma^2 \tag{5.25}$$

$$\Sigma^T\Sigma = \mathrm{diag}(\sigma_1, \sigma_2, \ldots \sigma_n) = \begin{pmatrix} \sigma_1 & & \\ & \ddots & \\ & & \sigma_n \end{pmatrix} \tag{5.26}$$

By inserting Eq. (5.25) into Eq. (5.24), Σ can be determined to be the square root of $\Lambda = \lambda I$, see Eq. (5.27).

$$\Sigma = \sqrt{\begin{pmatrix} \sigma_1 \\ \sigma_2 \\ \vdots \\ \sigma_n \end{pmatrix}} = \begin{pmatrix} \sqrt{\sigma_1} \\ \sqrt{\sigma_2} \\ \vdots \\ \sqrt{\sigma_n} \end{pmatrix} \tag{5.27}$$

There is, however, another way to allow for this mapping, by modifying Eq. (5.23) as follows.

$$\begin{aligned} A^T A &= V\Sigma^T\Sigma V^T && \text{Eq. (5.23), left multiply with } V^T \\ V^T(A^T A) &= V^T V\Sigma^T\Sigma V^T && V^T V = I, \text{ therefore dropped} \\ V^T(A^T A)V &= \Sigma^T\Sigma V^T V && V^T V = I, \text{ therefore dropped} \\ V^T(A^T A)V &= \Sigma^T\Sigma = diag(\sigma_1,\ \sigma_2,\ \ldots\ \sigma_n) \in^{n \times n} && \end{aligned} \tag{5.28}$$

The same applies for the $\Sigma\Sigma^T$ mapping, this can be derived from the left-singular eigenvectors U and A in the following way.

$$\begin{aligned} \mathrm{A} &= \mathrm{U\Sigma V^T} && \text{Eq. (5.17), right multiply with } \mathrm{A^T} \\ \mathrm{AA^T} &= \mathrm{U\Sigma V^T}\left(\mathrm{U\Sigma V^T}\right)^{\mathrm{T}} && \text{transposition reverses order of factors} \\ \mathrm{AA^T} &= \mathrm{U\Sigma V^T}\left(\mathrm{V^T}\right)^{\mathrm{T}}\mathrm{\Sigma^T U^T} && \left(\mathrm{V^T}\right)\left(\mathrm{V^T}\right)^{\mathrm{T}} = \mathrm{I}, \text{ therefore dropped} \\ \mathrm{AA^T} &= \mathrm{U\Sigma\Sigma^T U^T} && \text{left multiply with } \mathrm{U^T} \\ \mathrm{U^T AA^T} &= \mathrm{U^T U\Sigma\Sigma^T U^T} && \mathrm{U^T U} \text{ drops out, right multiply with U} \\ \mathrm{U^T AA^T U} &= \mathrm{\Sigma\Sigma^T U^T U} && \mathrm{U^T U} \text{ drops out} \end{aligned}$$

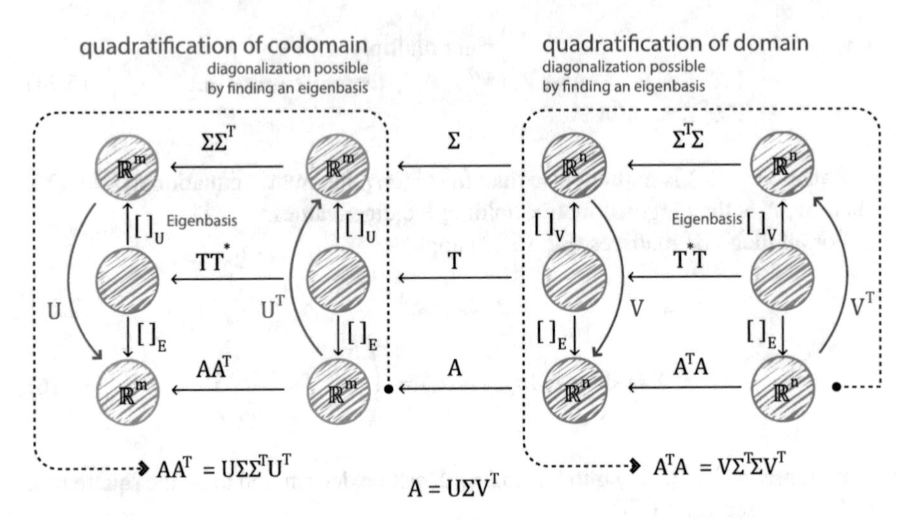

Fig. 5.5 Mappings in the singular value decomposition

$$U^T A A^T U = \Sigma\Sigma^T = \operatorname{diag}(\sigma_1, \sigma_2, \ldots \sigma_n, 0, \ldots 0) \in \mathbb{R}^{m\times m}$$

$$= \Sigma\Sigma^T = \operatorname{diag}(\sigma_1, \sigma_2, \ldots \sigma_n, 0, \ldots 0) = \begin{pmatrix} \sigma_1 & & & & \\ & \ddots & & & \\ & & \sigma_n & & \\ & & & 0 & \\ & & & & \ddots \\ & & & & & 0 \end{pmatrix} \tag{5.29}$$

Figure 5.5 depicts the relationships made possible through these equations. Note that on the left hand side, the involved transformation matrices reside in R^n, whereas on the right they reside in R^m.

For example, it is now possible to change basis from the originating basis $[]_E$ to the Eigenbasis$[]_V$ for A^TA by using V^T as a mapping. Moreover, it is possible to replace the mapping A^TA with $V\Sigma^T\Sigma V^T$ as indicated above in Eq. (5.23).

Similarly, U^T provides a mapping to transform AA^T from its originating basis $[]_E$ to the basis of the eigenspace$[]_U$. Moreover, again the transformation AA^T can be replaced by the alternative mapping U^TAA^TU [see Eq. (5.29)].

Virtually any combination of these mapping routes can be used. For example, the mapping A can be replaced with the full long route around Fig. 5.5:

$$A(= USV^T) = (A^TA)^{-1}V^T(V^T(A^TA)V)\,S\,(U^TAA^TU)\,U\,(AA^T)^{-1}$$

The ability of the underlying linear algebra to recombine these routes will become particularly relevant, when turning to the fold in routines that allow for ex post updating of the Eigenbasis conversion provided by the singular value decomposition.

To make an example of the actual implementation of this in R, these matrix mappings of the long route described above can be executed with the following code (see Listing 1). This line of code illustrates at the same time that the transformation chain produces the original result.

Listing 1 An example matrix mapping (results in A, the original document-term matrix[1]).

```
pinv(a %*% t(a)) %*%
u%*%
(t(u) %*% a %*% t(a) %*% u)[,1:ncol(u)] %*%
diag(s[1:ncol(u)]) %*%
((t(v) %*% (t(a) %*% a) %*% v) %*%
t(v) %*%
pinv(t(a) %*% a))[1:ncol(u),]
```

The singular value decomposition allows mapping the projection described by the document-term matrix from its originating base into the Eigenbasis. This allows to compare both row space and column space vectors in the same geometrical space, as illustrated in Fig. 5.6: in a classical vector space model, the document vectors $d_1, \ldots, d_5$ are coordinatised within their term-axes t_1, t_2, and t_3, while in the Eigenspace, both terms and documents are coordinatised along the orthogonal eigendimensions.

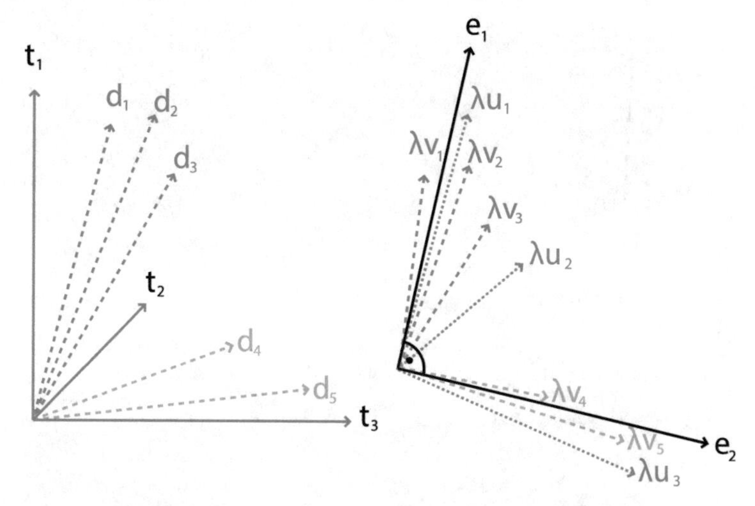

Fig. 5.6 Base transformation (from term-document space to Eigenspace)

[1] Minor rounding errors may happen, of course.

5.5 Stretch-Dependent Truncation

The eigenvalues of the eigenvectors of AA^T and A^TA are the same, since AA^T is symmetrical, while their eigenvectors—of course—differ. Geometrically, the eigenvalues are the 'stretch factors' with which the eigenvectors x are scaled to map to the image A^TAx.

Typically and not only for the sparse distributions of document-term matrices, the first few eigenvalues are rather big, whereas the last few eigenvalues tend to get very close to *0*, see Fig. 5.7. With the eigenvalues being the stretch factors of the eigenvectors, this effectively means that the first values are responsible for the bigger shares of the total transformation, whereas the latter eigenvalues are responsible for the smaller shifts.

Via the sum of eigenvalues Σ^2 it is now possible to calculate the total 'amount of stretch' accounted for by the number of eigenvalues at any given index position k. The total sum of all eigenvalues results in a mapping that reproduces the original matrix mapping, where as any smaller number of the first k eigenvalues results in producing a least square optimal approximation of the original matrix transformation (see Strang 1976, p. 123ff). In LSA, this is effect is used to lift the representation of the document-term matrix up to a 'latent' semantic level by loosing the smaller Eigendimensions, but a clear recommendation on what the desired number of dimensions should have is missing. Within this section, a novel, mathematically

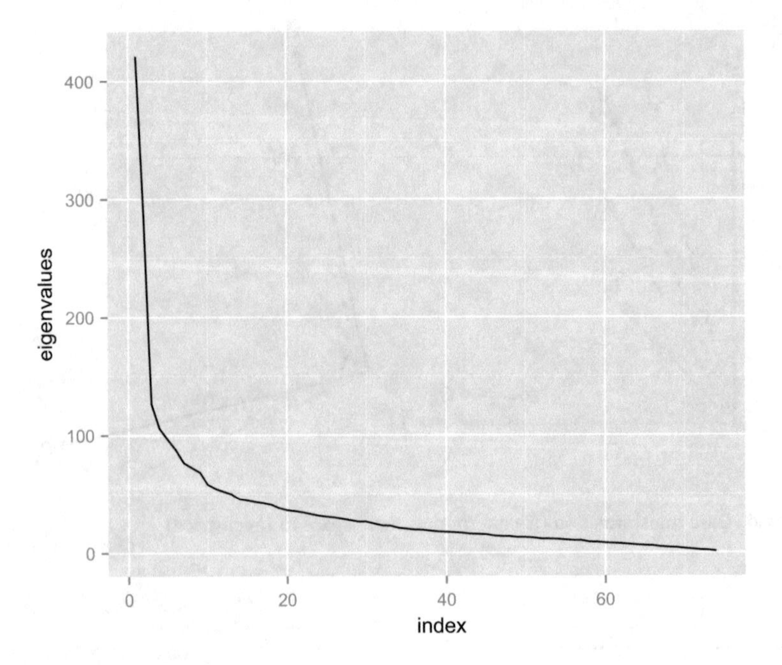

Fig. 5.7 Example eigenvalue distribution

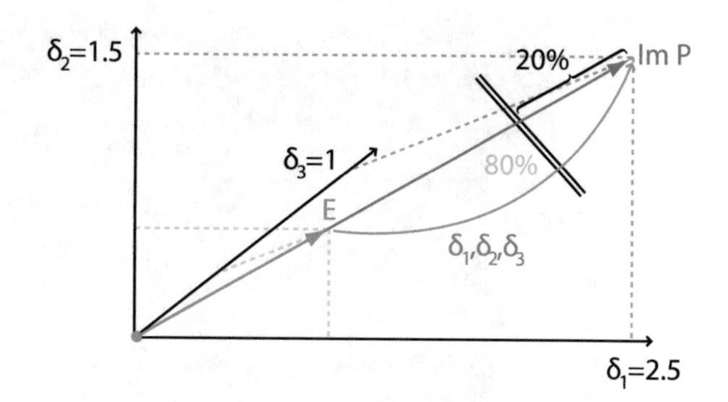

Fig. 5.8 Stretch-dependent truncation

grounded method for determining the number of Eigendimensions to retain will be presented.

Figure 5.8 illustrates this with an example using three eigenvalues. The eigenvector E is stretched to the image *Im P* by multiplying its components with the according eigenvalue δ_i. In the example, the first two eigenvalues already sum up to 80 % of the total stretch possible, so a useful truncation could be to two eigenvalues, *with the remaining 20 % missing to the full stretch being contributed by the third dimension*. Note that the 20 % share of the red vector from the origin to *Im P* is contributed by the eigenvalue δ_3 of the third dimension. For the singular value decomposition, this effectively means that the sum of squares of singular values sums up to 80 % [see also Eq. (5.27)].

Since the sum of the eigenvalues Σ^2 agrees with the sum of the trace of matrix, i.e. with the sum of values down the diagonal (Strang 2009, p. 288). This provides a shortcut for determining the threshold value up to which the eigenvalues (the squared singular values) of A^TA should be retained: 80 % of the trace $Tr(A^TA)$.

$$threshold = 0.8\ Tr(A^TA) \tag{5.30}$$

This is a significant computational advantage saving a vast amount of calculation time and memory: the cut-off value for Σ^2 can be determined already from the original A^TA matrix (or AA^T, for that matter) by iteratively calculating the next eigenvalue until the desired threshold is reached.

While there are many different options, how to implement this calculation of the trace and the chosen, application dependant cut-off point, a very efficient method in R for this is provided in Listing 2 (for this type of problem more efficient than using `Trace()` in package *pracma*: see Borchers 2014).

Listing 2 Efficient calculation of 80 % of the trace.

```
tr = 0.8 * sum(dtm*dtm)
```

The *LAS2* algorithm provided in *svdlibc* does not yet offer such a switch for terminating calculation after a desired threshold is exceeded, though its

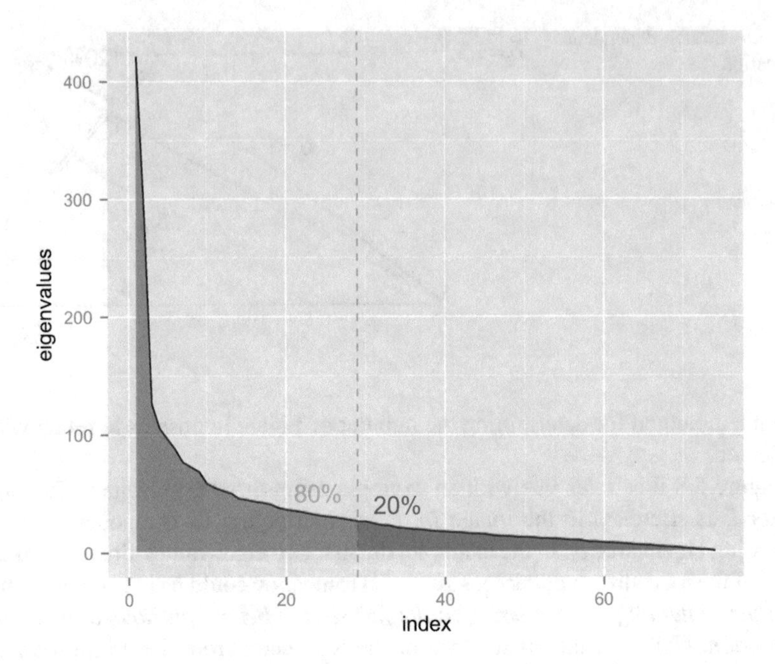

Fig. 5.9 The 80 % stretch in the eigenvalue example

implementation is now possible: basically through replacing or extending the switch for providing a cut-off point for the desired number of dimensions by a cut-off point for desired sum of eigenvalues.

Revisiting Fig. 5.7, such cut-off point can be identified from the trace of the original document-term matrix: $0.8Tr(A^TA) = 0.8 \times 2,671 = 2,136.8$ The last index position k for which the threshold value is not yet bypassed is $k = 29$ (with the sum of $\Sigma^2_{1\ldots29} = 2,123.9$, see Figure 5.9.

This allows breaking off the calculation of eigenvalues already at dimension number *29* instead of calculating all possible eigenvalues. Depending on the corpus size and variability in word use within the documents, deriving the full set of eigenvalues is usually a computationally intense endeavour.

For example, even in the with *769* terms and *74* documents comparatively small document-term matrix, all 74 possible eigenvalues are non-zero, thus effectively saving more than 45 iterations (>60 %) by voiding the calculation of the remaining eigenvalues through this new method of calculating a mathematically sound threshold.

Section 10.3.5 investigates the performance gains in more depth with more and realistic examples.

Same as in latent semantic analysis, the effect of the truncation can be inspected by re-multiplying the truncated Eigenvectors U_k and V_k with the truncated roots of the Eigenvalues Σ_k, see Fig. 5.10. Other than in LSA, there is now a clear

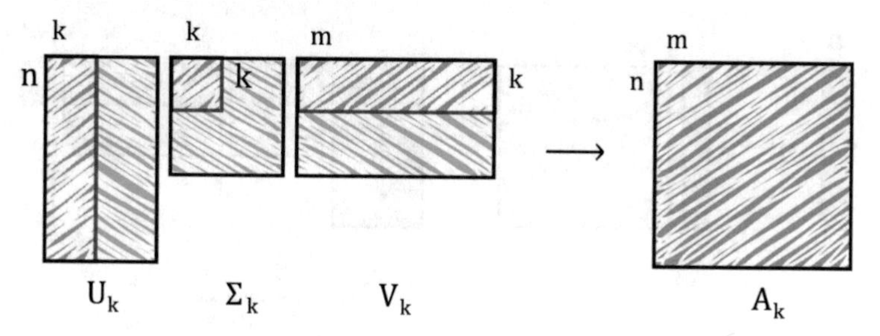

Fig. 5.10 Truncation of the Eigenbasis

explanation on what the amount of stretch means: setting it allows calibrating the right level of abstraction.

The singular value equation can be modified accordingly to reflect the novel method of determining the truncation index. This is shown in Eq. (5.31).

$$A_k = U_k \Sigma_k V_K^T, \; with \; k = 0.8 \; Tr(A^T A) \tag{5.31}$$

The general recommendation for the stretch is to aim for 80 %, but valid results can be achieved with other settings as well (see also Chap. 10).

5.6 Updating Using Ex Post Projection

Same as in latent semantic analysis, it is possible to insert additional column vectors into an existing Eigenspace through applying the following two mappings (see also Sect. 4.2: mathematical foundations). First, a projection of the add-on column vector a onto the span of the current row vectors U_k is required [see also Berry et al. 1994, Eq. (5.7) and page 16]:

$$v' = a^T U_k \Sigma_k^{-1} \tag{5.32}$$

Then in the second step, this new Eigenvector v' can be used to construct a new matrix column a' in the k-truncated Eigenspace$[]_k$:

$$a' = U_k \Sigma_k v'^T \tag{5.33}$$

Figure 5.11 illustrates this projection. The new eigenvector v′ is calculated from a using the row span U_k and the pseudoinverse of Σ_k. To create a column vector for A_k, a' is calculated subsequently using the three partial matrices.

This process is called ‘fold in’ (Berry et al. 1994, p. 5). Other than recalculating the singular value decomposition, this process is computationally much more

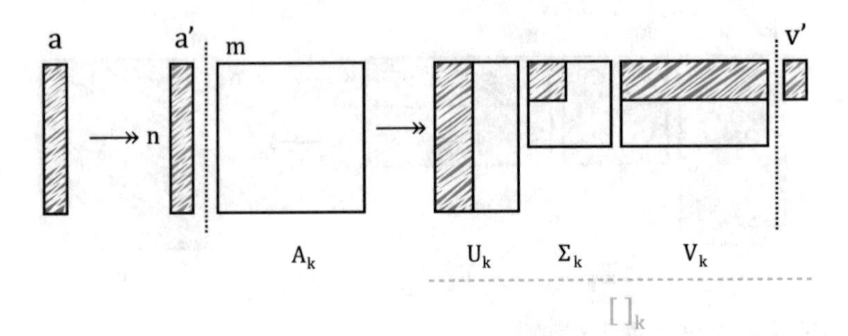

Fig. 5.11 Illustration of the fold in projection

efficient and it prevents unwanted changes in the latent semantics provided through the characteristics of a space (see also Sect. 4.2).

5.7 Proximity and Identity

Given the nature of analysis hidden in the singular value decomposition and given the nature of text corpora as such, it is better to speak of 'associative closeness', when interpreting any relatedness scores based on the closeness of vectors in a vector space. There are two basic types of associative closeness relevant to MPIA: proximity and identity.

Other than in latent semantic analysis, not a single, but two different measures of relatedness are required for MPIA: a measure to quantify proximity and a measure to determine identity.

'Proximity' thereby refers to the characteristic of two vectors being associatively close with a relatedness score above a certain (lower) threshold in the Eigenspace. 'Identity' then refers to the characteristic of two vectors being associatively very close with a relatedness score above a certain (higher) threshold in the Eigenspace.

Proximity captures weak to strong relatedness, whereas identity captures those relations where words or their bag-of-words combinations are used interchangeably—in the contexts provided through the text passages of the corpus.

While—similar to LSA—many different coefficients for measuring relatedness have been proposed over time (see Sect. 4.1), MPIA uses the Cosine measure for determining both proximity and identity, as it is insensitive to the number of zero values (Leydesdorff 2005, p. 769) and thus in general more appropriate to deal with largely sparse matrices. The findings of an experiment on calibration presented in the subsequent Chap. 7 support this choice.

Identity implements the meaning equivalence relation φ needed to establish information equivalence as introduced in Chap. 2. It should be noted that with more relaxed threshold values (such as the recommended 0.7), perfect transitivity of document identity is no longer given: three documents a, b, and c can be located in the Eigenspace in such a way that the cosine between a and b and between b and c is above the threshold, while the angle between a and c may already be too large. This is a shortcoming that can only be circumvented by choosing an appropriate stretch-factor truncation (as proposed in the previous Sect. 5.1.5) that is able to map identical meanings to the same location, so that a strict threshold of a cosine of one can be used for determining identity.

In practice, it is often not easy or possible to achieve perfect truncation over an ideal corpus. This may water down the accuracy in preserving transitivity of meaning equality relations, effectively requiring a fall back to, e.g., centroids of the transitive clusters rather than comparing their individual member vectors. This, however, does not affect the ability of the identity measure to ensure symmetry and reflectivity.

Proximity implements a weaker measure of associative closeness (with a recommended threshold of 0.3). It is used for the creation of the conceptual clustering on top of the individual terms, so as to layout the map projection of the high dimensional Eigenspace in a way that it preserves convergence and divergence as closely as possible.

Figure 5.12 illustrates the difference between the proximity and identity measure: only those angular distances above the higher threshold (in the example of a cosine of 0.7) are considered to be identical, whereas weak associative closeness is already established for those vectors above the lower threshold (in the example of 0.3). Typically, the Eigenspaces have a large number of dimensions (though not as large as the originating base), therefore requiring a certain amount of tolerance to reach the desired level of accuracy.

While the identity measure can be used to, for example, compare whether a learner-written essay closely enough resembles a tutor-written model solution or whether two learners share the exact same competence profile, the proximity measure can be used to evaluate what documents can be found in an MPIA

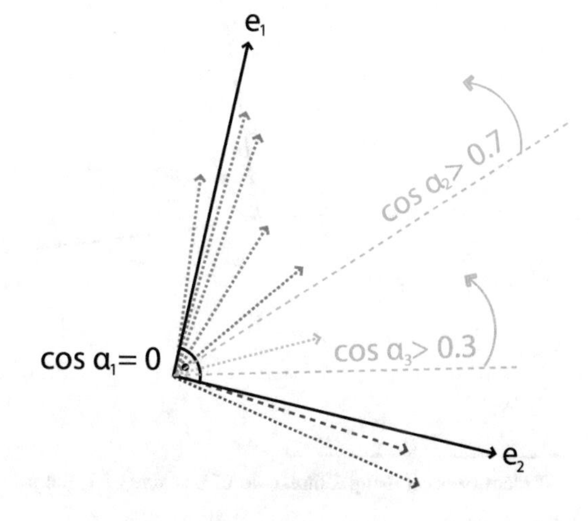

Fig. 5.12 Identity (*orange*) and proximity (*blue*) thresholds

implementation of the "zone of proximal development" (Vygotsky 1978, p. 86) or who the next best candidates with a similar competence profile would be.

Internally, MPIA calculates the cosine values between each vector pair, retrieving this relational data whenever required to determine if proximity or identity are given.

5.8 A Note on Compositionality: The Difference of Point, Centroid, and Pathway

The left- and right-singular eigenvectors in U_k and V_k map domain and codomain of the transformation A to their respective Eigenbasis. Both are an Abelian group, for which vector addition is commutative: if several vectors in—say—U_k are added, the resulting vector of such addition sequence points to one specific *location*, no matter in which order the addition has been executed.

For each set of vectors (e.g. document vectors), it is possible to calculate their centroid, i.e. the vector pointing to the balance point of the polygon created by the set, see Fig. 5.13. This centroid (aka 'balance point') can be calculated as the arithmetic mean of the composing vectors.

While the location changes when adding new vectors to a collection, centroids provide more stability: if, for example, an *identical* set of vectors is added to an existing collection, the centroid of the collection (roughly[2]) stays the same, while any single location resulting from their addition can—of course—be fundamentally different.

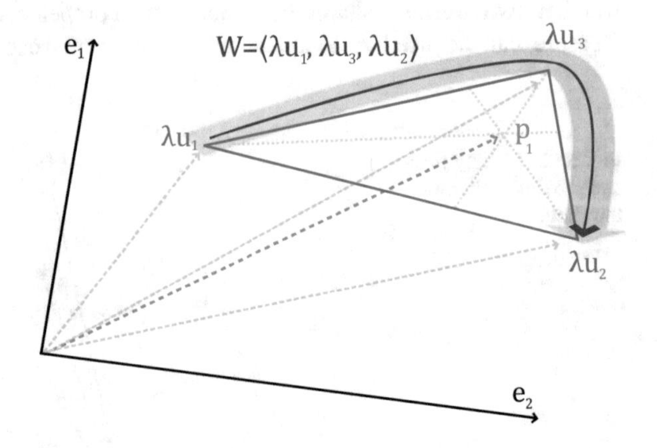

Fig. 5.13 Locations λu_n, position p_1, pathway W

[2] If identity is set using a threshold of a cosine of 1, the position even does not change at all.

The third noteworthy concept is that of a 'pathway': pathways stand for hyperedges, i.e. locations occupied by the component vectors of an addition *sequence* (see pathway *W* in Fig. 5.13). A pathway of vector locations in such ordered sequence is no longer an Abelian group, so commutativity is not given.

5.9 Performance Collections and Expertise Clusters

Proximity and identity relations allow deriving proposals for clustering together vectors in the Eigenspace—proposals for which the homogeneity can be quantified and well-defined cut-offs can be chosen.

There are many different clustering algorithms available: partitioning methods divide the dataset into a specified number of clusters, whereas hierarchical methods create (typically binary) trees through iterative agglomeration or division of the dataset (see Struyf et al. 1996, p. 2).

Complete linkage is one such agglomerative method. In each iteration complete linkage merges the most similar clusters of the n vectors, so that after (n-1) steps, a single, big cluster remains (Kaufman and Rousseeuw 1990). The decision on merging is based on the maximum distance between any two points in the two clusters, thereby preferring merging of close clusters over distant ones (Kaufman and Rousseeuw 1990; Lance and Williams 1967).

The result is a hierarchical tree, with the height indicating "the value of the criterion associated with the clustering method for the particular agglomeration" (R Core Team 2014). In case of complete linkage, this is the maximum distance between the clusters merged in each stage of the iteration. Figure 5.14 depicts such a cluster dendrogram.

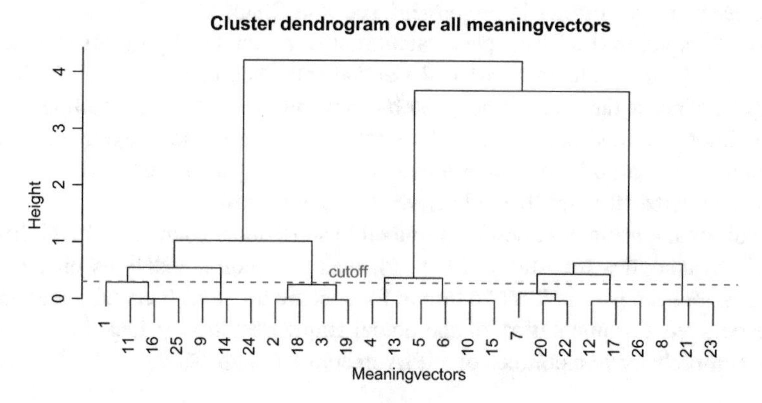

Fig. 5.14 Cluster dendrogram for a set of meaning vectors

In R, hclust (package *stats*) implements complete linkage (amongst other methods). Since the merging is based on distances, the proximity values have to be converted to distances first using as.dist((1+prox) / 2).

Listing 3 Agglomerative clustering using complete linkage.

```
a=hclust(as.dist((1+prox)/2), method="complete")
b=cutree(as.hclust(a), h=(1+identityThreshold)/2)
```

The tree can then be cut at the desired threshold level, thus retrieving a set of clusters with a defined homogeneity level (in our case: the identity threshold converted to distances to create crisp clusters, see Listing 3).

5.10 Summary

This chapter introduced the mathematical foundations of MPIA. This included the matrix algebra foundations used to create geometrical representations of texts, along the way of which a novel method to determine the number of Eigendimensions to retain *ex ante* from the trace of the original text matrix was proposed. This novel method allows for calculating a threshold value for the Eigenvalues before resolving the Eigenequations, thus reducing the computational efforts required in solving the Eigenvalue problem significantly compared to its predecessor LSA. The associative closeness relations known already from LSA were extended to differentiate between identity and proximity relations, thus paving the way for the use of projection surfaces. These fundamental identity and proximity relations allow quantifying their characteristics with accuracy.

Having proposed this new method, there are a lot of open questions emerging that offer room for further improvement, see also Chap. 11.

For example, in the Eigenspace calculations, the interesting question arises for the stretch factor problem of what the actual influence of the size of the bag-of-words is. What is the relation between bag size, number of words loading on each Eigendimension, and the amount of stretch expressed in their Eigenvalue? What would be the effect of applying grammar sensitive weights, for example, weighting nouns and verbs stronger than adjectives (or vice versa)?

Additional clarification will be provided in subsequent chapters, when introducing visual analytics for MPIA (Chap. 6), making recommendations on the additional calibration parameters and tuning for specific domains (Chap. 7), and adding more detailed documentation of the actual implementation (Chap. 8) as well as more comprehensive examples of MPIA in action (Chap. 9).

References

Barth, F., Muehlbauer, P., Nikol, F., Woerle, K.: Mathematische Formeln und Definitionen. J. Lindauer Verlag (Schaefer), Muenchen (1998)

Berry, M., Dumais, S., O'Brien, G.: Using Linear Algebra for Intelligent Information Retrieval, CS-94-270. Preprint (1994)

Borchers, H.W.: pracma: Practical Numerical Math Functions. R Package Version 1.6.4. http://CRAN.R-project.org/package=pracma (2014). Accessed 11 Aug 2014

Golub, G., van Loan, C.F.: Matrix Computations. North Oxford Academic, Oxford (1983)

Kaufman, L., Rousseeuw, P.: Finding Groups in Data: An Introduction to Cluster Analysis. Wiley, New York (1990)

Lance, G.N., Williams, W.T.: A general theory of classificatory sorting strategies: 1. Hierarchical systems. Comp. J. **9**(4), 373–380 (1967)

Leydesdorff, L.: Similarity measures, author cocitation analysis, and information theory. J. Am. Soc. Inf. Sci. **56**(7), 69–772 (2005)

R Core Team: R: A Language and Environment for Statistical Computing. R Foundation for Statistical Computing, Vienna, Austria (2014). http://www.R-project.org/. ISBN 3-900051-07-0

Strang, G.: Linear Algebra and Its Applications. Academic, New York (1976)

Strang, G.: Linear Algebra and Its Applications. Thomson Learning, Belmont, CA (1988)

Strang, G.: Introduction to Linear Algebra, 4th edn. Wellesley Cambridge Press, Wellesley, MA (2009)

Struyf, A., Hubert, M., Rousseeuw, P.: Clustering in an object-oriented environment. J. Stat. Softw. **1**(4), 1–30 (1996)

Vygotsky, L.: Mind in Society: The Development of Higher Psychological Processes. Harvard University Press, Cambridge, MA (1978)

Chapter 6
Visual Analytics Using Vector Maps as Projection Surfaces

High dimensional vector spaces and the linear equation systems of its matrix representation are not processed by human analysts with ease, particularly once the number of dimensions exceeds a certain limit. Visual representations can help to remediate this, but ultimately cannot help overcome complexity completely.

Visual analytics therefore aim wider then mere visualisation and additionally set focus on supporting analytical reasoning: "Visual analytics is the science of analytical reasoning facilitated by interactive visual interfaces" (Thomas and Cook 2005, p. 4).

Visual Analytics tap into four areas: the study of "analytical reasoning techniques", "visual representation and interaction techniques", "data representations and transformations", and techniques for "presentation, production, and dissemination" (Thomas and Cook 2005, p. 4; Thomas and Kielman 2009, p. 310).

This chapter focuses on the foundations underlying the representation and interaction techniques.

Data handling aspects are partially already covered in the preceding chapter, more—together with further presentation and sharing aspects—will be added in the subsequent Chap. 8, when turning to the actual software implementation (see especially Sect. 8.3.1 and Annex A for materialisation and storage aspects).

Together Chap. 8 (with the generic workflow examples) and Chap. 9 (with the application cases in learning analytics) will add substance on the use of *mpia* for analytical reasoning.

This chapter tends especially to the third and final objective of this work, namely '*to re-represent this back to the users in order to provide guidance and support decision-making about and during learning*' (Sect. 1.2). It sets focus on the foundations of such visualisation algorithm and its connected interaction interfaces that allow manipulating the created displays further during analysis.

Many existing visualisation methods fall short in creating the required projection stability needed to explore and further (visually) analyse high volume data. The proposed method provides a way out: In this chapter, the process to create knowledge maps with MPIA using a cartographic metaphor is presented.

F. Wild, *Learning Analytics in R with SNA, LSA, and MPIA*,
DOI 10.1007/978-3-319-28791-1_6

Since any visualisation of high dimensional data to two or three dimensions means loss in accuracy, Chap. 10 includes an investigation of how big such error introduced is for different-sized spaces (Sect. 10.3.4).

Juenger and Mutzel (2004) list the five archetypes 'tree layout', 'layered layout', 'planarization' (not to be mixed up with planar projection!), 'orthogonal layout', and 'force-directed layout' for automatic graph drawing. While tree layouts are well known through their metaphor, layered layouts form the special case in which "all vertices are drawn on parallel horizontal lines" or 'layers' (ibid, p. 24). Planarisation then refers to finding a "maximum planar subgraph" (ibid, p. 32) through edge removal, followed by subsequent edge insertion. In orthogonal layouts "each edge is represented as a chain of horizontal and vertical segments", introducing so-called bends into the edges (ibid, p. 37), while force-directed layouts resort to the idea of a physical system with edges modelled as attracting springs and "charged particles with mutual repulsion" (ibid, p. 44). Thereby, "due to their general applicability and the lack of special structural assumptions as well as for the ease of their implementation, force-directed methods play a central role in automatic graph drawing" (ibid, p. 41). Moreover, as Gronemann et al. (2013) add, network visualisations using topographic maps can be added to this list among the more recent methods.

While tree, planarization, layered, and orthogonal layouts are no longer preserving distance and proximity in the projection of multidimensional graphs to two- or three-dimensional displays, force-directed layouts and topographic maps do.

Since pure force-directed layouts, however, cannot provide the location-stability needed for overlays and further graphical interaction as required for *visual analytics* (see also Sect. 1.6), this chapter will introduce a new form of generating topographic maps from force-directed layouts.

Such maps then can serve as projection surface for setting focus of analysis, adding overlay information, reasoning, and the like, thereby ensuring topographic persistence. Moreover, they provide overview on the Gestalt of the space in the background, while overlaying details in the foreground.

There has been extensive research (Fabrikant et al. 2010, p. 253) on the use of the landscape metaphor in information visualization, often for the unsubstantiated claim that the "everyone intuitively understands landscapes". This seems not to be the case and understanding of the "landscape metaphor is not as self-evident as information designers seem to believe" (ibid, p. 267).

Fabrikant et al. (2010) conclude that both users and information designers are not aware that terraforming of mountains is mostly the absence of activity and that the equivalent of 'information' is therefore to be found rather in valleys and rivers, thus leaving them wanting to try new metaphors for which this natural interpretation of 'higher is more' is valid—e.g. using cityscapes rather than landscapes. Even though, several common-sense geomorphology interpretations hold in their investigation: besides the 'higher is more' concept, this was found to be the case for relative location and—surprisingly—for the notion of centrality versus periphery.

Natural understanding of cartographic visualizations is not a given and a certain amount of effort in developing the visual literacy required by a novel visualization format can be expected.

The calculation method and its design decisions will be documented here, supported by examples. This chapter will also provide insight into how the visualization and the according analytical processes work in practice. Such design decisions in the processing and choice of visual variables, however, aim to provide the optimal trade-off between accuracy and oversight. They are documented here, while—where possible—indicating potential alternatives.

6.1 Proximity-Driven Link Erosion

The cosine between vectors in the Eigenbasis[1] as well as the cosine between row or column vectors of the resulting mapping A_k can be used to create an affiliation matrix, just as typically required in (social) network analysis—with one significant difference: Since the cosine is defined for any angle between vectors, the result is always a complete network with every relation quantified. Analysing and visualising a full network typically results in a giant glob, where the human analyst no longer can discern associative closeness from associative distance.

To ease analysis, those relatedness scores that are not in proximity can be removed from the network matrix, hence this step is called 'proximity-driven link erosion'. Figure 6.1 further illustrates this process: in step 1, those relations below the proximity threshold are removed (the dotted, red lines) and replaced with missing values. This effectively removes the edges in the network visualisation, resulting in an affiliation matrix (and network visualisation) such as the one depicted in step 2 to the right hand of Fig. 6.1.

Since such link erosion, however, can introduce isolates and micro components to the subsequent force-directed layout, a minimum connected component size is defined and eventual outliers are reattached with an strong edge weight (the proximity cell value) of the proximity threshold. Such outliers otherwise create sort of 'meteorite belts' around the big connected components, thereby not only hindering clipping of the coordinate data, but typically also pushing together the connected components in the centre of the display.

Reattaching them with the proximity threshold ensures tight binding to the nearest node they can be attached to (though introducing loss). In case it is not a single isolate, but a component below the defined minimum component size of the square root of the number of terms, the node(s) with the highest betweenness is chosen and the removed link to its closest node in the other component(s) is added with the value of the proximity threshold.

[1] Proximity between all eigenvectors times the eigenvalue stretch factor.

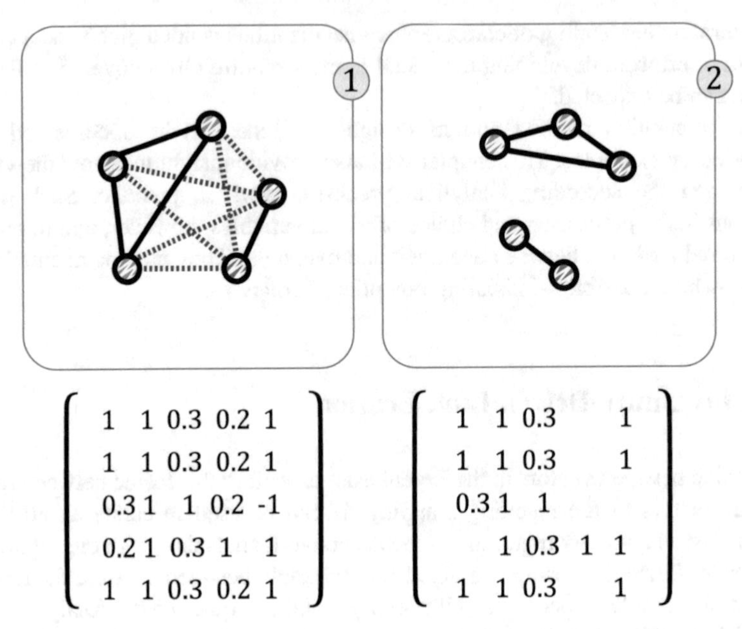

Fig. 6.1 Link erosion based on proximity

Typically, such number of isolates and stray micro components is very small with hardly any influence on the overall amount of error introduced in the planar projection of the underlying multidimensional graph data. Its effect on the layout of the display, however, is enormous and 'meteorite belts' around a giant prong are a thing of the past.

6.2 Planar Projection With Monotonic Convergence

A force-directed placement algorithm is used to create an injective mapping onto a 2D planar surface that maps the proximity matrix created from the closeness relations of the vectors in the multidimensional Eigenspace. An example of such mapping is depicted in step 2 in Fig. 6.1 above.

Such force-directed placement algorithms—also known as spring-embedders (Fruchterman and Reingold 1991, p. 1129)—try to minimise the difference between theoretic (i.e. proximity values) and Euclidian (geometric) distance in the planar placement of vertices and edges.

Often, 'energy minimisation' is used as a metaphor to explain the working principle of such placement algorithm coming to effect for reaching an equilibrium within a given number of iterations: A stable state with minimal energy consumption of the network is sought with edges being springs that attract the vertices they are connected to, while minimising the number of edge crossings required, and with the vertices typically behaving like electrically charged particles, so as to repulse each other. An optimal state is then "the state in which the total spring energy of the system is minimal" (Kamada and Kawai 1989, p. 7). Spring-embedders are known for providing aesthetically pleasing layouts, which can be processed by the human visual system with ease. One of the reasons for this aesthetic pleasure is argued to be in its ability to preserve symmetry (Kamada and Kawai 1989, p. 13; Fruchterman and Reingold 1991, p. 1139).

Moreover, spring-embedders have been shown to converge monotonically, i.e. with each iteration the energy required will be reduced, thus providing a more optimal layout (see De Leeuw 1988, for a proof).

The algorithm used in MPIA for planar projection is the one proposed in Kamada and Kawai (1989) as implemented in the *network* package for R (Butts 2008; Butts et al. 2012). A possible alternative is the one proposed in Fruchterman and Reingold (1991), which is provided in the *network* package as well.

The total energy required by the system is formulated by Kamada and Kawai (1989, p. 8, Eq. 6.1) as:

$$\sum_{i=1}^{|V|-1} \sum_{j=i+1}^{|V|} \frac{1}{2} k_{ij} (|p_i - p_j| - l_{ij})^2 \tag{6.1}$$

Thereby, p_i and p_j signify the positions of all pairs of vertices, k_{ij} stand for the stiffness factor of the spring between them, and l_{ij} is their optimal (theoretic) distance multiplied with the desirable length of the edge in the plane.[2] For vertices that are not directly connected, this derivate of Eades (1984) original spring-embedder proposal uses the geodesic between them.

Equation 6.1 is a reformulation of Hooke's law about the force (and thus energy) needed to extend or compress a spring (cf. Fruchterman and Reingold 1991, p. 1130, footnote):

$$F = -kx \tag{6.2}$$

$$E = \frac{1}{2} kx^2 \tag{6.3}$$

The force constant of the springs k is set to be normalized to unit length, such that (Kamada and Kawai 1989, p. 8, Eq. 6.4):

[2] Kamada and Kawai (1989) propose to use $L = L_0 / \, max(d_{ij})$ to calculate the desirable length in the plane, with L_0 being the length of a side of the square (!) plane. Since hardly any digital display surfaces are quadratic, this would offer room for further improvement, e.g., by determining L_0 through the length of the display diagonal and subsequently using adapted perturbations for x and y coordinates.

$$k_{ij} = \frac{K}{d_{ij}^2} \tag{6.4}$$

The energy equation [Eq. 6.1] allows computing local minima using the Newton–Raphson method, resulting in Eq. 6.5 [see Kamada and Kawai (1989, p. 9f), for proof]:

$$\Delta_m = \sqrt{\left\{\frac{\partial E}{\partial x_m}\right\}^2 + \left\{\frac{\partial E}{\partial y_m}\right\}^2} \tag{6.5}$$

Thereby, the unknowns ∂x_m and ∂y_m can be found iteratively by using the Newton–Raphson method, aiming to minimize the energy required by the overall system.

In the R implementation,[3] this is done in each local minimization step in the sum of potential energy saved Δ_m by changing the position of a vertex v_j:

$$\Delta_m = \sum \Delta_{mj}, \quad \text{for all m} \neq \text{j} \tag{6.6}$$

with

$$\Delta_{jm} = \hat{k}\, \frac{(|p_j - p_m| - l_{jm})^2 - (|p_j' - p_m| - l_{jm})^2}{{l_{jm}}^2} \tag{6.7}$$

using a constant $\hat{k}$ [4]

$$|p_j - p_m| = \sqrt{(x_j - x_m)^2 + (y_j - y_m)^2} \tag{6.8}$$

and

$$|p_j' - p_m| = \sqrt{(x_j' - x_m)^2 + (y_j' - y_m)^2} \tag{6.9}$$

The potential energy to be saved by changing positions through Gaussian perturbation[5] of the vertex position of p_m thereby is reduced in each iteration with a constant cooling factor. Per default the amount of energy to be saved is set to

---

[3] Equations 6.6–6.9 were taken from the C code implemented in Butts et al. (2012).

[4] The R implementation uses $|V|^2$ as k.

[5] Instead of the originally proposed Newton–Raphson method, the R implementation uses Gaussian perturbation with (per default): $y_j' = \text{rnorm}\left(y_j, \left(\frac{n}{4}\right) \cdot \left(10 \cdot \frac{0.99^{\text{iteration}}}{10}\right)\right)$ and $x_j' = \text{rnorm}\left(x_j, \left(\frac{n}{4}\right) \cdot \left(10 \cdot \frac{0.99^{\text{iteration}}}{10}\right)\right)$.

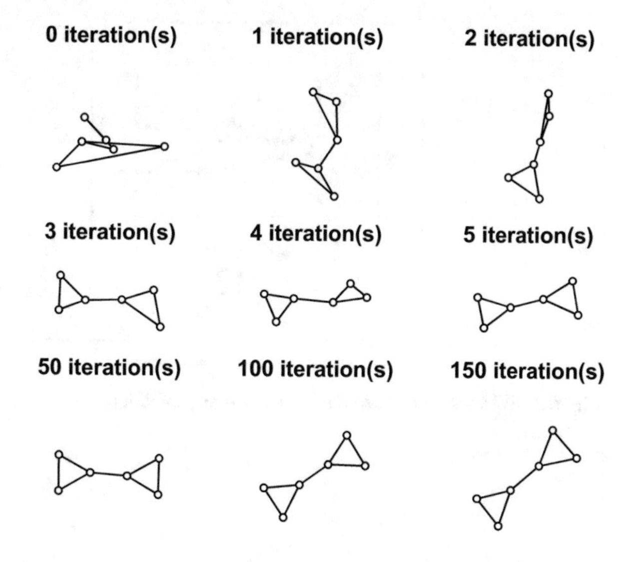

Fig. 6.2 The energy minimization process redrawn from Kamada and Kawai (1989, p. 11) using Butts et al. (2012) implementation

10 and the cooling factor with which it is multiplied is set to 0.99. The decision about keeping a position is thereby coupled to random.[6]

Figure 6.2 depicts the energy minimization process in action: in each iteration, local minima are further optimized, changing positions quickly in the first iterations, while reaching a more stable position with the higher number of iterations at the bottom of the figure.

The resulting matrix of placement coordinates in the Euclidian space shall henceforth be called network coordinates (short: 'netcoords').

Subsequently, this resulting planar surface placement can be mapped to a relief contour (a 'wireframe') to express density information, while internally storing the more precise positional data gained from the planar surface projection in order to still be able to precisely locate and distinguish vertices in any subsequent overlay visualisation.

This surface elevation is particularly useful in the visualisation of large volume data with the number of vertices exceeding the number of pixels available and the minimal positional difference noticeable by the human eye.

MPIA uses a grid-tiling algorithm for the calculation of this relief contour of—per default—a size of $n_r = 100$ rows $\times$ $n_c = 100$ columns. The wireframe segment each vertex position p_i falls into can be determined from its coordinates $p_i = (px_i, py_i)$ in the unit length as

$$x_i = \left\| \frac{px_i}{\frac{1}{n_c}} \right\| = \| px_i \cdot n_c \| \tag{6.10}$$

and

[6] See Butts et al. (2012), in the function network_layout_kamadakawai_R in layout.c for more details.

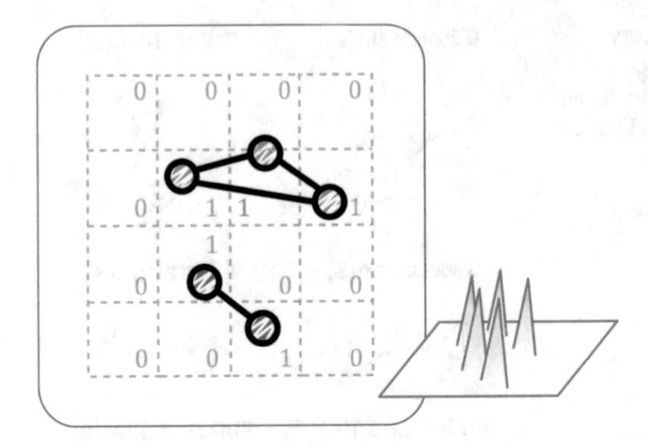

Fig. 6.3 Wireframe conversion of the planar projection

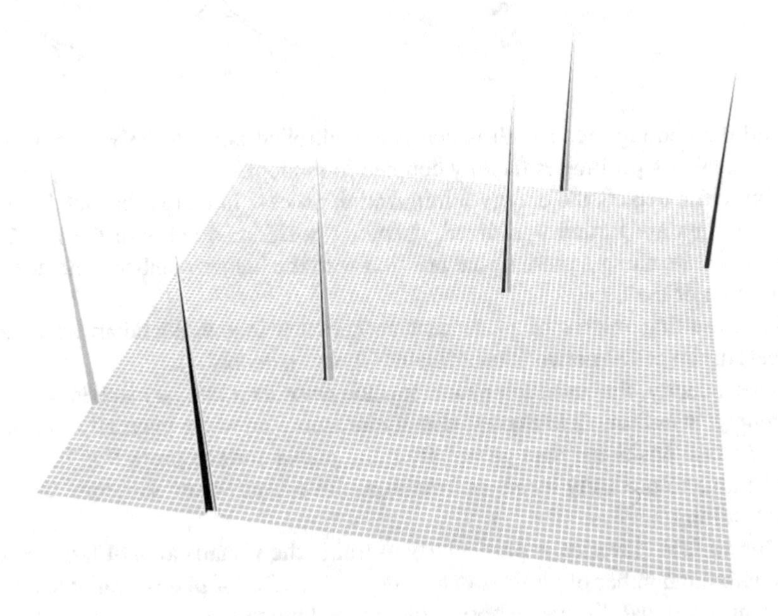

Fig. 6.4 Wireframe with elevation levels

$$y_i = \left\| \frac{py_i}{\frac{1}{n_r}} \right\| = ||py_i \cdot n_r|| \tag{6.11}$$

Any px_i or py_i coordinates being zero are put into the first column or first row respectively. Each wireframe segment is assigned the elevation level of the number of vertices contained (an integer in $\mathbb{N}$).

Figure 6.3 depicts this process: the vertices from the example from the link erosion in the previous section are converted to the elevation level of the grid sector they occupy.

As Fig. 6.4 illustrates, how this resulting wireframes suffers from abrupt changes in the elevation level: in the small example chosen (and used in Fig. 6.2), the elevation levels are spiking in the locations where vertices are placed. In more realistic examples, the available grid area is usually densely populated with elevation levels changing abruptly from one grid quadrant to another.

6.3 Kernel Smoothing

To avoid such effect, the surface wireframe can be smoothened using a Kernel smoothener provided in the *fields* package for R (Furrer et al. 2013), thus evening out the actual surface structure into an aesthetically more pleasing spline.

While this smoothening process reduces the information conveyed in the display to ordinal, the original precise quantitative positional information is still available in the network coordinates calculated with the algorithms provided in the previous section. The kernel smoothener maps the wireframe data from an integer in $\mathbb{N}$ to a real number in $\mathbb{R}$.

Figure 6.5 illustrates this process: The green numbers indicate the new, smoothened values that even out the spikiness of the surface wireframe shown earlier in Fig. 6.3. A visual example of what such surface mapping looks like with real data is presented in Fig. 6.6. What previously looked like stalagmites now starts to resemble a landscape.

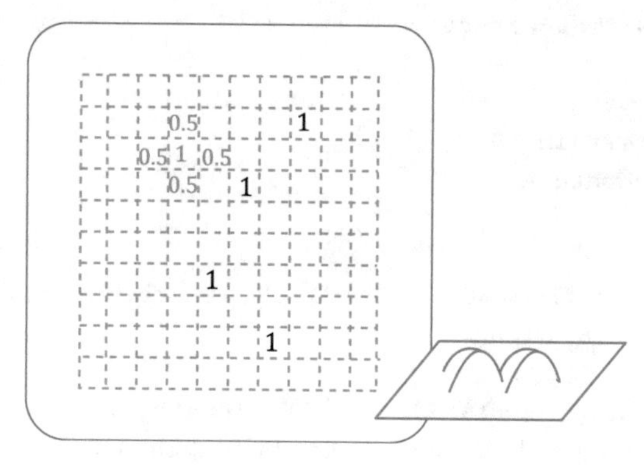

Fig. 6.5 Spline smoothening

Fig. 6.6 Smoothened surface wireframe

6.4 Spline Tile Colouring With Hypsometric Tints

In knowledge cartography, various techniques can be used to differentiate elevation levels. One such method is to use so-called hyposometric tints with "colors assigned to elevation zones" (Patterson and Kelso 2004, p. 34).

Listing 1 Generate colour palette with hyposometric tints.

```
topo.colors.pastel = function (n=21) {

 i = n - j - k # water
 j = n %/% 3 # terrain
 k = n %/% 3 # mountain

 cs = c(
  hsv(h=seq.int(from=38/60, to=31/60, length.out=i),
    s=0.5, v=0.8, alpha=1),

  hsv(h=seq.int(from=18/60, to=8/60, length.out=j),
    s=0.3, v=seq.int(from=0.6, to=1, length.out=j), alpha=1),

  hsv(h=seq.int(from=7.8/60, to=6/60, length.out=k),
    s=seq.int(from=0.3, to=0.1, length.out=k),
    v=seq.int(from=0.85, to=1, length.out=k), alpha = 1)
 )

 return(cs)

}
```

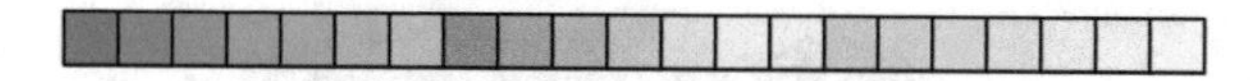

Fig. 6.7 The palette of hyposometric colours

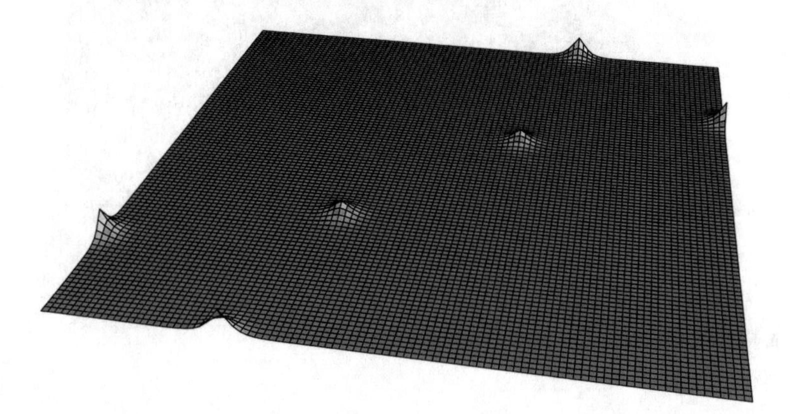

Fig. 6.8 Perspective plot using hyposometric tints

MPIA uses a default palette of 21 distinct colours for three 'terrain types' (water, land, mountain): see Fig. 6.7. They are generated using the cylindrical-coordinate representation of the hue-saturation-value colour specification with the method listed in Listing 1. The method generates a vector of three sets of colour shades, preferably with the same number of colours each: per default it creates seven blue tones ('water'), seven green tones ('terrain'), and seven brown/gray/white tones ('mountains'). Rotating the colour hue creates the blue tones, while the green tones vary both hue and brightness. The 'mountain' tones are generated varying all three: rotating the hue, while decreasing saturating and increasing brightness. The resulting palette of pastel tones is depicted in Fig. 6.7.

This palette is then used to colour the tiles of the wireframe mesh used in the perspective plot to visualize the projection surface, see Fig. 6.8: the six vertices of the example introduced in Fig. 6.2 rise as little island above the 'sea level'.

Conceptual projection surfaces are typically much more densely populated in practice. Even with relatively small evidence collections comprising just a few dozen of documents created or consumed by a handful of persons, the vocabulary contained easily can span several hundred if not thousand terms. In such case, aesthetically more interesting projection surfaces emerge, see Fig. 6.9: the depicted perspective plot of the projection surface was created from 33 business textbooks of the Open University, split into 741 chapters that contained 1193 terms. The 3D perspective plot depicts a projection surface created from the term-term proximity relations, using a proximity threshold of *0.3*.

The density, with which the MPIA conceptual space is populated, brings along the problem of labelling visually prominent landmarks in the landscape visualization in a meaningful way. Toponymy is the study of place names and MPIA provides several methods to automatically detect useful places and their names.

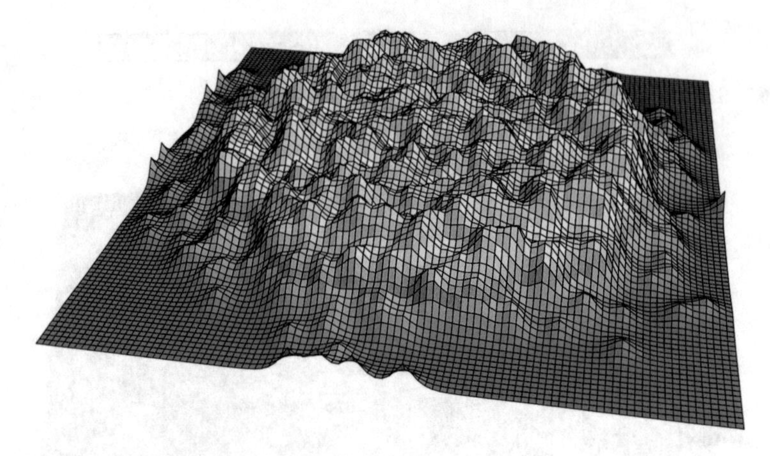

Fig. 6.9 Perspective plot of real-life example

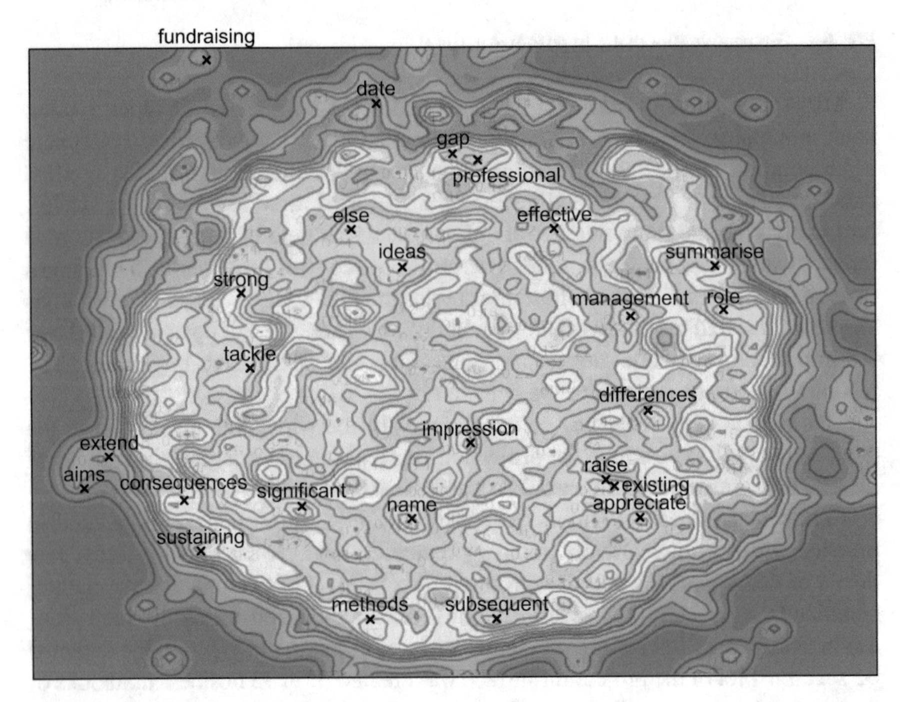

Fig. 6.10 Example topographic map projection

To not overload his section, their documentation will follow in the subsequent chapter.

MPIA comes with a set of visualization formats, as often for a particular case one visualization format is more expressive than another. To just show briefly another, two-dimensional one: Fig. 6.10 depicts a topographic map projection with contour lines indicating elevation levels in addition to the hyposometric colours. The toponymy was generated using the 'mountain' method: in each grid quadrant of a grid of default size 10 × 10, the term vector with the highest vertex prestige was identified and chosen as label. The location of the label thereby is precisely the net coordinate, while—visually—this typically coincides with the highest elevation level found in the quadrant.

6.5 Location, Position, and Pathway Revisited

As already indicated above in Sect. 5.1.8, there important differences to be made between the location of vectors, the position of collections, and the pathways an ordered sequence of vectors of a collection spans. These three basic types have their visual correlate: the location is a point in the projection surface, the position is the location of the centroid and balance point of a collection of vectors, and the pathway can be depicted as the curve connecting the locations of the vectors in the sequence provided by the collection.

The visual marker indicating the place occupied by a location or position vector is in MPIA implemented as a cross in the two dimensional visualization formats and as a flag in the three dimensional ones. The placement of the label thereby is slightly off, the marker, however, is put precisely to the location of the net coordinate for locations and to the centroid for positions.

The curves to connect pathways are implemented using x-splines (Blanc and Schlick 1995), as provided by `xspline()` in the R core package *graphics*. The x-spline thereby uses the locations of the vectors in the collection (with the control parameter s_k being 0 at the beginning and end and -1 in between). An example of the effect of the control parameter s_k is depicted in Fig. 6.11: $s_0 = s_6 = 0$, $s_1 = s_5 = 1$, $s_2 = s_3 = s_4 = 0$. Values of 0 create sharp edges, positive values (e.g. 1) confine within, negative values (e.g. -1) approximates the outer hull.

Figure 6.12 depicts an example of such x-spline visualization of a path. The path encompasses three vectors, i.e. textual performance demonstrations, leading from location 1 over location 2 to location 3.

The position occupied by this performance collection with three elements is depicted in the next Figure 6.13: the position is located at the weighted centroid of its component vectors. More details about its calculation will be provided in the

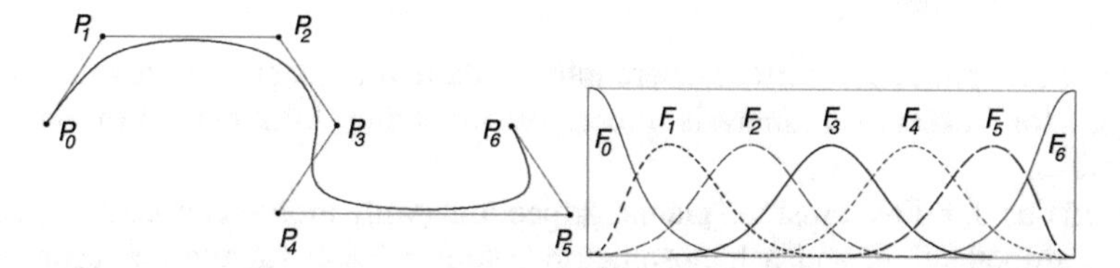

Fig. 6.11 X-spline and control points (Graphic: Blanc and Schlick 1995, p. 385: Fig. 6.10)

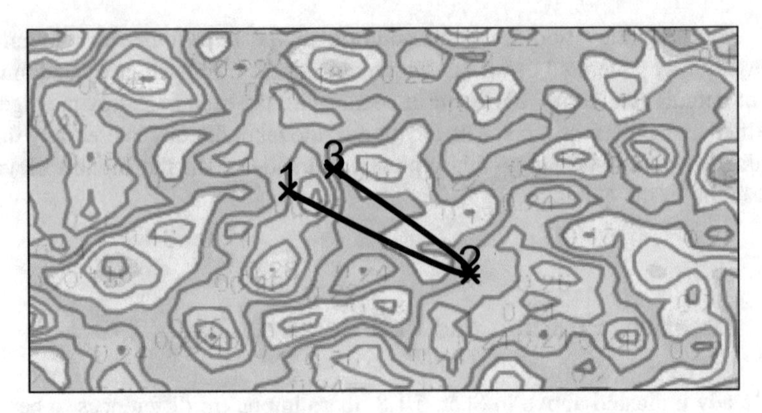

Fig. 6.12 Example ‘performance’ path

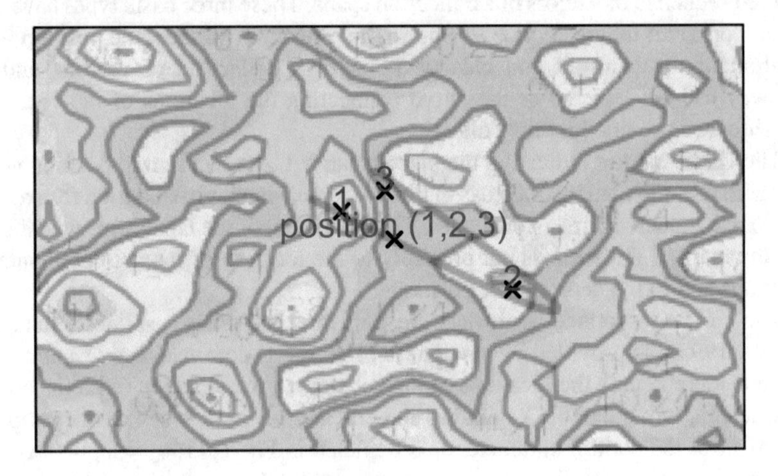

Fig. 6.13 Position occupied by the example path

documentation of the actual MPIA package implementation following in the next Chap. 6.

6.6 Summary

Geometrical projection surfaces were introduced in this chapter to provide a stable ‘stage’ for subsequent analysis processes involving locations, positions, and pathways.

Methods for link erosion, planar projection (with monotonic convergence!), kernel smoothening, and a hyposometric colour scheme for tile colouring were proposed that help in creating conceptual landscape visualizations.

It has been shown briefly (with the promise that more will come in the subsequent chapter), how these projection surfaces can be used to display positional information and performance collections, thus paving the way for manifold types of analysis: for example, this is allows investigating positioning of individuals and groups, based on the textual evidence held in their performance collections.

The means presented here complement the analytical processes introduced in the previous chapter with a—as we will see—powerful visual instrument. In their combination, the two provide means to analyse social semantic performance networks.

There is room for improvement, still: For the force-directed placement algorithm, the resulting placements typically prefer vaguely circular ('island') layouts due to the built in preference for symmetry (this is the same for the alternative proposed by Fruchterman and Reingold 1991). This may be wasteful of space, as today's displays typically are rectangular with unequal length and width. What other algorithms could be used instead of this—and what impact would this have on the accuracy of the plot?

Subsequently, the next Chap. 7 will further investigate, how calibration operations such as sanitising operations influence effectiveness of spaces. Chap. 8 will then describe the actual software implementation into the R package *mpia*. Examples of learning analytics with MPIA follow then in Chap. 9, leading over to the evaluation Chap. 10.

References

Blanc, C., Schlick, C.: X-Splines: a spline model designed for the end-user. In: Proceedings of the SIGGRAPH'95, pp. 377–386 (1995)

Butts, C.T.: network: a package for managing relational data in R. J. Stat. Softw. **24**(2), (2008)

Butts, C.T., Hunter, D., Handcock, M.S.: network: classes for relational data. R package version 1.7-1, Irvine, CA. http://statnet.org/ (2012)

De Leeuw, J.: Convergence of the majorization method for multidimensional scaling. J. Classif. **5** (2), 163–180 (1988)

Eades, P.: A heuristics for graph drawing. Congressus. Numerantium. **42**, 149–160 (1984)

Fabrikant, S., Montello, D., Mark, D.: The natural landscape metaphor in information visualization: the role of commonsense geomorphology. J. Am. Soc. Inf. Sci. Technol. **61**(2), 253–270 (2010)

Fruchterman, T., Reingold, E.: Graph drawing by force-directed placement. Softw. Pract. Experience **21**(11), 1129–1164 (1991)

Furrer, R., Nychka, D., Sain, S.: Fields: Tools for Spatial Data, R Package Version 6.8. http://CRAN.R-project.org/package=fields (2013)

Gronemann, M., Gutwenger, C., Jünger, M., Mutzel, P.: Algorithm engineering im Graphenzeichnen. Informatik. Spektrum. **36**(2), 162–173 (2013)

Juenger, M., Mutzel, P.: Technical foundations. In: Juenger, M., Mutzel, P. (eds.) Graph Drawing Software. Springer, Berlin (2004)

Kamada, T., Kawai, S.: An algorithm for drawing general undirected graphs. Inf. Process. Lett. **31** (1989), 7–15 (1989)

Patterson, T., Kelso, N.: Hal Shelton revisited: designing and producing natural-color maps with satellite land cover data. Cartogr. Perspect. **47**, 28–55 (2004)
Thomas, J., Kielman, J.: Challenges for visual analytics. Inf. Vis. **8**(4), 309–314 (2009)
Thomas, J.J., Cook, K.A.: Illuminating the path: the research and development agenda for visual analytics. IEEE Computer Society Press, Los Alamitos, CA (2005)

Chapter 7
Calibrating for Specific Domains

Eigenspace-based models were shown to create greater effectiveness than the pure vector space model in settings that benefit from fuzziness (e.g., information retrieval, recommender systems). In settings that have to rely on more precise representation structures (e.g., essay scoring, conceptual relationship mining), better means to predict behaviour under certain parameter settings could ease the applicability and increase efficiency by reducing tuning times.

One significant problem in analytical processing of conceptual and social interaction is that for good topical coverage of spaces a large amount of evidence (and background material) is required to feed into calculation.

In this chapter, a comprehensive investigation of the influencing parameters, their potential settings, and their interdependencies is reported in order to enable a more effective application.

The systematic variation of the influencing parameters along the settings described in Sects. 7.1 and 7.2 created 1,370,961 different spaces for the nine essay collections studied. The average processing and calculation time for each space was 12.56 s.

This chapter reports experiences and experiment results from this systematic variation—indicating how to reduce the amount of training material necessary by using intelligent filtering operations. Furthermore, it investigates the influence of additional sanitising operations such as vocabulary filtering, sampling of the document collection, changing dimensionality, and altering the degree to which evidence can be replaced with generic documents (aka 'degree of specialisation').

Trends indicate that the smaller the corpus, the more specialisation is required. Moreover, recommendations for vocabulary filtering can be derived, depending on the size of the corpus. These results are picked up in the implementation, a report on which follows in Chap. 8. Furthermore, they are followed in the demos and examples described in Chap. 9. It should be noted, that the essay collections used for this experiment were also used for evaluation, the results of which are described separately in Chap. 10.

F. Wild, *Learning Analytics in R with SNA, LSA, and MPIA*,
DOI 10.1007/978-3-319-28791-1_7

Typically, proposals on how to create effective subspaces of the Eigenspace stay either agnostic about the ideal size of a corpus, while at the same time warning not to use too small ones (Quesada 2007), or they recommend the use of large spaces with an ill-conceived 'bigger is better' assumption. Contradicting this, several authors report positive results with relatively small corpora and spaces, thus encouraging research on their composition: amongst others the inventors of latent semantic analysis provide in their seminal paper (Deerwester et al. 1990) a convincing micro example utilising only nine documents, twelve terms, and two dimensions.

Within this section, results of an experiment are reported that shed more light on the deployment and deployment conditions of small corpora and small spaces as such and on relevant influencing parameters in their calibration. As will be shown, small spaces as such work equally well as bigger ones—given the right sampling method and configuration.

The underlying problem is a problem of creating a space from a corpus that is big enough to cover the desired target domain and apt to separate the relevant senses, connotations, and meanings from the—in this field and for this application—irrelevant, but at the same time small enough to allow for computational efficiency. This is at the same time a question of validity: can convergent and discriminant validity be found in small spaces.

Using a small corpus as input to the singular value decomposition shown in Sect. 5.4 can be used to produce a small, lower-dimensional vector space, which allows evaluating texts with respect to this limited domain (i.e. limited coverage of meanings). The number of documents, the number of terms, and the number of dimensions chosen determine space size. The advantage of using a smaller space is the reduced working memory consumption and increased speed while retaining the same or comparable level of accuracy as bigger ones.

The rest of this chapter is organised as follows. First, the corpus model is outlined and the set-up of the investigation is described, which varies parameter settings and sampling methods. Second, the gathered data are investigated and results analysed. Last but not least, the findings are discussed against their implications and an outlook on unresolved problems is given.

7.1 Sampling Model

A corpus is a collection of documents that contain terms with certain frequencies in a specific order. Vector space models neglect this order to focus with a bag of words approach on the distribution of the contained terms.

With Eigenspace-based methods, an input vector space is converted to a lower-order abstraction that ideally comes closer to the meaning structures exposed by the texts constituting the corpus. This is accomplished by pruning the number of dimensions extracted, skimming terms, or dropping documents.

Normally, the justification for dropping particular dimensions, terms, or documents is that they express variation that is too small to be considered useful or interesting.

The upper limit of the rank of the matrix is either the number of terms or the number of documents—whichever is smaller. In this uninteresting upper limit case, no reduction of the underlying data is performed and the original matrix reconstructed. The optimal rank of the data matrix reflects the intercorrelation of terms in the data, i.e. if there is little shared meaning, the effective rank is higher and it is lower, when there are relatively less connections in the texts.

To the extent that some terms co-occur preferentially in certain documents and not in others, the optimal rank is smaller than the number of documents. A method for determining the optimal rank has been proposed in Chap. 5.

The computational efficiency of the calculation is, however, not completely independent of additional factors.

Regardless of the format in which the input corpus is represented,—as a sparse or non-sparse text matrix—, the size of the resulting Eigenspace representation is constrained largely by three factors: the size of the vocabulary, the number of documents, and the number of dimensions retained after the stretch-dependant truncation (see Sect. 5.5).

The vocabulary can be restricted in several ways. By concentrating on the medium frequent terms, the most semantically discriminant terms can be selected. This can be achieved with frequency thresholds, either absolute or normalised in relation to vocabulary size. Alternatively, stopword filtering can serve as a simplified removal method for highly frequent terms. The vocabulary size for the resulting space can be assessed by calculating the number of terms, inspecting their (global) frequency distribution, and frequency means plus standard deviation.

The document base composition is driven by the number of documents selected and by the relation of domain-specific to generic documents. The document number can be linearly extended to investigate its influence. The degree of specialisation can be linearly varied to create different mixes of domain-specific versus generic documents. The document base selection can be assessed by measuring the resulting number of documents, the mean document length and standard deviation, and by the degree of specialisation (share of domain-specific to all documents).

The number of dimensions to be retained from the decomposition can be varied. As already stated, a mathematically sound prediction method has been introduced in Chap. 5. Applying, however, the above filtering and selection methods, affects the originating basis and it is unclear in how far this distorts prediction and resulting spaces.

The experiment presented here aims to unveil the interdependencies between these influencing factors.

Frequency-filtering the terms or varying the degree of specialisation amongst the documents, for example, can be assumed to influence the optimal number of dimensions. Sampling the number of dimensions retained linearly in steps should ensure that the experiment is not negatively influenced by the prediction methods failing to provide a useful number. This is a limitation. Since, however, aim of this analysis is not to validate the stretch-truncation (or any other method of prediction), but rather to uncover interdependencies of the overall influencing factors, this will not affect the findings.

There are further limitations. Different proximity measures can be used for the similarity assessment in spaces. Some of them are sensitive to zeros, thus influencing performance. In the experiment described below, several different measures have been tested and the best working measure was selected in each experiment run to rule out influence of choosing the wrong similarity measure (see Leydesdorff 2005, for the advantages and disadvantages of different proximity measures: depending e.g. on sparseness of the matrix, the one or other can be advantageous). Moreover, local and global weights as well as stemming were not tested. Term frequency weighting schemes further influence effectiveness of spaces. They, however, offer no computational advantages for their reduction, although they could serve as alternatives for calculating term frequency thresholds.

7.2 Investigation

In this study, the configuration settings for this processing model were varied and externally validated against human judgements in order to allow for the identification of successful sampling strategies.

To avoid artefacts from the document selection and order, the 'run' of the investigation was repeated five times while the base corpus (generic and domain-specific documents) was random sampled.

The external validation deployed nine essay collections in German in the wider subject areas of business administration and information systems (from the areas of programming, knowledge representation, e-commerce, business processes, marketing, software procurement, requirements engineering, software testing). Human raters scored the essays beforehand.

Students in management information systems, for example, were given the instruction to provide an overview of the most relevant developments in the area of 'business process intelligence'. A good student would respond to this instruction with an essay similar to the one listed below (translated from German):

> Organisations automate their business processes in order to increase efficiency, reduce costs, improve quality, and reduce errors. Graphical tools for the display of business processes and simulation tools are standard, but analytical tools rather the exception. It is possible to display business processes with a sort of petri net. Another approach is to use formal logic. All these approaches share that they aim at checking the processes. They are not analysed or optimised. In 'business process intelligence', data warehousing and data mining technologies are used to analyse and optimise the processes. Process management systems collect a multitude of information and save them in data bases, for example: being/end, input/output, errors, exceptions, resource allocation. Reports can be generated from these data using own tools. But these have restricted analysis facilities. A data warehouse that imports data can conduct analyses with the help of OLAP and data mining. These provide insights into matters of performance, quality of resources, understanding and

prediction of exceptions, information on process definition and optimisation of process sequences. 'Process discovery' tries to identify the structure of processes from log-data of workflows. Identified structures can subsequently be implemented as processes.

Altogether, nine collections of essays amounted to 481 essays plus an additional $3 \times 9 = 27$ model solutions (also called 'gold standard solutions'). In average the essays had a length of 55.58 words.

For evaluating of the effectiveness of each space calculated, the Spearman rank correlation between the human and machine judgements of the essays in each collection was chosen. The machine judgements were calculated as the average closeness of an essay to three 'gold standard' model solutions (three essays removed from each essay collection, which had been evaluated by the human raters with maximum scores).

It is well known that this scoring method of an essay against three model solutions is not the best method available to score essays.

Aspect methods, for example, which break down the model solution into partial aspects to be covered, have been shown to outperform this simple scoring method (Landauer et al. 1998). As more complex scoring methods would introduce additional influencing factors, this simple scoring method was considered adequate.

The experiment was wrapped in an execution script and distributed across the 152 computation nodes (64-bit) of the cluster of the Vienna University of Economics and Business, Austria, with a total of 336 GB of RAM. Its overall execution time was 199.29 days, run within the time frame of a few days thanks to the distribution across the cluster nodes. The number of successful calculations conducted was 1,370,961. A random sample of 100,000 experiment runs had to be drawn from this in order to reduce the complexity of analysis.[1]

The base corpus from which the sampling was conducted consisted of 1600 generic documents collected from news in two newspapers ('Presse' and 'Standard'): 400 thereof random-sampled from the one and 1200 from the other that were spread over ten sections of the newspaper and 12 months in 1 year with ten articles each. This was complemented by 746 domain specific documents (a textbook from the target domain information systems, split into paragraph sized units and an additional sample created from Google hits to the exam question posed). The essays were not included in the corpus, as this would violate requirements of the scenario of quick, dynamic evaluations mentioned above.

The number of input documents increased by steps of 50 (up to 410) and 100 (from 410 on) starting with 10. The degree of specialisation, i.e. the percentage of domain-specific documents in the corpus, was varied in steps of 20 from 0 to 100.

The resulting input corpus characteristics were assessed, thereby measuring the number of documents retained, the number of terms contained therein, and several measures about the vocabulary distribution (number of terms in the domain-specific documents, size of the vocabulary expressed in the generic documents, mean document length of the domain-specific texts and of the generic documents).

[1] It turned out that the collected data was too big to analyse on a machine with 32 GB memory.

Subsequently, upper-bound and lower-bound vocabulary filters were applied: 'fixed', 'flexible', or 'stopwords' for the upper-bound, 'fixed' or 'flexible' for the lower bound. Alternatively, no threshold was used ('none'). The minimum word length was set to two. The upper boundary was thereby set to 100 for the fixed and according to Listing 1 for flexible boundaries.

Listing 1 Flexible upper boundary for the frequency filter.

```
trunc(100 x log(nrow(M), base=100))
```

For the lower-boundary fixed was set to 3 and flex according to the calculation provided in Listing 2.

Listing 2 Flexible lower boundary for the frequency filter.

```
trunc(3*log(nrow(M), base=100))
```

To make sure that a good number of dimensions to keep was identified, the number of dimensions was varied from 2 in steps of 30 up to a maximum of 300. As this introduces a lot of weak results, the results had to be reverse filtered for the best performing number of dimensions.

Each resulting space was measured with respect to the number of documents, terms, size (sum of frequencies), the mean document length and its standard deviation, and—similarly—the mean term frequency and its standard deviation.

Into any such space, the essay collections were folded in and evaluated. This offers the advantage of very efficient evaluation of small numbers of documents. It is a scenario typically occurring in essay scoring and similar applications, where end-users need to be served quickly with results, thus avoiding the time-consuming space calculation by pre-calculation of the space. Although folding is a non-lossless projection, it is "generally appropriate to fold documents in only occasionally" (Berry et al. 1999, p. 355).

The score for each essay was thereby evaluated three times with the different closeness measures Spearman, Pearson, and Cosine. It was determined as the average correlation to the three model solutions in each of the essay collections.

The correlation of each essay collection to be evaluated against the human ratings was calculated using Spearman's rank correlation test, as the metric of the human ratings was in points and the closeness values of the space were evaluated using the closeness measures. The resulting Spearman's rho, S, and p-value were returned.

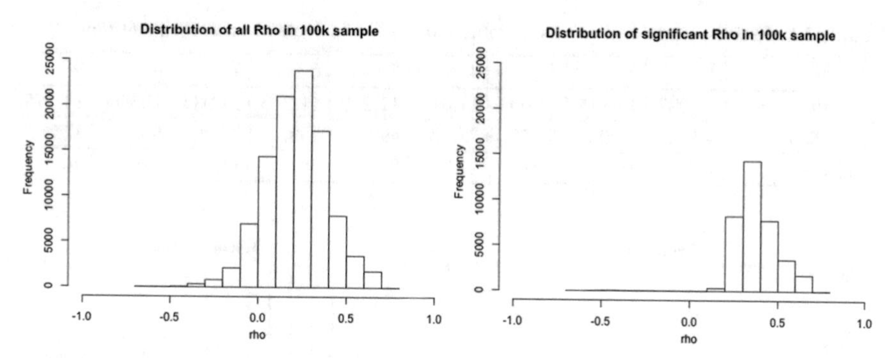

Fig. 7.1 Distribution of the rank correlations

7.3 Results

Not all experiment runs produce significant correlations between the human and machine judgements on a p-value below 0.05, nor do they all provide a correlation high enough to indicate dependency between the human and machine scores. In the sample of 100,000 experiment runs, the rank correlations are distributed as depicted in Fig. 7.1. The mean Spearman Rho of all 100,000 results was 0.22 and had a standard deviation of 0.17.

Looking at the significant results only (significant on a p-value below 0.05), the rank correlation was found to be in mean of 0.38 and had a standard deviation of 0.12. Filtering for the significant results eliminated the majority of results with a rank correlation below 0.5 and did not drop any above. 36,100 results were found to be significant of which 5766 had a correlation higher than 0.5. This shows that—as expected—the parameter settings tested brute-force created a lot of ineffective results.

As especially the number of singular values retained in the stretch truncation of the matrix decomposition was varied extensively, the size of number of ineffective test runs is not very surprising. To clean the result set from these irrelevant test runs, a reverse look-up of the best performing number of singular values for each number of documents in each essay collection was identified. Table 7.1 shows that each essay collection did receive attention in an almost equal number of test runs. The significance, however, varied between the collections. Filtering for the best performing test runs per document collection size per number of singular values, however, selected again a similar range of experiment runs. Altogether, 1065 experiment runs were selected.

Looking more closely at the resulting numbers of documents, the discrete steps from the experiment set-up can still be seen: Fig. 7.2 shows that the steps do approximate a fitted line (left-hand side), but the steps are retained and vary only slightly. The small variations in each staircase step can be explained with training documents dropped from the corpus for being empty after the term frequency filters were applied. As the histogram to the right of Fig. 7.2 shows, each of the classes is

Table 7.1 Results, significant results (p-value < 0.05), # of selected best experiment runs

Essay	1	2	3	4	5	6	7	8	9
Results	11.082	11.145	11.095	11.017	11.248	11.210	11.084	10.974	11.145
Significant	4.145	9.900	5.720	2.022	499	198	8.090	5.201	325
Selected	131	164	150	113	88	58	152	139	70

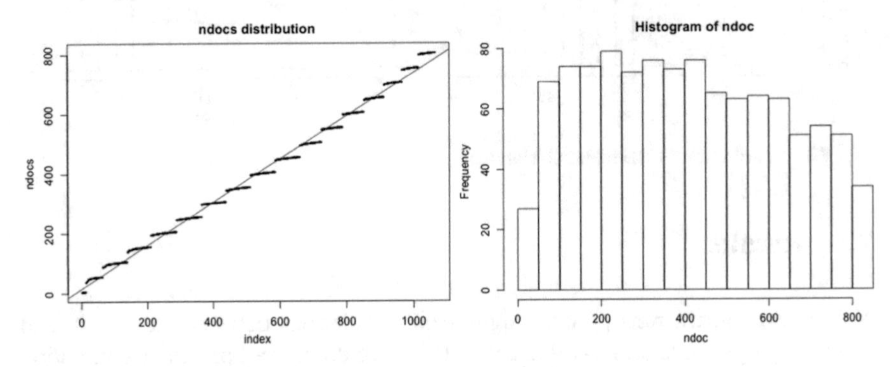

Fig. 7.2 Distribution of the number of documents among the selected results

almost equally populated. The rather low number of very small document sizes can be explained with such spaces failing in accuracy, indicated by a higher number of non-significant and low-effect experiment runs.

The situation is different, when looking at the number of terms. Figure 7.3 displays the—for display reasons minimally jittered—number of terms. A large number of test-runs have a moderate vocabulary size: The first quartile reaches a maximum of 551 terms, whereas the third quartile reaches 2270. Only 107 experiment runs resulted in a vocabulary size greater than 5000 terms (marked by the horizontal line in Fig. 7.3). The red curve in Fig. 7.3 depicts the overall growth of the number of terms. As the number of documents had been increased almost linearly in steps starting with small numbers, this distribution was expected: only a small number of corpora that consist of a large number of documents will actually result in a large vocabulary; additionally, only the settings in which the vocabulary is not subjected to frequency threshold filtering can result in large numbers of terms.

Turning to the number of singular values retained from the stretch truncation, again the steps of the increments by 30 starting from two singular values as given in the experiment set-up are visible among the best performing results. Although this trend is visible, the number of high performing experiment runs is inversely correlated with the number of singular values retained: the higher the number of singular values, the lower the number of successful experiments (see Table 7.2). This seems to show that the incremental variation of the number of singular values successfully identified good numbers of singular values to keep: the mean of 104 dimensions at a standard deviation of 88 is below the maximum number of 272 allowed in the experiment.

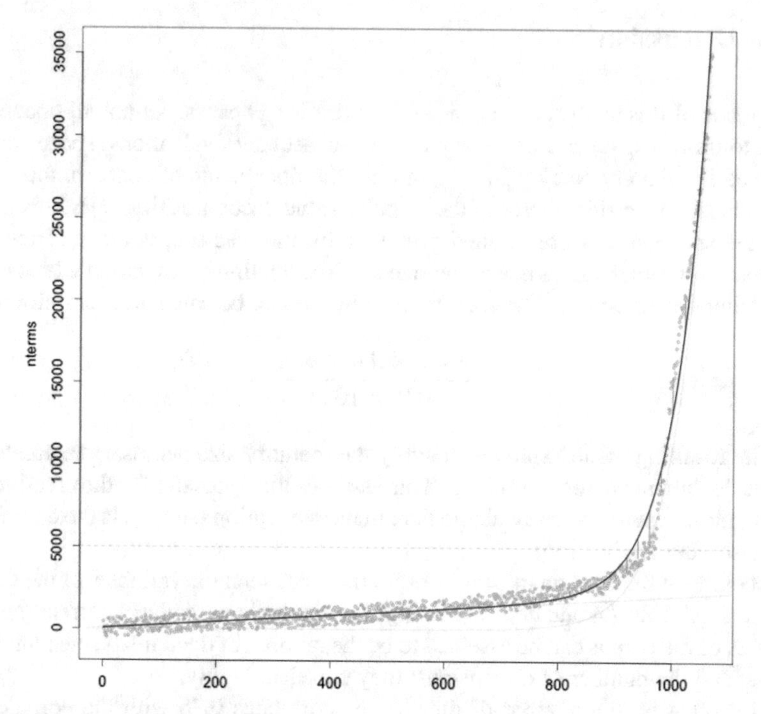

Fig. 7.3 Distribution of the number of terms

Table 7.2 Distribution of the number of singular values

Dimensions	2	32	62	92	122	152	182	212	242	272
Frequency	197	186	132	109	77	68	74	78	77	67

Table 7.3 Specialisation and lower/upper threshold filters

Specialisation	**0 %**	**20 %**	**40 %**	**60 %**	**80 %**
	60	208	265	220	312
Lower filter	**Fixed**	**Flex**	**None**		
	291	665	109		
Upper filter	**Fixed**	**Flex**	**None**	**Stopwords**	
	287	197	240	341	

The degree of specialisation in the corpus and the term frequency filters with upper and lower thresholds resulted in the distribution summarised in Table 7.3. Noteworthy is that a specialisation of 0 %, i.e. with no domain-specific documents, created clearly less performing results. The lower frequency filter seems to be set to ‘flex’ best, as it creates significantly more results. The upper frequency filter does not seem to drive the effectiveness very much, its impact on size, however, remains to be shown.

7.4 Discussion

The focus of this analysis is the size of the resulting spaces. As not all documents used to train a space are necessary for the subsequent evaluations, spaces can be reduced in memory size by folding in only the documents of concern, thus using only the left-hand side matrix of the singular value decomposition (T_k). This means that the space size is constrained primarily by the resulting number of terms n, number of singular values d, and the size w of the floating-point variable type on the operating system in use. The size (in megabytes) can be calculated as follows:

$$\frac{w \times n \times d + w \times d}{1024 \times 1024} \tag{7.1}$$

The resulting value expresses roughly the memory size necessary to create one single document vector in a fold in. This assumes that typically for the creation of a space, more resources are available than in its application (such as is the case for an essay scoring application).

Looking at the Spearman's rank correlations between the variables of the model, as listed in Table 7.4 and depicted in Fig. 7.4, the basic parameters correlating with the size of the corpus can be assessed to be the number of documents, the number of terms, and the number of dimensions: they correlate highly.

Most clearly, the number of dimensions correlates 0.78 with the corpus size. Similarly, the number of terms and documents correlate 0.66 and 0.61 respectively. The corpus size is the result of these three influencing parameters: they causally drive the size. Figure 7.4 underlines this graphically: in the row with the scatter plots against 'size', all three 'ndoc', 'nterm', and 'dims' exhibit the underlying relationship with 'spray patterns'.

Additionally, the number of terms and number of documents correlate with 0.74 significantly high with each other. Astonishingly, neither the number of terms, nor the number of documents correlates with the number of dimensions (and the other way round). Besides 'impossible' areas (where the number of documents is smaller than the number of singular values) in the scatter plot dims versus ndoc, there is no trend that can be spotted visually in addition to the Spearman rank correlation. The degree of specialisation is with −0.47 slightly negatively correlated with the

Table 7.4 Spearman's rank correlations of the variables

	ndoc	nterm	dims	spec	size	rho
ndoc	1.00					
nterm	0.74^{***}	1.00				
dims	0.31^{***}	0.14^{***}	1.00			
spec	-0.19^{***}	-0.47^{***}	0.24^{***}	1.00		
size	0.61^{***}	0.66^{***}	0.78^{***}	-0.14^{***}	1.00	
rho	0.10^{***}	0.21^{***}	0.11^{***}	−0.02	0.21^{***}	1.00

***p-value < 0.001, **p-value < 0.01, *p-value < 0.05

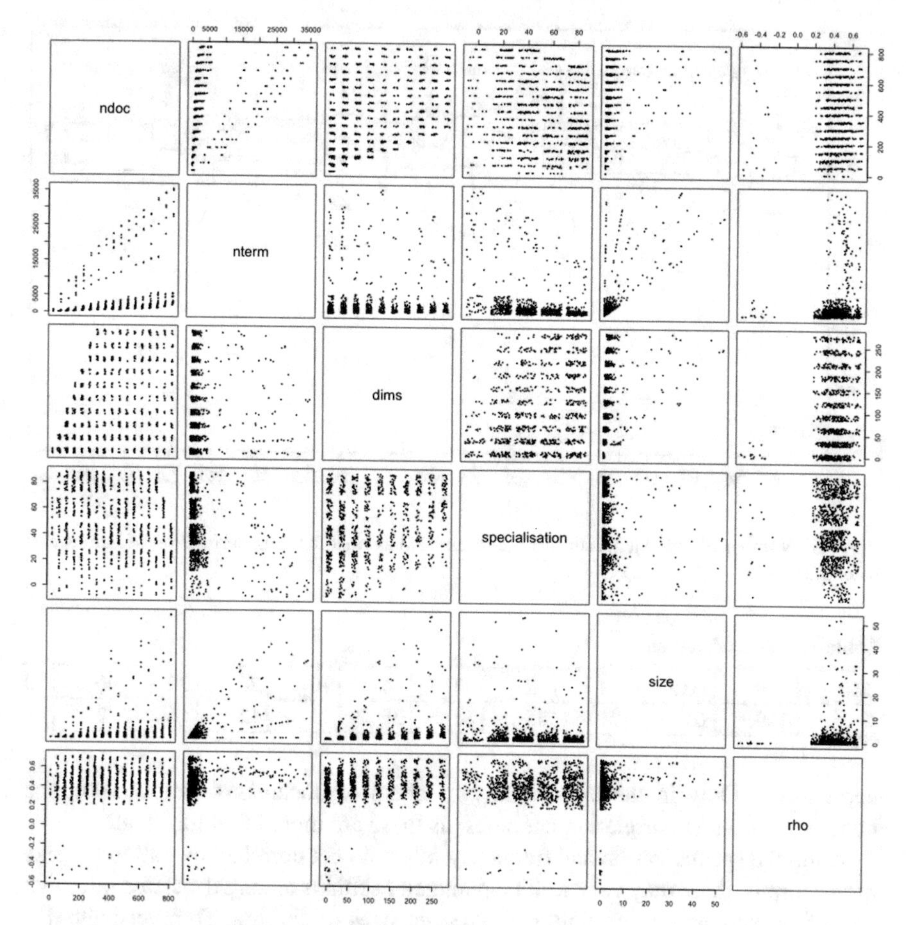

Fig. 7.4 Overview on the correlations between the variables

number of terms: higher degrees of specialisation reduce the number of terms; this effect, however, does not break through to the size.

Most notably, there is no correlation of the investigated variables with the human-machine agreement, measured with Spearman's rho. Specifically the size is only very weakly correlated with the human-machine agreement: their rank correlation is 0.21: all space sizes create effective and ineffective essay evaluations.

Focusing back on the number of documents that could be easily mistaken for the size of a corpus, Figure 7.5 provides further insight: the notches of the bars indicate the median Spearman's rho of human to machine scores, the columns of the diagram partition the data into histogram classes. The lower and upper hinges display the first and third quantile, the whiskers the extrema. Outliers in the data have been depicted with small circles. The figure underlines again that there is no overall relation between the number of documents and the human-to-machine

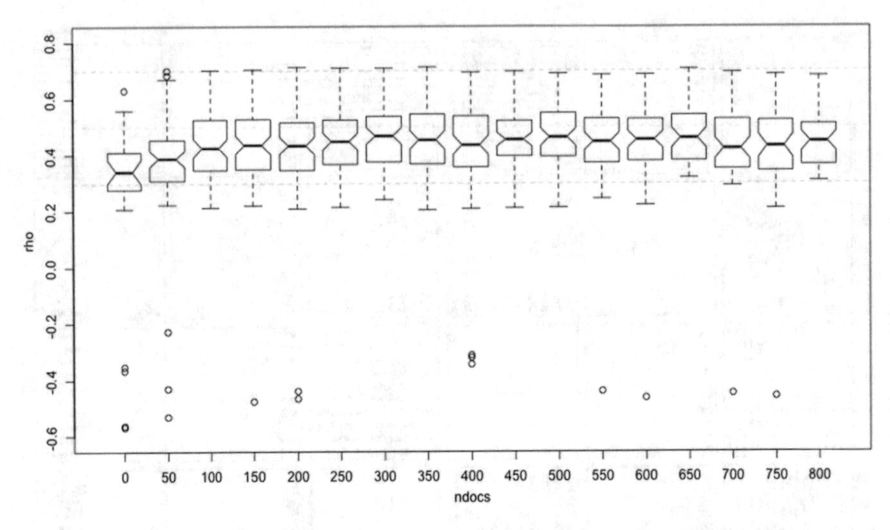

Fig. 7.5 Number of documents (in histogram classes) against the rank correlation of human and machine scores

Table 7.5 Size tabulation

MB	0	0.5	1	2	3	4	10	20	30	40	50
#	109	65	59	35	17	36	12	12	6	2	1

correlations. Only in the area of tiny numbers of documents (below 100), the human-to-machine correlation degrades, as these are more often too small.

Although specialization and frequency filters do not correlate overall to the size of the corpus, there may very well be parameter settings among the tested ones that do in fact correlate or correlate with specific sizes of corpora. To investigate this further, a principal component analysis was conducted over the test results. The principal component analysis thereby aimed to investigate which variable settings occur in combination. Thereby, the size was partitioned into the bandwidths listed in Table 7.5.

Specialisation and the frequency filters were already partitioned from the experiment setup. Analysing the principal components of the frequency filters shows the following. Very small corpora with a size between 0 and 0.5 MB appear often in combination with stopword filtering. Small corpora from 1 up to 4 MB are served with no upper boundary better and with a fixed or flexible lower threshold more frequently. Larger corpora of 4–20 MB (and 40–50 MB) show the trend to be in favour of no lower threshold filter and slightly in favour of no and flexible upper threshold filters (Fig. 7.6, Table 7.6).

Turning to the degree of specialization, i.e. the share of domain specific to generic documents, the following tendencies can be read from the results of the second principal component analysis depicted in Fig. 7.7. Very small corpora of up to 1 MB seem to co-occur with a high degree of specialization. Smaller corpora

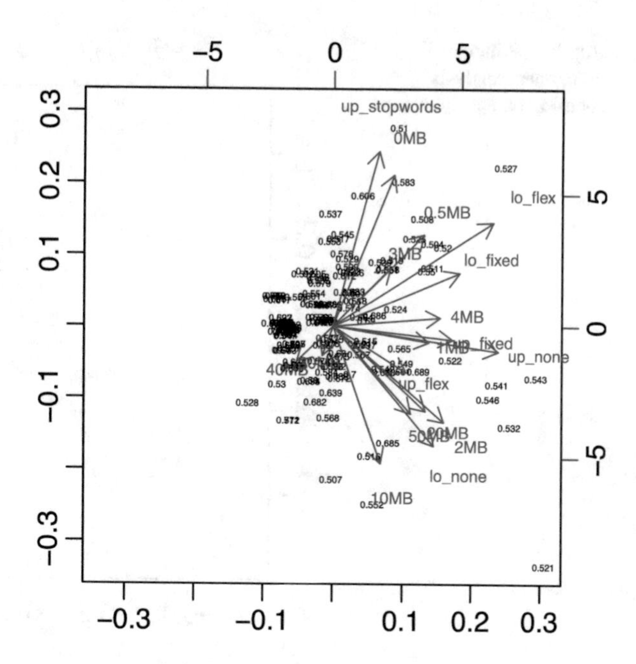

Fig. 7.6 PCA focusing on frequency filters

Table 7.6 Spearman correlations between vocabulary filter settings and size

	Upper threshold				Lower threshold		
	Fixed	Flex	None	Stopw.	Fixed	Flex	None
0 MB	0.32***	0.12	−0.12	0.49***	0.17*	0.29***	0.05
0.5 MB	0.11	0.23**	0.12	0.17*	0.16*	0.33***	0.02
1 MB	0.09	0.20**	0.25**	−0.04	0.02	0.39***	0.01
2 MB	0.14	0.04	0.29***	−0.19*	0.26***	0.19*	0.02
3 MB	−0.01	−0.12	0.19*	0.11	0.16*	0.08	−0.06
4 MB	0.08	0.02	0.34***	0.10	0.18*	0.10	0.27***
10 MB	0.12	0.14	0.07	−0.10	0.02	−0.15	0.43***
20 MB	0.10	0.14	0.15	−0.04	−0.03	0.11	0.42***
30 MB	−0.07	−0.03	0.05	−0.01	−0.08	0.01	0.23**
40 MB	−0.09	0.05	0.03	−0.09	−0.01	−0.15	0.13
50 MB	0.09	0.18*	0.10	−0.06	0.06	0.13	0.15

***p-value < 0.001, **p-value < 0.01, *p-value < 0.05

seem to co-occur more often with 20–40 %. Larger corpora seem to tend to address a specialization of 40 %. There is a trend to require less specialization for larger corpora (see also Table 7.7).

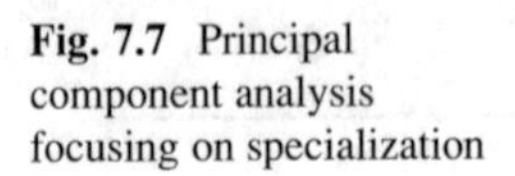

Fig. 7.7 Principal component analysis focusing on specialization

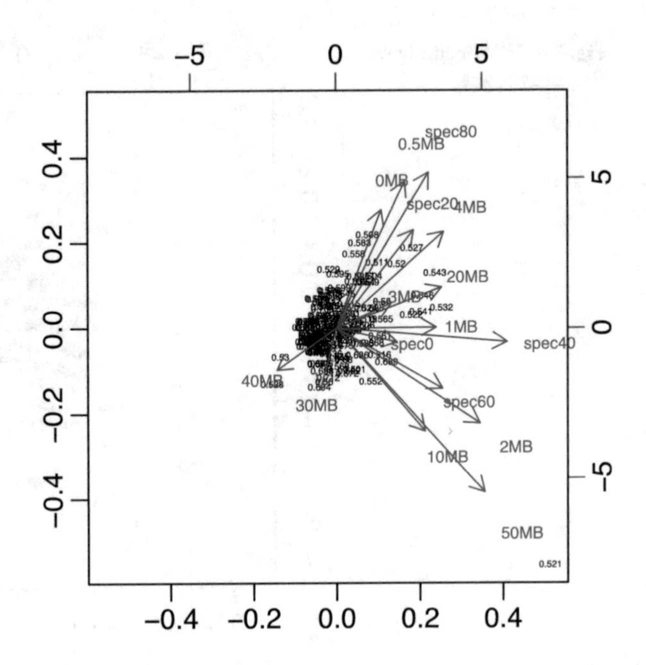

Table 7.7 Spearman correlations between specialization settings and size

	0 %	20 %	40 %	60 %	80 %
0 MB	0.13	0.17*	0.19*	0.09	0.16
0.5 MB	0.19*	0.14	0.10	−0.01	0.25***
1 MB	−0.14	0.04	0.21**	0.13	0.26***
2 MB	0.05	0.08	0.01	0.24***	0.14
3 MB	−0.03	0.06	0.19*	0.16*	0.01
4 MB	0.12	0.23**	0.14	0.09	0.09
10 MB	0.21*	0.01	0.08	0.02	0.00
20 MB	0.10	0.02	0.18**	−0.02	0.21**
30 MB	−0.03	−0.06	0.06	0.13	−0.02
40 MB	0.07	−0.07	−0.09	0.03	−0.03
50 MB	0.15	−0.05	0.16*	0.15	−0.08

***p-value < 0.001, **p-value < 0.01, *p-value < 0.05

7.5 Summary

Does size matter? Is bigger better? ‘No, it’s the technique’, is a conclusion that can be drawn from the presented results. It has been shown that the size of a space does only weakly drive its effectiveness. Effective spaces can be created with all sizes, no matter whether big or small. The number of documents, the number of terms, and the number of dimensions chosen controls the size.

Whereas overall the degree of specialisation has no or only weak effects, there are weaker tendencies indicating how well it might suit particular corpora sizes. Similar tendencies could be identified for upper and lower thresholds of vocabulary filters, some of them rather more than a mere tendency.

It might be argued that due to the experiment set-up, systematic bias might have been introduced to exclude specifically tiny corpora—as random sampling of the domain and background corpus might very often result in a situation where the necessary vocabulary is lacking and the target essays cannot be evaluated at all or only with very low success. Through the random sampling loop, however, this effect should be reduced in its influence, though it may very well be possible that more intelligent sampling methods pave the way for success of tiny corpora.

The results found in this experiment have shown that sampling a smaller corpus from a larger one to serve particular domain needs is possible. The role of seed documents and vocabulary growing procedures deserve special attention in the future in order to better support the retention of small corpora for a specific context from an open corpus.

The subsequent Chap. 8 now turns to the actual implementation of meaningful, purposive interaction analysis into the R package *mpia*. Comprehensive usage examples of learning analytics with MPIA will follow then in Chap. 9 and a critical evaluation in Chap. 10.

References

Berry, M., Drmac, Z., Jessup, E.R.: Matrices, vector spaces, and information retrieval. SIAM. Rev. **41**(2), 335–362 (1999)

Deerwester, S., Dumais, S., Furnas, G., Landauer, T., Harshman, R.: Indexing by latent semantic analysis. J. Am. Soc. Inf. Sci. **41**(6), 391–407 (1990)

Landauer, T., Foltz, P., Laham, D.: An introduction to latent semantic analysis. Discourse. Processes. **25**(2–3), 259–284 (1998)

Leydesdorff, L.: Similarity measures, author cocitation analysis, and information theory. J. Am. Soc. Inf. Sci. **56**(7), 69–772 (2005)

Quesada, J.: Creating your own LSA space. In: Landauer, T.K., McNamara, D.S., Dennis, S., Kintsch, W. (eds.) Handbook of Latent Semantic Analysis. Lawrence Erlbaum Associates, Mahwah, NJ. (2007)

Chapter 8
Implementation: The MPIA Package

The previous two chapters introduced the mathematical foundations in matrix theory of MPIA and visual analytics for MPIA. The first part thereby focused on algebra and geometry of vector spaces created from performance collections, whereas the second part introduced a method to create stable projection surfaces for visual analytics. For both, important concepts and relations were defined: most notably, identity and proximity relations as well as the concepts of location, position, and pathway.

This chapter will provide an introduction to the actual implementation, i.e. the *mpia* R package, its class system as well as key data manipulation and visualization methods. It describes the software package delivered to overcome the key challenge of this work, formulated in all three objectives described in Sect. 1.2.

It provides the instruments to automatically represent and further analyse the conceptual development evident from the performance of learners, while at the same time introducing re-representation mechanisms for visual analytics to guide and support decision-making about and during learning.

Use case (Sect. 8.1) and activity diagrams of an idealized analysis workflow (Sect. 8.2) will thereby foster understanding of how MPIA analysis processes can be conducted.

The implementation (Sect. 8.3) hides a lot of the mathematical and functional complexity, encapsulating it in high-level functionality, while providing configuration options via parameters where required.

Finally, a summary (Sect. 8.4) concludes this chapter.

8.1 Use Cases for the Analyst

Other than the implementations for *sna* and *lsa*, the *mpia* package targets facilitating analysis on an abstraction level closer to its application in learning analytics. The use cases depicted in Fig. 8.1 thus fall into the following five distinct groups:

F. Wild, *Learning Analytics in R with SNA, LSA, and MPIA*,
DOI 10.1007/978-3-319-28791-1_8

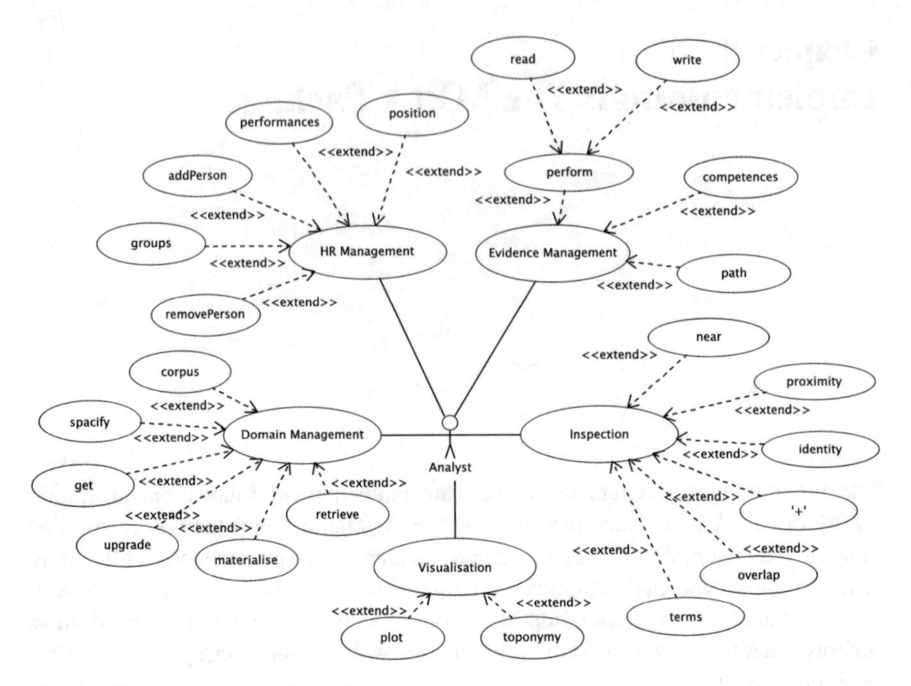

Fig. 8.1 MPIA use cases of the analyst

domain management, human resource management, evidence management, further inspection, and visualization.

Thereby, domain management subsumes those human to system interactions that deal with calculating, storing, and retrieving social semantic performance networks.

The human analyst typically adds a *corpus*, from which the initial conceptual vector space in its Eigenbasis is derived—a process called '*spacification*'.

Analyst thereby refers to the persona of an investigator or researcher of learning, for example a learner, tutor, teacher, faculty administrator, system designer, or researcher.

Since processing the Eigenspace representation is computationally expensive, a persistence layer is added, comprising the cases of *materializing*, *finding ('get')*, and *retrieving* space objects to a storage device and back into memory. The use case of *upgrading* covers the routines required to update the classes of persistently stored space objects (while retaining their pre-calculated data), thus ensuring downwards compatibility of the package routines with eventual future releases.

The cases grouped together in human resource management deal with *adding* and *removing* representations of learners and their meta-data. This includes retrieving evidence of *performance* previously associated with them as well as calculating competence *positions* they hold. Moreover, an interface for identifying *groups* with shared or similar competence positions.

Evidence management bundles those use cases dealing with the administration of the digital records representing human action: basic performance records hold a representation of meaning and a human-readable label for the purpose of the action. The case '*perform*' creates such record, while *read* and *write* extend the case for modelling reading and writing actions. Underlying *competences* and learning *paths* can be extracted.

The group of inspection cases refers to basic operations the analyst might want to conduct: testing for proximity and identity (see Sect. 5.7), finding performance records and persons *near* to a specified one, *combining* performance records ('+'), and describing records in form of their highlight loading *terms* or with respect to shared terms *overlapping* between records.

The final group of visualisation cases groups the fundamental use cases required for depicting visually the domains, locations, positions, and paths of persons, groups, and individual performance records. Plot thereby refers to depict the involved objects visually in form of a cartographic representation and toponymy refers to the use cases needed for describing the planar projection introduced in Chap. 6.

While this use cases overview may be overwhelming, more clarity will emerge when the basic analysis workflow is introduced in the subsequent section with the help of activity diagrams. This shallow overview, moreover, shall be complemented with additional detail in the description of the class system implemented in the package. Other ways to group together the use cases are possible: particularly the separation of the inspection cases may look somewhat arbitrary. As these cases, however, can operate across person and performance records they justify separate treatment.

8.2 Analysis Workflow

The basic analysis workflow follows the steps depicted in Fig. 8.2. An analyst first selects or creates a Domain model, typically using the DomainManager to instantiate, create, and materialise the basic language model trained for a particular domain. This step is similar to creating a latent semantic network, with the notable differences that the dimensionality for the Eigenbasis is determined automatically using the stretch-dependant truncation proposed in the previous Chap. 5.

Moreover, the subspace of the Eigenspace chosen serves as a representation space for the actual analyses added subsequently in the steps of adding evidence and conducting measurement, filtering, and inspection operations. Adding evidence thereby refers to using the HumanResourceManager to add representations of learners (instantiating Person objects) and inserting Performance objects that hold both meta-data (such as data about intended purposes, source texts, scores, and labels) as well as meaning vectors that represent the evidence mapped into the subspace of the Eigenspace. Such broken down workflow of adding evidence is depicted to the left of Fig. 8.3.

Not least because such meaning vectors are typically of high dimensionality, inspection routines are used to describe the concepts activated by the meaning

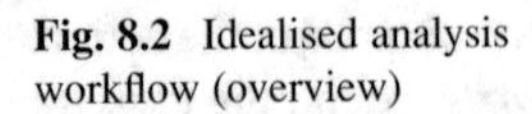

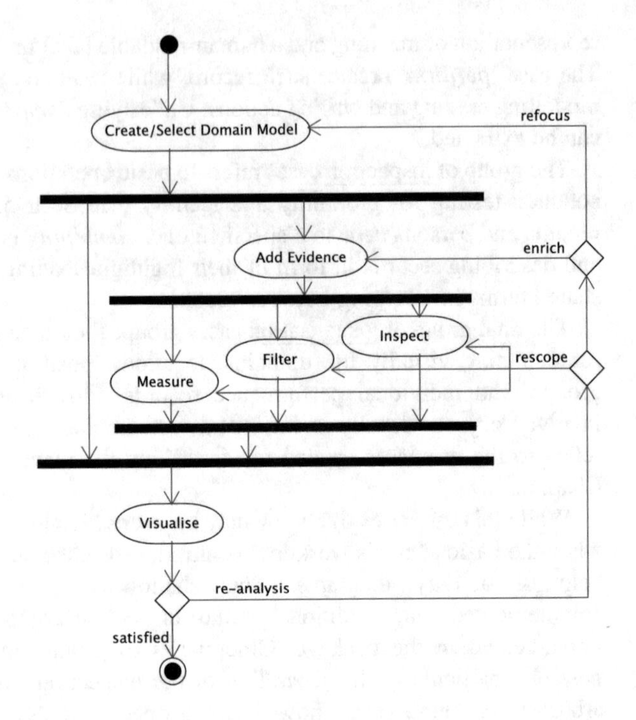

Fig. 8.2 Idealised analysis workflow (overview)

vector in the Eigenspace representation. Filtering operations help with selecting particular paths, topical areas, or groups of persons. Measurement operations allow exploring structural properties of the evidence investigated: for example, identity and proximity measurements can be conducted. Moreover, position information can be derived, uncovering the competences underlying the performances mapped.

If the information need of the analyst is satisfied, the analysis workflow concludes. If this, however, is not the case and re-analysis is required, the analyst can refocus and turn to a different domain. Evidence can be collected for multiple domains at the same time, while the analysis focuses on a single domain at a time. In an inner loop, the scope of the analysis can be changed and additional evidence can be mapped. This is depicted with the decision points to the right of Fig. 8.2: the analysis is either enriched, thereby adding new data, or it is re-scoped, thereby changing filters and changing measurements and inspection. This allows for exploratory data manipulation.

The activity diagram depicted in Fig. 8.3 adds more detail to this basic analysis workflow. The steps of getting/setting domains via the *DomainManager* and of adding evidence with the help of the *HumanResourceManager* and *Person* plus *Performance* objects have already been introduced above.

The activity diagram, however, adds more detail to the filtering, measurement, and inspection actions at the bottom of the illustration (in between the diagram-spanning split bars): nine essentially different actions can be distinguished in the analysis.

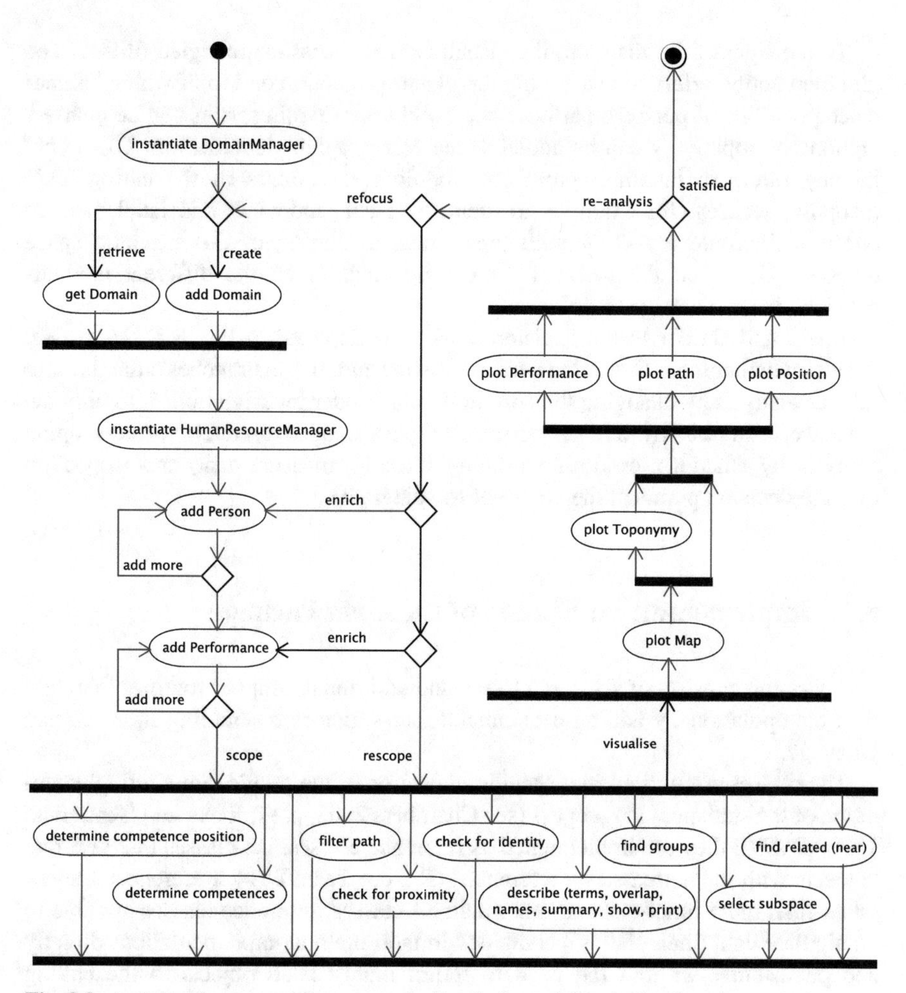

Fig. 8.3 Activity diagramme of a prototypical analysis workflow in detail

At the heart of it (bottom, middle) lie the tests for identity and proximity. To the left, actions supporting the identification of competences underlying the performances can be found. The analyst needs to determine current or past competence positions of learners represented as well as the competences underlying the performance of individual persons, individual performances, or groups thereof. For this, filtering for learning paths is required, for example, in order to restrict analysis to a specific time interval or to restrict analysis to specific purposive contexts, such as a particular assessment. To the right, actions are shown that deal with further, explorative selection and filtering: finding groups and finding persons with similar profiles or closely related performances or competences can be counted amongst these. A bigger group of actions required to inspect persons, performances, and competences are listed as well as subspace selection routines for basic data manipulation with the array accessors ‘[]’.

To the right of the diagram, the visualization process is untangled further. The plot map action refers to visualizing the planar projection on top of which competence positions of persons, performances, and their learning paths can be charted. Optionally, toponymy can be added to the cartographic visualization. Toponymy thereby refers to labelling prominent topological features of the cartographic mapping, seeking, for example, to identify visual landmarks and label them in order to facilitate a sort of with topological anchoring of the semantic space depicted. See Sect. 8.3 below for more information on the different methods provided in the package to do this.

Again and similar to the decision work flow depicted in Fig. 8.2, the analyst decides whether the information need is satisfied and, if not, branches either back to refocus analysis by changing the current domain under investigation, enriching the evidence data set (by adding persons or performance records), or re-scoping analysis by changing or further refining filtering, measurement, and inspection (see the decision points at the centre of the diagram).

8.3 Implementation: Classes of the *mpia* Package

The package consists of six core classes and additional wrapper routines that ease data manipulation by adding a command-like structure resembling more human language.

The classes are written in a specific dialect of R, the public domain implementation of the statistical language S (see Chambers 2008, p. 9f; Ihaka and Gentleman 1996, p. 299): they are implemented as reference classes, i.e. classes that generate "objects with fields treated by reference" (R Core Team 2014, ReferenceClasses). Other than mere functions and other than S4 classes, reference classes are able to manipulate data (their 'fields') contained in their method implementations directly and persistently without the need to return manipulated objects to the calling environment.

The wrapper routines implement so-called generic functions, i.e., they dispatch calls to ordinary functions and methods based on the classes of the objects handed over in the arguments[1] (Chambers 2008, p. 396ff).

The core classes and their dependencies are depicted in Fig. 8.4 (see also Annex A for individual diagrams for each class and the full methods and fields tables): The *HumanResourceManager* holds objects of class *Person*, which again holds the record objects of class *Performance*.

[1] The generic registers a function to be called if the arguments handed over match the so-called signature of the method. This way, for example, different 'print' function implementations can be called for objects of class 'numeric' and class 'character, while to the user a single 'print()' generic function is exposed.

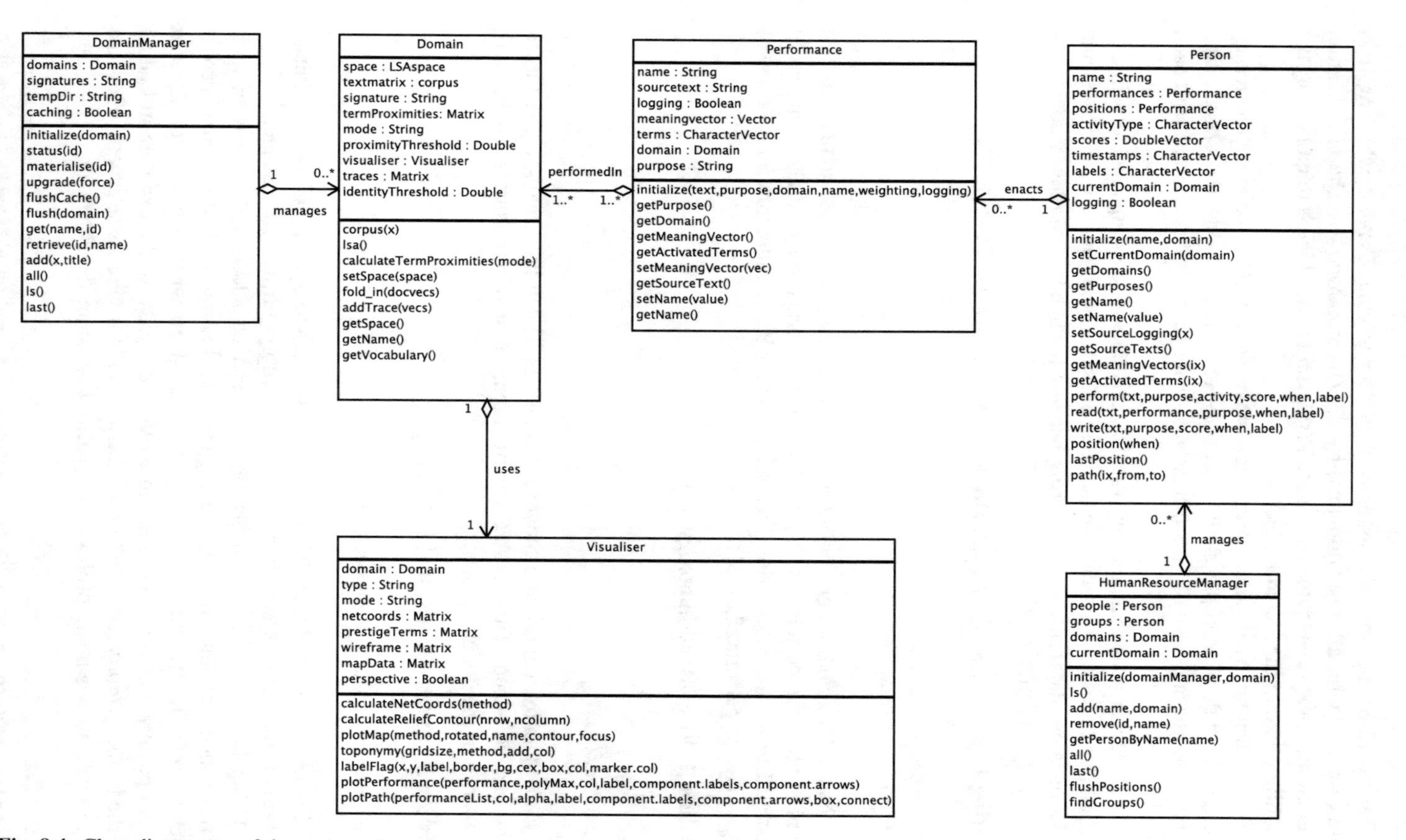

Fig. 8.4 Class diagramme of the *mpia* package

Each performance record is associated with a particular *Domain*, which again is held (together with all other domains) by the *DomainManager*. The visualisation routines are encapsulated into a separate *Visualiser* class, which is referenced by the according field in the *Domain* objects.

Before turning to a detailed description of the classes and associated generics, an analysis workflow example shall be made—to illustrate what the classes sign responsible for and how objects instantiated from such classes interact with each other.

As all earlier code examples provided in this book, the listings included in this section aim to foster re-executable (same code, same data) and reproducible (same code, different data) research, lowering barriers for uptake and reuse through this tutorial style of presentation.

Listing 1 Instantiate the *DomainManager* and fetch a *Domain*.

```
dmgr = DomainManager()
d = dmgr$get("generic")
```

First, a *DomainManager* object is instantiated and one of the domain models coming with the package is retrieved from the storage device. The example in Listing 1 features a generic domain model within which evaluations for a variety of topics can be performed.

Listing 2 Visualise terminology map.

```
plot(d)
toponymy(d)
```

Subsequently, the domain (now in memory as d) can be visualised and toponymy can be added. This creates a cartographic plot of the conceptual network contained in the domain model.

Listing 3 Instantiate a virtual HR manager.

```
ppl = HumanResourceManager(dmgr)
```

The next step is to instantiate a HumanResourceManager in order to administer person objects. The object *ppl* now holds an 'empty' virtual HR manager.[2]

Using this virtual HR manager, learners can be added and records of their performance can subsequently be logged, see Listing 4. Performance records thereby refer to evidence of learning such as given when a learner produces a forum posting, answers a question in writing, or reads. In the example in Listing 4, the learner *fridolin* has written short answers to four micro assessments about the topics risk management, profits, risk control, and donor development.

[2] For a user story about a human resource manager see the foundational examples of Chaps. 3, 4, and 9.

Listing 4 Add a learner and performance records.

```
fridolin=ppl$add(name="fridolin")

fridolin$write("risk management is crucial for
  preventing a business from failure.", label="risk")

fridolin$write("shareholders can maximize their
  profits, if the price level is increased:
  most shareholders appreciate such performance
  management oriented approach to business",
  label="profits")

fridolin$write("in management, it is the role of
  the ceo in business to appreciate the contributions
  of the panel members and to balance risk",
  label="control")

fridolin$write("donor development in fundraising is
  all about ensuring that you and your donors get
  the most you can from your relationship in ways
  which are mutually agreeable and beneficial",
  label="donors")
```

The method *write* thereby instantiates a new Performance object, which cares for mapping the source text to its *mpia* representation in the chosen subspace of the Eigenbasis. As this representation is not necessarily in a form easy to process for the human analyst (due to its usually large number of dimensions), inspection methods and visualisation methods can be used to support their investigation.

Listing 5 Inspect and plot competence positions and learning paths.

```
terms(fridolin)
terms(position(fridolin))
plot(position(fridolin), col="red")
plot(path(fridolin), col="red")
```

Listing 5 provides examples for this: a call of the generic *terms* over the performance records and over the aggregated position of *fridolin* is used to describe which were the most highly activated terms in the space. Moreover, the *plot* methods are used to enrich the cartographic visualisation created above in Listing 2 with a visual marker of the competence position of *fridolin* as well as with a path spline.

Listing 6 Add a second learner, add performance records, visualize.

```
max=ppl$add(name="maximilian")

max$write("advertising includes the activity in
  the media such as running newspaper ads, direct
  marketing, cold calling", label="advertising")

max$write("marketing management, however, is a
  wide field: this basically boils down to comparing
  investment into advertising with the return on
  investment and making decisions", label="marketing")

plot(position(max), col="yellow")
```

The HR manager object is then used to add a second learner to the learning network, map two performance records (about marketing and advertisements), and again enrich the existing plot with an additional marker representing *max*.

Listing 7 Test for competence profile identity, proximity, and extract underlying competences.

```
max == fridolin
near(max, fridolin)
competences(fridolin)
```

The final Listing 7 adds examples of further data exploration. Line one checks for identity of the two learners with respect to their competence position, whereas line two tests for proximity. The final line returns a list of competences underlying the performances of one of the learners.

8.3.1 The DomainManager

The *DomainManager* is used to administer *Domains*, i.e. those objects holding corpus, space, traces, and additional meta-data used in investigating learning conversations. It provides methods to add, retrieve, upgrade, materialise, and remove domain objects.

Some of the calculation processes involved in creating such domain representations are computationally rather costly. The *DomainManager* assists in efficient persistence management: it provides methods to materialise to a local cache directory and it supports the user in finding, loading, and upgrading domain objects (and their visualizers) to new class releases.

8.3.2 *The* Domain

The *Domain* class encapsulates fields and methods for processing and storing social semantic vector spaces: a given corpus of texts is analysed to derive an Eigenbasis such that—per default—their first Eigenvalues explain 80 % of the total stretch needed to expand its eigenvectors to the mapping provided by the raw document-term matrix (constructed over the input corpus).

Through this approximation, the input corpus is lifted up to a more semantic representation, thus allowing to investigate the nature of the associative closeness relations of its term vectors, document vectors, or any of their combinations used in representing competence positions and performance locations of a *Person* or groups.

8.3.3 *The* Visualiser

Every *Domain* uses a *Visualiser* object for the creation of cartographic plots of the social semantic network it holds: the conceptual graph can be visualized to be then used as a projection surface in order to display locations, positions, and pathways of learners or individual performance records.

The *Visualiser* is responsible for all plotting activities: all other objects interface with it to output to the plot. The *Domain* uses the *Visualiser* to render a cartographic map and label its landmarks. The *Person* uses the *Visualiser* to mark position, path, or location of individual *Performance* objects.

The *Visualiser* class implements several variations for plotting cartographic representations: 'topographic', 'persp', 'wireframe', and 'contour'. Thereby, topographic and contour relate to two-dimensional, flat visualisations (using contour lines and shading to indicate elevation) and perspective and wireframe refer to three-dimensional plots (using central perspective with light sources and tile shading to depict elevation).

Depending on whether the intended plot is *perspective* in nature, an internal mapping of points on the wireframe surface needs to be constructed, thereby using the viewing transformation matrix mapData that results from plotting the map. For such mapping of any point, first its original position in netcoords is calculated (see above, Sect. 6.2), then mapped to the wireframe coordinates (after grid-tiling as described at the end of Sect. 6.2). The resulting wireframe position (row, column, height) is then transformed to the x, y, and z coordinates of the plot (and the x, y coordinates of the viewing device).

The *Visualiser* class additionally implements a method to support the identification of interesting topographic features (places) in the chosen cartographic representation, thereby inspecting which of the conceptual right-singular Eigenvectors cause them in order to select ideally a smaller number of labels. Currently, four such methods are implemented: 'mountain', 'gridprestige', 'gridcenter', and 'all'. While the latter simply displays all term vector labels at their exact location, scaled and shaded by their prestige scores (Butts 2010, function 'prestige'; Freeman 1979, p. 220; see also Chap. 3, Listing 14) in the graph, the other three try to select a smaller, but characteristic number of labels.

The method 'mountain' thereby overlays a grid with the default size 10×10 over the projection surface to then identify the highest point in the wireframe quadrant each grid tile spans. Once the highest point is found, the first term vector below this point of maximal elevation is selected as label for the mountain. This way, preference in labelling is given to points of elevation, while at the same time, the maximum number of labels is restricted to 100 (or smaller, given that there usually are empty quadrants on the periphery).

The method 'gridprestige' follows a similar approach using a per default 10×10 grid overlay. The method, however, gives preference not to elevation level, but to prestige scores: it will select the term vector location and label for the term vector with the highest prestige score in the subspace inhabiting the quadrant.

The method 'gridcenter' follows the same approach (though with a smaller default grid size of 5×5), with the small difference that it selects the three terms with the highest prestige scores in the grid quadrant. Also it places them at the centre of the grid quadrant (and not at the exact netcoords location of the term vector).

8.3.4 *The* HumanResourceManager

Just like in a company, the *HumanResourceManager* cares for the people working in it (see also Sects. 3.2, 4.3, and 9.2): it holds pointers to the *Person* objects, provides an interface for mining groups, and helps with identifying experts with a particular profile. The HR manager act in a particular *Domain*.

The HR manager is used to add or remove people, to cluster together those persons into groups that share a similar competence profile, and to find persons with known experience in a particular area.

When the HumanResourceManager is used to identify groups of people with similar positions, the found groups are stored as a *Person* object and added to the field *groups*. The name for the group thereby is the collated list of names of the persons it contains (coma separated).

The groups, competences, and near generics will be described below in Sect. 8.3.7.

8.3.5 *The* Person

Person implements a representation of the learner: persons have names, competences, and positions, calculated from the activity traces that are stored about them. Such traces are implemented as so-called *Performance*.

Persons should best be added using the *HumanResourceManager*, rather than instantiating them directly. This way, the HR manager maintains a record of all persons of interest in the analysis.

8.3.6 *The* Performance

The *Performance* holds a digital record representing a particular human action: it stores a representation of meaning and a human-readable label for the purpose of the action. The purpose thereby is a character string, whereas the meaning vector is constructed by mapping the descriptive source text to its social semantic vector space representation of the *Domain*.

Several class methods (and generics) support in inspecting the result of such mapping: terms, for example, returns all terms activated above threshold in the vector space through the fold in of the source text. Each *Performance* object is bound to a particular *Domain*.

When mapping the source text to the meaning vector in the Domain, the constructor calls the Domain's *addTrace* method to append another vector to the *traces* matrix and storing a reference to the index position rather than the full vector in its local field *meaningvector*. This way, memory use is managed more efficiently: all Performance records are appended to the trace matrix, thus enabling the use of matrix operations in further calculations.

The introspection routines provided in *getActivatedTerms* take a threshold value (per default this is the *proximityThreshold* of the *Domain*) to return all term labels where the cell value exceeds the threshold. Choosing different threshold values depends on intended use (which may favour recall over precision, for example) and the properties of the space.

8.3.7 *The Generic Functions*

As already mentioned in the introduction of this section, several generic functions provide wrappers to functionality of the package in order to ease the syntax of data manipulation.

All generics are already described already in brief above, listing what functionality they provide in each signature implementation for the different classes. There are, however, a few generics the functionality of which is a little bit more complex: groups, competences, overlap, terms, near, and addition (+). They shall be described subsequently, thereby adding more detail on how they work internally and what they effectively do.

Listing 8 Determining competence positions.

```
positions=NULL
for (p in .self$people) {
  positions=c(positions, p$position())
}
```

The generic *groups* command provided by the *mpia* package implements a method for identifying expertise clusters in the personnel administered by a given *HumanResourceManager*, following the proposal in Sect. 5.9. It therefore first calculates the competence position of every person (see Listing 8).

Once positions are determined, the proximity table of the closeness of each position to each other is calculated (using *proximity*, i.e., the cosine similarity, see Listing 9).

Listing 9 Calculating the proximity table.

```
prox=matrix(0, nrow=length(positions),
            ncol=length(positions))

for (i in 1:length(positions)) {
  for (l in 1:length(positions)) {

    if (l==i) prox[l,i]=1
    else {
     prox[l,i]=proximity(positions[[l]],
                positions[[i]])
    }

  } # for l
} # for i
```

Using this proximity table, agglomerative clustering (using complete linkage with hclust, see Kaufman and Rouseeuw 1990; Lance and Williams 1967) is applied, resulting in a hierarchical cluster dendrogram. This dendrogram indicates in its height the level of dissimilarity within each clusters at a given agglomeration level. Such measure is implemented as maximal distance of any two members of two clusters merged: the cosine proximities are converted to distances as described in Listing 10 (see also Sect. 5.9).

Listing 10 Agglomerative nesting and cluster selection.

```
a=hclust(as.dist((1+prox)/2), method="complete")
b=cutree(as.hclust(a), h=
      (1+.self$currentDomain$identityThreshold)/2)
```

The dendrogram is then cut off at a level of the *identityThreshold* (converted to distance) in order to obtain homogeneous clusters. Depending on the competence profiles of the *Person* objects managed, this may result in a smaller or greater number of clusters.

The generic *competences* method is working with complete linkage as clustering algorithm in quite a similar way: it uses hclust over the proximity table of the performances handed over or retrieved from the persons managed by a human resource manager. Once they have been retrieved, complete linkage clustering is applied and again the resulting dendrogram is cut into the desired clusters at the height of the *identityThreshold*.

The generic *terms* is used to describe the meaning vectors created for each performance in the social semantic space. Therefore, the basis of the meaning vector *mv* is mapped back from the Eigenbasis $[]_V$ into the originating basis $[]_E$—following the proposal of Sect. 5.4.

The decision on which terms shall be used to describe the performance is then made based on whether the frequency values of the $[]_E$-coordinatised performance are above a given threshold (per default, this uses the same threshold of *0.3* as the *proximityThreshold*, see Sect. 5.7). Listing 11 provides the essential lines of code for this: first *dtm* is resolved as the mapping of *mv* back to the originating basis. Then, *ixs* determines which of the cell frequencies of the resulting vector are above the given activation threshold.

Listing 11 Mapping the meaning vector back to the originating basis and determining which term frequencies are above threshold.

```
dtm = crossprod(
        t(crossprod(
           t(.self$domain$space$tk),
           diag(.self$domain$space$sk)
        )),
        t(mv)
        )
ixs = which(dtm > threshold)
```

The generic *overlap* provides enhanced functionality, acting on top of the generic *terms*. It determines which terms are good descriptors of two or more performances to then return only those terms that are shared by the performances investigated: it returns only those that ‘overlap’ between the given incidences.

The generic *position* returns the centroid vector (non-weighted average, aka ‘balance point’) of a set of meaning vectors. When executed over a Person object, position first retrieves the complete (or filtered) *path*, to then calculate the position underlying the performance records contained in the path. Listing 12 shows how the centroid is calculated from a given number of meaning vectors *mvecs*.

Listing 12 Position centroid calculated as non-weighted average of its meaning vectors.

```
meaningvector = colSums(mvecs) / nrow(mvecs)
```

The generic ‘+’ method acts on two Performance objects. It allows recombining performance records by calculating their centroid position, see also Sect. 5.8.

The generic *near* signs with several signatures: it acts on *Person* objects, *Performance* objects, and combinations of the two. It returns those persons or performance records that are close to a given one above the *proximity* threshold.

Several of the generic functions sign for multiple signature classes. Table 8.1 provides a comprehensive overview on which generic acts on what classes.

Table 8.1 Overview on generics and their signatures

Generic	HumanResourceManager	Person	Performance	DomainManager	Domain	Visualiser
performances	×	×				
path		×				
+-method			×			
[-accessor		×				
competences	×	×	×			
cosine		×	×			
identity		×	×			
proximity	×	×	×			
near	×	×	×			
terms		×	×			
overlap			×			
position		×	×			
groups	×					
plot		×	×		×	
toponymy					×	
names	×	×	×			
summary		×	×		×	×
print	×	×	×	×	×	×
show	×	×	×	×	×	×

8.4 Summary

The community knowledge expressed digitally in from of unstructured text such as provided in a collection of messages, essays, articles, or books can be used to derive its specific 'conceptual graph' of both generic as well as professional language terms and their associative proximity relations.

Such conceptual graph provides a projection surface on top of which it is possible to investigate social as well as semantic relations of persons, their performance, and underlying competence.

The package provides all routines required to construct and investigate representations of social semantic performance networks. It additionally provides visual analytics to support the analyst.

There are, however, limitations to the implementation. Most notably, the only processing mode available currently for the *Visualiser* class is 'terminology', though alternatives can be thought of: focus of the visualization could alternatively emphasise 'incidents', i.e. using the left-singular Eigenvectors over the A^TA pairwise document incidences. In such visualisation, it would be the document proximity in the stretch-truncated Eigenspace that would govern the layout of the planar projection.

The second alternative would be to focus on both Eigenvectors U and V simultaneously, possibly even including the traces mapped ex post into the truncated subspace of the Eigenspace.

Both alternatives, however, come with the disadvantage of providing less stable projection surfaces: placement of traces is then no longer (fully) determined by their conceptual representation, but rather (or 'additionally') by their incidence similarity.

The individual class diagrams and detailed fields and methods tables are included in the Annex A, while the full package documentation is attached in Annex B to this book.

References

Butts, C.T.: sna: Tools for Social Network Analysis, R Package Version 2.2-0. http://CRAN.R-project.org/package=sna (2010)

Chambers, J.: Software for Data Analysis: Programming with R. Springer, Berlin (2008)

Freeman, L.: Centrality in Social Networks. Conceptual Clarification. Soc. Networks. **1**, 215–239 (1979)

Ihaka, R., Gentleman, R.: R: a language for data analysis and graphics. J. Comput. Graph. Stat. **5**, 299–314 (1996)

Kaufman, L., Rouseeuw, P.: Finding Groups in Data: An Introduction to Cluster Analysis. Wiley, New York (1990)

Lance, G.N., Williams, W.T.: A general theory of classificatory sorting strategies: 1. Hierarchical systems. Comput. J. **9**(4), 373–380 (1967)

R Core Team: R: A Language and Environment for Statistical Computing, R Foundation for Statistical Computing, Vienna, Austria. ISBN 3-900051-07-0. http://www.R-project.org/ (2014)

Chapter 9
MPIA in Action: Example Learning Analytics

This chapter presents comprehensive application examples of *MPIA*, thereby contributing an extended, more holistic approach to learning analytics, as the review of the state of the art will show.

Already the previous Chap. 8 (Sect. 8.3) illustrated the general analysis workflow with the *mpia* software package with the help of a parsimonious-code example. This chapter now aims to extend this and showcase *MPIA in action*, using increasingly more complex learning analytics applications and—in the two final ones—real-life data.

The chapter is lead in by a brief review of the relevant state of the art in applications of learning analytics (Sect. 9.1) in this subarea it roots in of content analysis and social network analysis (see Sect. 1.5).

Then, the first foundational example (Sect. 9.2) revisits the introductory demonstrations of the *SNA* (Sect. 3.2) and *LSA* (Sect. 4.3) chapters in order to illustrate, where the benefits of *MPIA* lie and how *MPIA* goes beyond these predecessor technologies.

Two more complex examples follow: Example two (Sect. 9.3) revisits the automated essay-scoring example presented back in Sect. 4.5. Example three (Sect. 9.4) provides an application implementing a more open learning-path scenario.

In the tradition of the R community, the examples of this chapter are also included as demos (commented code only) in the *mpia* software package. This chapter provides the narrative to these 'naked code' demos. Together, re-executable (same code, same data) and reproducible (same code, different data) research is sought, not only to ensure quality of ideas and their implementation, but also to facilitate impact. Consequently, the one or other line of code may appear trivial, though then it typically hides the complexity described above in Chaps. 5, 6, and 7. Moreover, all demos provided are exhaustive, the code reduced to its bare minimum possible without compromising completeness.

F. Wild, *Learning Analytics in R with SNA, LSA, and MPIA*,
DOI 10.1007/978-3-319-28791-1_9

Throughout the examples, clarity is sought how *MPIA* can be used for inspecting social and conceptual structure of learning. The chapter is wound up with a summary (Sect. 9.5).

9.1 Brief Review of the State of the Art in Learning Analytics

Following the differentiation brought forward in the introduction, the state of the art in Learning Analytics—in the relevant classes of social network analysis and content analysis—can be summarized as follows.

The prevalent approach in the area of analysing content (and the user's interaction with it) is transaction log analysis of the web servers involved, sometimes combining the findings with historical assessment data of the learner.

The Signals[1] project, for example, predicts learner performance with a traffic light coding of how likely they are to pass a course, based on the time spent in the learning management system and previous summative (plus current formative) assessment scores (Arnold 2010).

Govaerts et al. (2011) propose a Student Activity Meter (SAM), rendering a visualisation of how many web resources learners use over time, additionally providing a visualisation of the logged time spent on them. Learners can compare their performance along these two dimensions with the performance of their peers.

Retalis et al. (2006, p. 2) propose a system called cosyLMSAnalytics, which supports "in automatically gathering and analysing data concerning learners' access patterns" with the help of cluster analysis applied over the user learning paths, i.e. sequences of web document references, as tracked by the learning management system (in their case: Moodle).

Crespo García et al. (2012) and Scheffel et al. (2012) use contextualised attention meta-data to transcribe the web usage and software development activity of learners recorded on personal virtual machines handed out to computer science students in their experiments. They develop several different types of presenting the data visually. Their work, however, is—at this stage—descriptive, focusing rather on data collection than facilitating individual or group feedback and subsequent performance improvement.

Transaction log based techniques are insufficient in providing content-based feedback. Trausan-Matu et al. (2010) therefore propose a system called Pensum, which supports learners in writing summaries of literature. Using latent semantic analysis, the system analyses the learner written summary and matches it sentence by sentence to the papers to be summarised. The system helps in identifying whether all relevant passages in the literature are actually reflected in the text

[1] http://www.itap.purdue.edu/tlt/signals/

written by the student, through measuring the associative closeness as expressed by the cosine similarity of the sentence vectors.

With respect to analysing social networks, Rebedea et al. (2011) propose a system called PolyCAFe, which monitors and visualises conversational collaboration through watching utterance reply structures as exposed in chat rooms or forum discussions. Tracking on-topic postings against given assignments, the system is able to calculate a so-called interanimation score: this score reflects the degree to which participants contribute to an on-topic discussion. In the shared roadmap together with Pensum (and the system Conspect, which is further described in this book, see Chap. 10), Gerdemann et al. (2011) express the need for improved accuracy and validity, as well as improved processing and visualisation facilities.

Social Networks Adapting Pedagogical Practice (SNAPP) is a system proposed in Dawson et al. (2010). It monitors the social interaction of students by visualizing the social graph expressed in forum reply structures. Analogously, Crespo García et al. (2012) use sociogrammes in the tools they propose in order to visualise the social network hidden in forum postings.

With the notable exception of PolyCAFe, none of these Learning Analytics try to bring together content analysis with social network analysis to solve the shortcomings of each analysis when done in isolation. PolyCAFe was developed in the same EC-funded project 'language technologies for lifelong learning (LTfLL)' in parallel to Conspect, one of the prototypes further described and developed in this book (see Chap. 10). PolyCAFe, however, focuses on conversation feedback in a structured online chat system.

9.2 The Foundational SNA and LSA Examples Revisited

Section 3.2 above introduced a foundational social network analysis example with a user story, in which the human resource manager of a multinational business seeks an equally skilled employee who can fill in for a person on sick leave.

Utilising SNA, the manager derives such information by analysing the affiliation of the learners with particular courses and with more informal online learning groups. The affiliation data is mapped to an association matrix, which lead to the following visualisation depicted in Fig. 9.1. The three employees Alba, Joanna, and Peter participated in similar contexts as the person on sick leave (Christina).

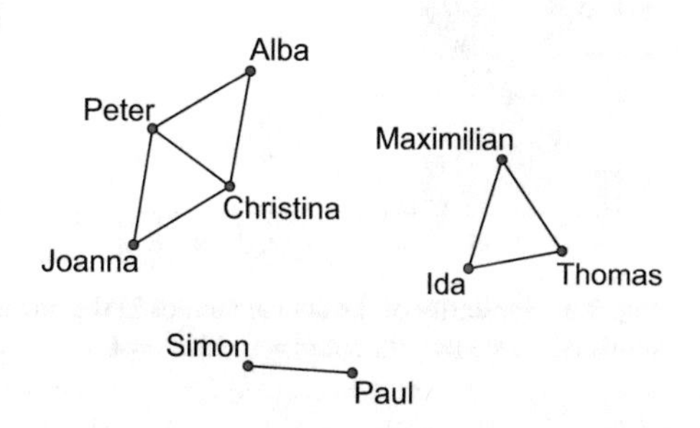

Fig. 9.1 Learner network visualization (from Sect. 3.1, Fig. 3)

It is possible to further aggregate and filter the underlying data in order to shed a bit more light on the 'why' and 'in what way' learners may or may not be similarly skilled, but the whole analysis process overall is agnostic of the content covered.

Latent semantic analysis, on the other hand, is well up for supporting the analysis of such contents—as demonstrated in the foundational example provided in Sect. 4.3 above. In this example, the human resource manager now analyses the short memos written by the employees about their learning experiences. Fourteen different memos describe 14 different learning opportunities of both formal and informal nature.

The result is a similarity table, such as the one shown earlier in Sect. 4.3 (Table 4.8) and depicted here visually in Fig. 9.2. The table to the left shows that each memo is in its vector space representation (Sect. 4.3, Table 4.7)—naturally—identical to itself.

The three expected clusters of computing, math and pedagogy memos, however, become salient only for the latent semantic space, the results of which are depicted to the right of Fig. 9.2. The identity is visible in both cases along the black diagonal.

Using the factor loadings on the three factors of this constructed example, the term and document vectors can be visualized as shown earlier in (Fig. 4.8)—see Fig. 9.3. For spaces with more relevant dimensions, this perspective plot method is no longer valid, as the distortion gets too big. Moreover, larger dimensional spaces tend to group the more generic vocabulary together in the first factors, thus rendering this method useless in bringing out the differences (not commonalities) of a corpus under investigation.

Although LSA enables the human resource manager to inspect contents and the semantic relationships expressed in and across the memos, the relationship of the employees to these semantic relations is lost. It is possible to discover that the

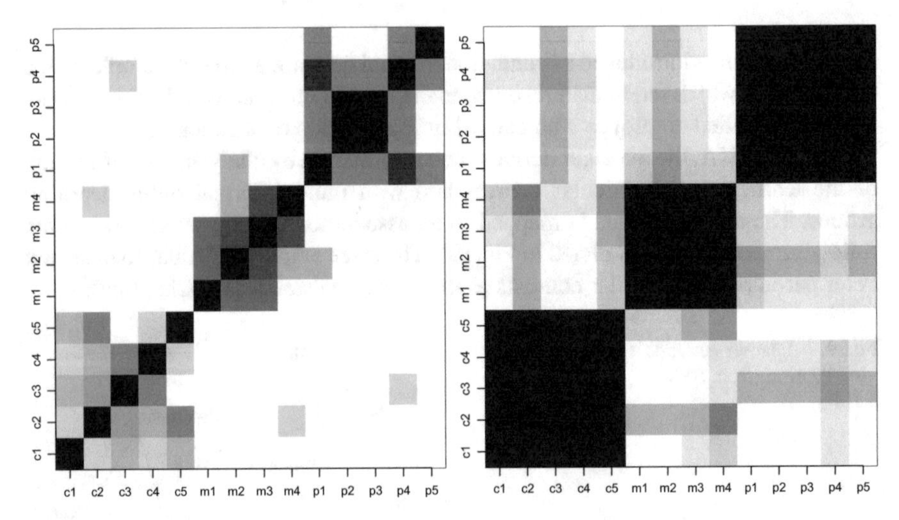

Fig. 9.2 Similarity of the course memos in the document-term vector space (*left*) and in the latent semantic space (*right*): *white* = 0, *black* = 1

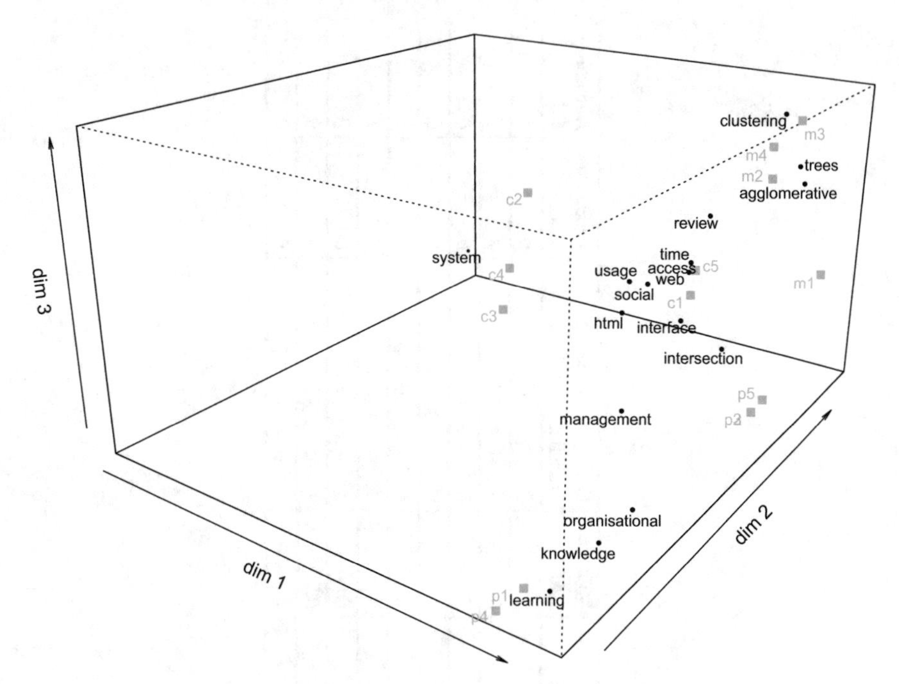

Fig. 9.3 Visualization of the document and term vector loadings on the first three factors

training measures undertaken by the employees fall into three groups—computing, math, and pedagogy—, but it is no longer possible to directly relate the attendance data with these findings.

In the following, the two data sets of attendance data and memo corpus shall be brought together in an MPIA.

The incidence matrix from Sect. 3.2 (Table 3.9) captures the attendance information of the employees: the nine employees participated in twelve courses or online learning groups, see the reprinted Table 9.1.

The memos describing these learning experiences are already provided above in Sect. 4.4 (Table 4.2): each course and online group is described with a short text.

Table 9.2 relists these memos and the learning opportunities they describe (in the same order as in the incidence matrix).

The space calculated above in Sect. 4.4 (Listing 5) is reused and added via the domain manager, as shown in Listing 1. The first line thereby instantiates a Domain Manager object to then add and subsequently assign a Domain object to *d*.

Listing 1 Adding a domain with a precalculated space.

```
dmgr = DomainManager()
id = dmgr$add(space, title = "memospace")
d = dmgr$get("memospace")
```

Table 9.1 Incidence matrix *im* of learners and their participations

	OU-CS	UR-Informatik	MOOC-PED	MOOC-TEL	MOOC-Math	OU-PED	MOOC-ocTEL	MOOC-LAK	OU-Statistics	Facebook-Statistics	Facebook-TEL	Linkedin-CS
Paul	*1*	*1*	*0*	*0*	*0*	*0*	*0*	*0*	*0*	*0*	*0*	*1*
Joanna	*0*	*0*	*1*	*0*	*0*	*1*	*0*	*0*	*0*	*0*	*1*	*0*
Maximilian	*0*	*0*	*0*	*0*	*1*	*0*	*0*	*0*	*0*	*1*	*0*	*0*
Peter	*0*	*0*	*0*	*1*	*0*	*1*	*1*	*1*	*0*	*0*	*0*	*0*
Christina	*0*	*0*	*1*	*1*	*0*	*1*	*1*	*1*	*0*	*0*	*0*	*0*
Simon	*0*	*1*	*0*	*0*	*0*	*0*	*0*	*0*	*0*	*0*	*0*	*1*
Ida	*0*	*0*	*0*	*0*	*0*	*0*	*0*	*0*	*1*	*1*	*0*	*0*
Thomas	*0*	*0*	*0*	*0*	*0*	*0*	*0*	*0*	*1*	*1*	*0*	*0*
Alba	*0*	*0*	*0*	*1*	*0*	*0*	*1*	*1*	*0*	*0*	*0*	*0*

Table 9.2 Learning opportunities and their description in *thedocs*

	Learning opportunity	Memo
c1	OU-CS	A web interface for social media applications
c2	UR-Informatik	Review of access time restrictions on web system usage
p1	MOOC-PED	The intersection of learning and organisational knowledge sharing
c3	MOOC-TEL	Content management system usage of the HTML 5 interface
m1	MOOC-Math	The generation of random unordered trees
p2	OU-PED	A transactional perspective on teaching and learning
c4	MOOC-ocTEL	Error spotting in HTML: social system versus software system
m2	MOOC-LAK	A survey of divisive clustering along the intersection of partial trees
m3	OU-Statistics	Width and height of trees in using agglomerative clustering with Agnes
m4	Facebook-Statistics	Agglomerative clustering algorithms: a review
p3	Facebook-TEL	Innovations in online learning: moving beyond no significant difference
c5	Linkedin-CS	Barriers to access and time spent in social mobile apps

In the next steps, the evidence data provided through Tables 9.1 and 9.2 is added with the help of the instantiated Human Resource Manager object, see Listing 2.

Listing 2 Instantiating the Human Resource Manager.

```
ppl = HumanResourceManager(domainmanager = dmgr, domain = d)
```

Each assignment thereby follows the form shown in Listing 3: first, a new Person object is instantiated and added to the regime of the Human Resource Manager, then the memo record is submitted to describe the performance. This can be repeated manually for all employees and memos.

Listing 3 Example assignment Person object & Performance record.

```
paul = ppl$add(name="Paul")
paul$write( ".. memo text .. " )
```

Since, however, the data is already available, this can be added in a more automated way: As the incidence matrix is already provided in *im* and the documents are available in the same sort order via *thedocs* (see above), the following Listing 4 can be used to add the employee and evidence records en bulk.

The outer for loop thereby traverses the incidence matrix row by row, handing over the name of the employee. The second line within this outer loop adds a Person object and assigns it subsequently to the variable name (e.g. ‘paul’). The inner for loop then adds the memo for each incidence as a Performance record.

Listing 4 Adding the performance records for all employees.

```
for ( p in rownames(im) ) {
  assign( tolower(p), ppl$add(name=p) )
  for (pf in which(im[p,]>0)) {
    get(tolower(p))$write(
      thedocs[pf,3],
        label=thedocs[pf,1],
        purpose=colnames(im)[pf]
    )
  } # for each incidence
} # for each person
```

The result is that the human resource manager object cares for nine people, for which a total of 26 performance records were created.

The next step is to visualize the projection surface capturing the conceptual proximity relations, as done in Listing 5. In this case, the perspective plot is chosen as visualization variant, whereas for the toponymy all labels shall be plotted. In more complex examples plotting all possible labels may not be as useful (though often it helps in gaining an initial overview).

Listing 5 Plotting the projection surface.

```
plot(d, method="persp", rotated=TRUE)
toponymy(d, method="all", add=TRUE, col="black")
```

The resulting visualization of the knowledge cartography is depicted in Fig. 9.4. Already here it is visible that the three different topical clusters are laid out into

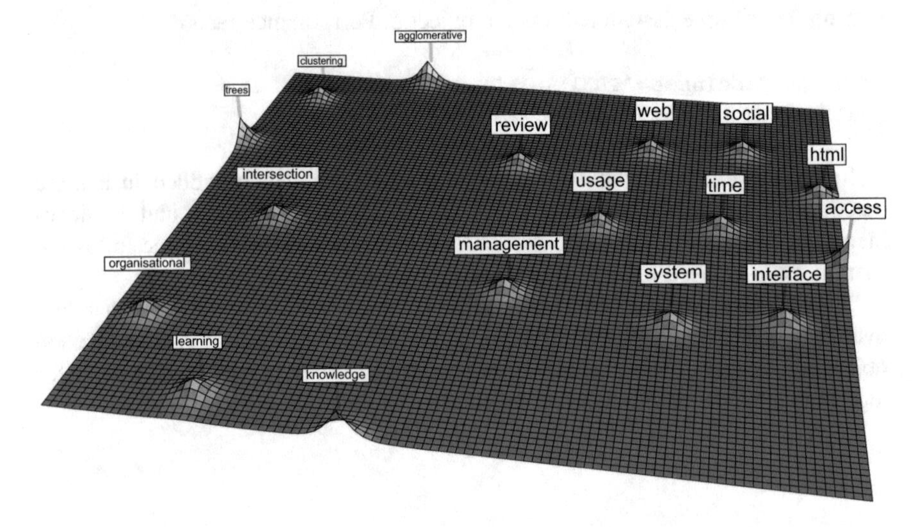

Fig. 9.4 Projection surface

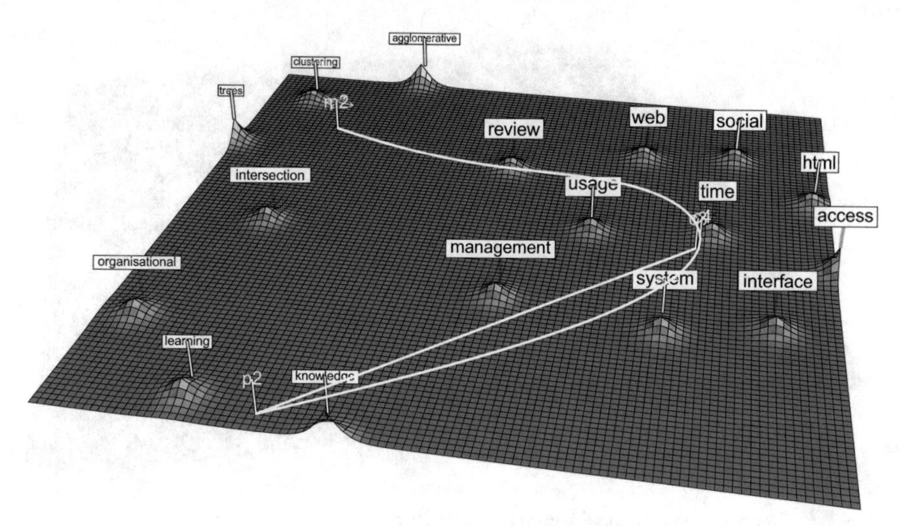

Fig. 9.5 Learning path of Peter

different areas of the plot: pedagogy bottom left, math top left, and computing top right—with management forming sort of a bridging concept between them.

This projection surface can then be used in order to visualize the performance of the employees. For example, Peter's learning path can be depicted using the command shown in Listing 1. The path shows the location of each of the performance records and connects them—in chronological order of adding—with a line. The result of this operation is depicted in Fig. 9.5.

Listing 6 Plotting a learning path of an individual.

```
plot(path(peter), col = "white")
```

As indicated earlier in Sect. 5.8, the pathway provided through the locations of the individual performance records of a learner differs from the overall position of the leaner in the space. Moreover, Sect. 5.9 already introduced that commonly appearing competence positions can be extracted from the underlying performance data by looking into the proximity. Figure 9.5 helps to understand this with a practical example: two of the performance records lie in close proximity to each other. To be precise, the location of performance record one and three is even identical, which means they provide evidence of the same underlying competence. Listing 7 shows how to conduct such test: it uses the identity operator '==' to compare the two performance records, yielding the result 'true'.

Listing 7 Testing two performance records for identity.

```
peter[1] == peter[3]
```

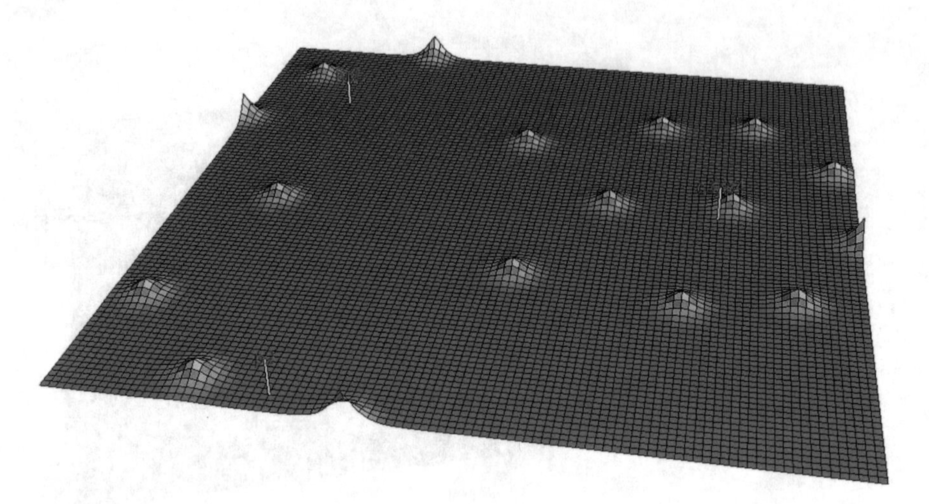

Fig. 9.6 Competences extracted from the path of Peter

Extracting the underlying common competence positions (as proposed in Sect. 5.9 using agglomerative nesting) results in three distinct competence positions, as depicted in Fig. 9.6: the three positions are marked up in red by calling the code of Listing 8.

Listing 8 Plotting the competence positions underlying the performance records of Peter.

```
plot(
   competences(peter),
   col = "red",
   component.labels = FALSE,
   connect = FALSE,
   alpha = 1
)
```

The two performances to the right of Fig. 9.5 above are now conflated to the single red marker 'c3/c4' to the right of Fig. 9.6.

This allows for comparing the learners under scrutiny. Figure 9.7 shows such visualization, using a flat, topographic plot (without the wireframe visualization for clarity). Listing 9 provides the code with which this visualization is generated: first, a new, empty plot is prepared (lines 1–2). Then the toponymy analysed and plotted, using a black grid. Subsequently, the learner positions and paths are plotted in different colours.[2]

[2] Note that two of the learners have been suppressed from this plot (Paul and Thomas), since their position and label is overlapping with existing learners, thus cluttering the display. Paul's position is the same as Alba's and Thomas's position the same as Ida.

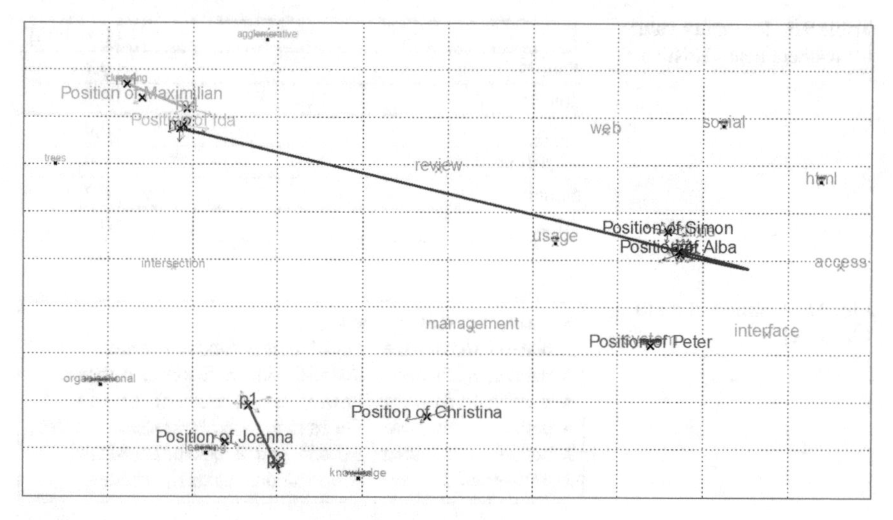

Fig. 9.7 Positions of several learners, three paths marked

Listing 9 Visualizing path and positions of several learners.

```
par(mar=c(0,0,0,0))
plot.new()
toponymy(d, method="all", add=FALSE,
   col="darkgray", grid.col="black")
plot(peter, col="darkgreen")
plot(christina, col="darkgreen")
plot(path(maximilian), col="orange")
plot(maximilian, col="orange")
plot(path(joanna), col="red")
plot(joanna, col="red")
plot(path(alba), col="purple")
plot(alba, col="purple")
plot(ida, col="lightgreen")
plot(simon, col="brown")
```

As can be seen in one glimpse from the visualization, there are several learners that demonstrated performance only in a single competence group (Max and Ida in math, Simon and Alba in computing, and Joanna in pedagogy). The two remaining learners Christina and Peter obtained competence positions in between the clusters, since their performance demonstrations touched into all of them. Their focus points seem to be different, however, as Christina is drawn closer to pedagogy and Peter closer to computing. This is the case, because Christina shows more activity in the pedagogical area (and Peter in computing).

What can be derived visually, can also be tested formally by utilizing the *proximity* and *near* generics.

Table 9.3 Proximity values for learners near Christina

	Proximity
Paul	0.7
Joanna	0.6
Peter	0.9
Christina	1.0
Simon	0.7
Alba	0.9

Fig. 9.8 Learners close to Christina

```
> near(ppl, christina)
A person with name 'Paul' and 3 textual traces.
A person with name 'Joanna' and 3 textual traces.
A person with name 'Peter' and 4 textual traces.
A person with name 'Christina' and 5 textual traces.
A person with name 'Simon' and 2 textual traces.
A person with name 'Alba' and 3 textual traces.
```

For example, the *proximity (simon, alba)* of Simon and Alba is quantified in the space with 0.96, thus yielding *true* for the *near (simon, alba)* evaluation. The *proximity (simon, joanna)* between Simon and Joanna, on the contrary, is only −0.11, thus yielding *false* for the evaluation with *near (simon, joanna)*.

Seeking a worthy replacement for Christina, the person who is on sick leave, the human resource manager evaluates the statement provided in Listing 10.

Listing 10 Finding a replacement for Christina.

```
near(ppl, christina)
```

Six names pop up (including the self reference to Christina). These are the persons occupying positions close (above proximity threshold) to Christina. To further inspect, how close these persons are, the manager can first look into the actual proximity values, using the code provided in Listing 11. This results in the proximity table provided in Table 9.3: the two closest persons are Peter and Alba. All persons are in visible proximity to Christina: Alba, Paul, Simon in the cluster to the right,[3] Peter and Joanna to the left and right of Christina.

While Peter is a very obvious choice with a high proximity value of 0.9, it is surprising that Alba has a higher proximity value compared to Joanna (0.9 compared to 0.6, see Table 9.3 and Fig. 9.8). Further inspection can uncover the differences and provide an explanation.

[3] Note again that Paul is not depicted in Figure 9.7 above, as his position is essentially the same as Alba's (thus cluttering the label).

```
> overlap(path(christina), path(alba))
[1] "agglomerative" "clustering"    "knowledge"     "learning"
"review"        "trees"
> overlap(path(christina), path(joanna))
[1] "access"         "html"           "interface"      "organisational"
"social"         "system"
[7] "time"           "usage"          "web"
> overlap(competences(christina), competences(alba))
 [1] "access"         "agglomerative" "clustering"    "html"
"interface"     "review"        "social"
 [8] "system"        "time"          "trees"         "usage"
"web"
> overlap(competences(christina), competences(joanna))
[1] "intersection"   "knowledge"     "learning"        "management"
"organisational"
```

Fig. 9.9 Overlapping terms of paths and competence positions

Listing 11 Extracting proximity values for learners near Christina.

```
replacements = sapply( near(ppl, christina), names )
proximity(ppl)['Christina', replacements]
```

Such further exploration can, for example, be conducted by inspecting the overlapping descriptor terms of both the paths of two persons and their underlying competences. This can be conducted by executing the code provided in Listing 12.

Listing 12 Inspecting the difference between two close learners.

```
overlap(path(christina), path(alba))
overlap(path(christina), path(joanna))
overlap(competences(christina), competences(alba))
overlap(competences(christina), competences(joanna))
```

It turns out that while the performance records of Christina and Joanna share more overlapping strong loading descriptors than Christina and Alba (see Fig. 9.9), their competence positions do not: the competence positions of Alba and Christina have more terms in common than for Joanna and Christina. This is the reflection of the centroids calculated from the competence positions being different from the centroids calculated from the path locations, thus resulting in a lower proximity value and differing descriptors.

The human resource manager can now check through these inspection methods the system recommendations who to replace Christina with—Peter or Alba (first choice), or Simon, Paul, or Joanna (second choice)—and who is most suited for the competence needs of the customer project.

9.3 Revisiting Automated Essay Scoring: Positioning

Within this section, as promised above (Sect. 4.5), essay scoring is revisited in order to illustrate the improved analysis facilities provided by MPIA. This demonstration thereby will put emphasis on *positions* of learners in the social semantic network and its map projection.

Even though the quality of the essays introduced in Sect. 4.5 varies—the human graders scored these essays with as low as 1¼ to as high as 4 points—, all of them try to answer to the same question, thus at least the better ones can be expected to be located in relative vicinity to each other, possibly occupying a rather small number of (if not a single) competence positions.

Moreover, eight additional collections of essays will be considered in this essay-scoring example. The full collection is written in German and holds 481 learner essays with an average length of 55.58 words. All originate the wider area of economics and business administration, but they vary in collection size (see Table 9.4). The nine collections have already been used in the previous Chap. 7, where they facilitated the analysis of how to train specialised spaces. Collection 4 was dropped because of the low interrater agreement between the two human raters (Spearman Rho of –0.07).

Using this full set allows to show how learners and their activities within the wider subject area of business administration and information systems differ and how this can be unveiled in the analysis. An accuracy evaluation for the full super set will follow in Chap. 10.

First, an MPIA space has to be created from the essays in the collections. The collections are stored in single coma-separated values (CSV) file, see extract presented in Table 9.5.

The data set provided in this coma-separated values file has to be converted to a corpus, cleaned and sanitised to then be mapped to a text matrix. The following steps have to be applied (see Listing 13). First, data is read in using readLines. Then, a Corpus object is established over the vector source, thereby setting basic

Table 9.4 The essay collections

#	Topic: Assignment	Essays
1	Programming: define 'information hiding'	102
2	E-Commerce: define 'electronic catalogue'	69
3	Marketing: argue for one of two wine marketing plans	40
5	Information systems: define 'meta-data'	23
6	Information systems: define 'LSA'	23
7	Programming: define 'interface'	94
8	Procurement: argue reengineering of legacy system versus procurement of standard software	45
9	Information systems: explain 'requirements engineering'	46
10	Information systems: explain component/integration/system test	39
		481

Table 9.5 Example essays (collection 1, maximum human scores, translated from German to English)

Essay filename	Essay text
data1_40_064.txt	No direct acces to variables, only via method calls isLoggedIn() and getSurname(); declaration of the variables as private is called 'data encapsulation'. With this access can be restricted. The methods may be public.
data1_40_077.txt	'Information Hiding' prevents direct referencing of instance variables from other classes. In this example, the instance variables surname and loggedIn of the class student are declared private, i.e. the can only be changed (referenced) from within the class. The variable firstname, however, can be accessed also from outside of the class (default value of the package). The methods isLoggedIn and getSurname are declared as public, i.e. they can be accessed from within the class, but also from all other classes. Both methods return the value of loggedIn and surname to the calling object. I.H. prevents data manipulation, but also data changes by mistake. Moreover, it renders possible access control.
data1_40_088.txt	Data encapsulation. Hiding of the implementation behind methods. No access from outside possible. Applied by declaring variables as private. Access only via method calls possible. For this get and set methods declared public are needed. Via method get, the value of the variable can be read, while set is used to change or reassign.
data1_40_100.txt	'Information Hiding' helps restricting visibility and access control of variables. For this, the variables have to be declared 'private' in the class (e.g. private String surname) and they have to be returned via methods that are declared public. Outside of the claa, the variables can then only be accessed via this method.

configuration options, such as dropping stop words,[4] in German, removing punctuation, and setting the minimum character length for terms to three. The third line adds a mapping function that is able to remove special characters not caught by the *tm* package's internal routines (such as the "..." character often used in MS Word texts), while preserving all alpha-numeric characters (German, hence the added Umlauts). Word with less than the minimum number of characters are eliminated and all numbers are removed. Finally, everything is converted to lower case.

Listing 13 Creating the corpus from the CSV file.

```
essays.content = read.csv2(file= "essays.content.csv")
tm = Corpus(
  VectorSource(essays.content[,2]),
  readerControl = list(
    reader = readPlain,
    language = "de",
    load = TRUE,
```

[4] Using the German stop word list provided by the tm package, which originates from the author's lsa package.

```
        removePunctuation = TRUE,
        stopwords = TRUE,
        minWordLength = 3,
        removeNumbers = TRUE
    )
)
tm = tm_map(tm,
    function(e) return(gsub("[^a-zA-Z0-9äöüÄÖÜß]", " ", e))
)
tm = tm_map(tm, tolower)
```

The object *tm* now holds a slightly cleaned corpus with 481 text documents. In preparation of the stem completion, the original vocabulary is saved as a dictionary *dict* (Listing 14). It has a size of 4154 words.

Listing 14 Save full dictionary for stem completion (see below).

```
dict = Terms(DocumentTermMatrix(
    tm,
    control = list(
        removePunctuation = TRUE, stopwords = FALSE,
        minWordLength = 1, removeNumbers = TRUE
    )
))
```

In the next steps, stemming is applied (Listing 15). Stemming reduces the original 'raw' vocabulary of 4154 words to 3115 word stems.

Listing 15 Apply stemming.

```
tm = tm_map(tm, stemDocument, language = "ger")
dtm = TermDocumentMatrix(tm, control = list(
    removePunctuation = TRUE,
    removeNumbers = TRUE,
    stopwords = TRUE,
    minWordLength = 3,
    bounds = list(global = c(1, Inf))
))
```

To improve readability, stem completion is applied (Listing 16). Therefore, it uses the previously created dictionary *dict* to complete each word stem to its shortest possible raw form in dict. It then in the 2nd group of lines restores the original (stemmed) name for those terms for which no completion could be identified. The third line assigns the restored term labels *sc* to the matrix.

Listing 16 Reverse stemming: stem completion for improved readability.

```
sc=as.character( stemCompletion(
rownames(dtm), dictionary=dict, type="shortest"
))
sc[which(is.na(sc))]=rownames(dtm)[which(is.na(sc))]
rownames(dtm)=sc
```

This operation of stemming and stem completion, however, is not idempotent. For example, the word 'aktualisieren' is stemmed by the Porter stemmer to 'aktualisi', while 'aktualisierten' is stemmed to 'aktualisiert'. Both, however, are expaneded to 'aktualisiert', when using shortest word forms as a setting for stem completion.

Therefore, Listing 17 is used to conflate those (duplicate) row vectors to a single entry, whose stem form resulted in the same completed stem. This conversion is necessary, since stemming and completion deploy different algorithms, so their mappings are not reversible. This step finds 154 duplicates and thus reduces the matrix to 2961 rows.

Listing 17 Dupe cleaning for stem completion.

```
if (any(duplicated(rownames(dtm)))) {
   dupes=which(duplicated(rownames(dtm)))
   for (i in dupes) {
      target=hits[ which(! hits %in% which(duplicated(sc))) ]
      replvec=t(as.matrix( colSums(as.matrix(dtm[ hits, ])) ))
      rownames(replvec)=sc[target]
      dtm[ target,1:length(replvec) ]=replvec
   }
   dtm=dtm[!duplicated(rownames(dtm)),]
}
class(dtm)=c("TermDocumentMatrix", class(dtm))
```

The resulting text matrix of size 2961 terms against the 481 documents constructed from this corpus yielded the raw frequency distribution depicted in Fig. 9.10 (log-scaled on both axes).

The frequency spectrum follows the typical power law distribution: Few terms appear very often, the majority very rarely. The maximum frequency found is a term showing up in 451 documents. Only two terms show up in more than 50 % of the documents. 16 terms appear in more than 25 % of the documents and 59 in more than 10 %.

On the lower spectrum, more than half of the terms (50.6 %) appear only in one document in the collection (1498 terms). Only 567 appear more often than in 5, and 335 of these more often than in 10 documents.

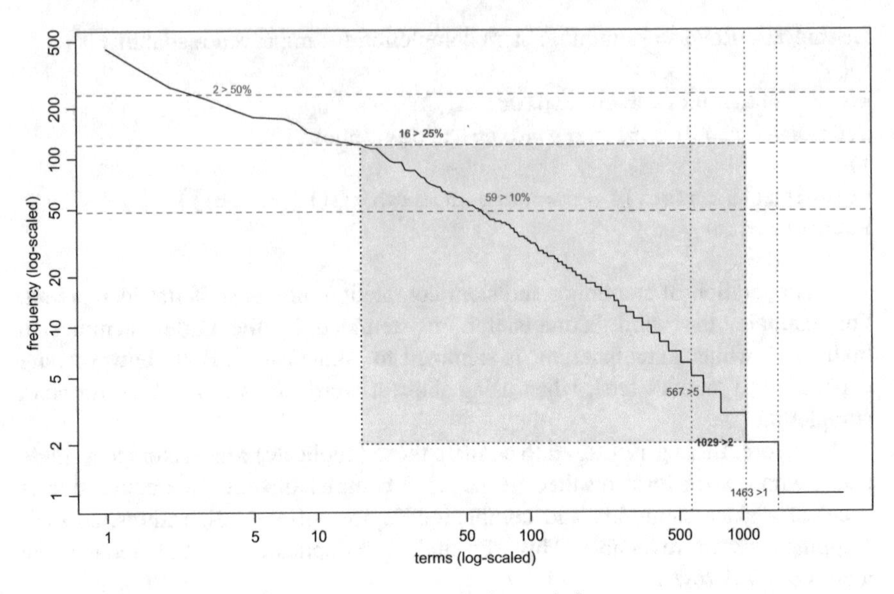

Fig. 9.10 Frequency distribution in the text matrix (log-scaled)

Listing 18 Cropping with lower and upper frequency threshold.

```
freqs = rowSums(as.matrix(dtm))
lower = which(freqs > 1)
dtm = dtm[lower,]
freqs = rowSums(as.matrix(dtm))
upper = which(freqs < ncol(dtm)/4)
dtm = dtm[upper,]
empty = as.integer( which(colSums(as.matrix(dtm)) == 0) )
dtm = dtm[,-(empty)]
dtm = as.matrix(dtm)
```

Cropping the matrix to the frequency spectrum of the medium frequent terms (see Listing 18) with a lower threshold of appearing at least twice in the collection and an upper threshold of appearing less often than in 25 % of the documents creates a text matrix of size 1447 terms and the 481 documents. This cropping operation effectively drops two documents that no longer have any terms associated with them (two essays scored with zero point by the human raters that contained a single word as the answer).

Listing 19 Determining the threshold value for the Eigenspace trunction.

```
tr = sum( dtm*dtm )
tr * 0.8
```

The cut off value of Eigendimensions to retain is determined using the trace of the text matrix, calculated before the spectral decomposition: the value of the trace is 20,564, with 16,709.6 equalling the chosen amount of 80 % stretch.

Listing 20 Space calculation and stretch truncation (by hand).

```
class(dtm) = "textmatrix"
space=lsa(dtm, dims=dimcalc_raw())
threshold=tr * 0.8
r=0
for (i in 1:length(space$sk)) {
  r=r+(space$sk[i]^2)
  if (r>=threshold) {
    cutoff=i-1
    break()
  }
}
space$tk=space$tk[, 1:cutoff]
space$dk=space$dk[, 1:cutoff]
space$sk=space$sk[ 1:cutoff ]
```

Such value is reached at dimension 109 (see Listing 20), allowing to stop calculation after the first 109 dimensions, using the iterative singular value decomposition routines provided in e.g. the svdlibc implementation (Rhode 2014, based on Berry et al. 1992). Truncating the Eigendimensions to 80 % of the full stretch, results in the Eigenvalue distribution depicted in Fig. 9.11.

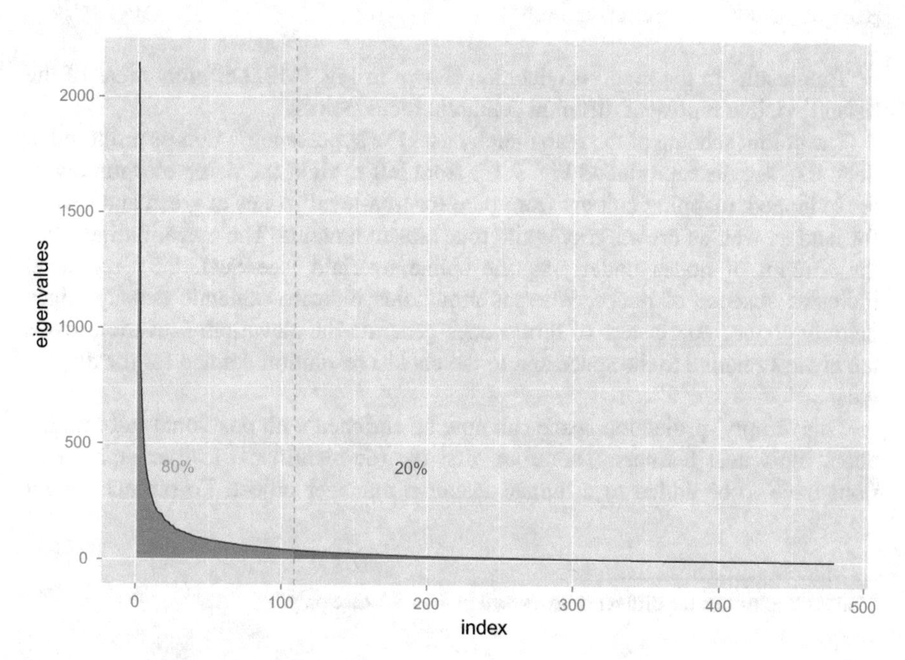

Fig. 9.11 Eigenvalue distribution of the sanitized text matrix

Subsequently, this manually calculated space is added to Domain and the domain to the DomainManager (Listing 21). The package's Domain class provides routines to do this (see the example in Sect. 9.3 below).[5]

Listing 21 Adding space to DomainManager.

```
d=Domain(name="essays")
d$setSpace(space)
dmgr=DomainManager()
dmgr$add(d)
```

For the visualization, first the term-to-term proximities need to be calculated (line 1 in Listing 22). The routine thereby already removed all proximity values below the Domain's proximity threshold (per default 0.3). Lines 2 and 3 then calculate the map data and wireframe.

Listing 22 Calculating term proximities and visualisation data.

```
d$calculateTermProximities()
d$visualiser$calculateNetCoords()
d$visualiser$calculateReliefContour()
```

The resulting projection plane can be plotted using the command shown in Listing 23.

Listing 23 Plotting the map projection.

```
plot(d, method="topographic")
```

This results in the map visualization shown in Fig. 9.12. Different areas of the 'island' visible represent different semantic focus points.

The colour scheme of the map thereby uses the hyposometric tints as proposed in Sect. 6.4, see the reproduced Fig. 9.13: from left to right the rising elevatin levels are indicated, mapping colours from blue for 'sea-level' zones to green and yellow for land as well as brown/gray/white tones for mountains. The elevation indicates the amount of nodes underlying the spline overlaid (see Sect. 6.3): sea level indicates absence of nodes, whereas mountains indicate semantic density, since densely connected clusters of term nodes generate the elevation. Elevation levels are always relative to the space and to the chosen resolution defined for the display area.

This 'empty' projection space can now be enriched with positional information about individual learners. Therefore, first the (de-identified!) learner representations have to be added to a human resource manager object. To protect learner

[5] This is to illustrate the difference to the svd in the LSA case only.

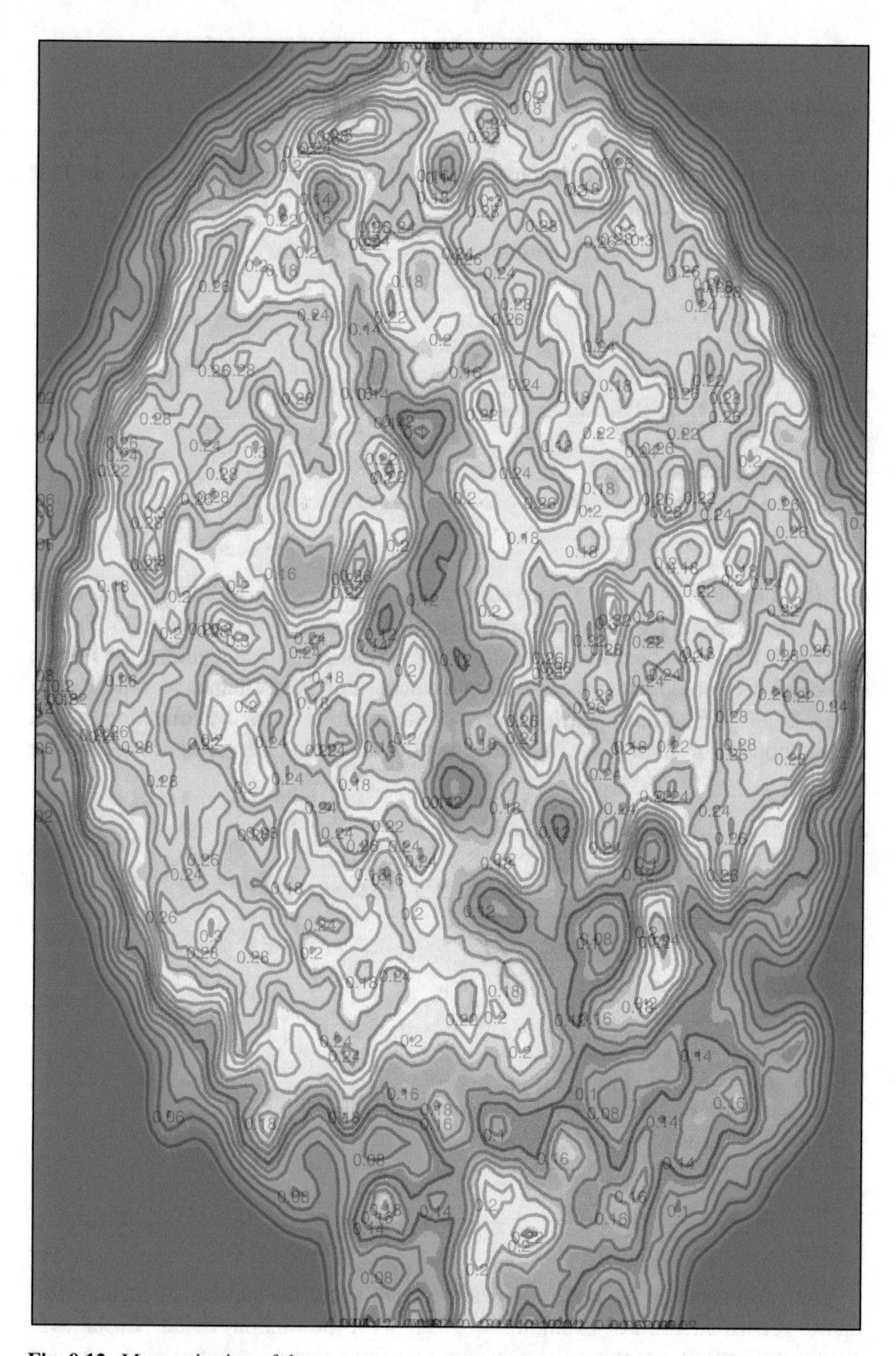

Fig. 9.12 Map projection of the essay space

Fig. 9.13 The palette of hyposometric colours (see Sect. 6.4)

identities and ensure learner anonymity, each learner is assigned a random first name, as shown in Listing 24.

Listing 24 Deidentification: random first name selection.

```
set.seed(22031977)
firstnames=c(
  readLines("male.txt", warn=F),
  readLines("female.txt", warn=F)
)
student.names=firstnames[
  sample(length(firstnames), nrow(essays.content))
]
```

Moreover, it is possible to store with each performance record the human scores assigned. To allow for this, the scores are filed in as shown in Listing 25. Subsequently, both *student.names* and *humanscores* are utilized, when adding the Person object for each student and when adding the according Performance record for each essay: Listing 26 iterates (starting line 4) through the essays in the collection, adds a Person object for each entry, and attaches the according Performance record for each essay.

Listing 25 Loading the human scores.

```
scorefiles=dir("essays.scores/", full.names=TRUE)
humanscores=NULL
for (i in 1:length(scorefiles)) {
  humanscores=rbind(
    humanscores,
    read.table(
      scorefiles[i],
      col.names=c("file","score"),
      row.names= "file"
    )
  )
}
```

The *mpia* package thereby provides feedback about the success of each Performance record added. For example, for essay 474, the random name "Lorin" was chosen and, when attaching the according performance record, the most strongly activated keywords are listed (see Listing 27).

Listing 26 Creating Person objects and Performance records for all students and their essays.

```
ppl = HumanResourceManager(domainmanager = dmgr, domain = d)
essays = NULL
essays.scores = NULL
for (i in 1:nrow(essays.content)) {
  essays.scores[i] = humanscores[essays.content[i,1],1]
  p = ppl$add( name = student.names[i] )
  essays[i] = gsub(
    "[^a-zA-Z0-9äöüÄÖÜß]", " ", essays.content[i,2]
  )
  p$perform(
    essays[i],
    activity = "essay",
    purpose = "exam",
    score = essays.scores[i]
  )
}
```

Listing 27 Feedback for adding a Performance record (success).

```
Person 'Lorin' performed a meaningful activity ('exam').
about: spezifikation, erwartet, anforderungsspezifikation,
verhandlung, prozess, ergebnis, anforderungsdefinition,
repraesentation, definition, anforderung, kommuniziert,
mitarbeiter, softwarespezifikation, entwickler,
benoetigt, klar.
```

The performance records produce warnings, where a low number of terms results from the projection operation that maps the source terms to the Eigenspace. An example of such warning is included in Listing 28 (note that the numbers in the warning message are estimates—and only a proper analysis can unveil which terms were actually dropped by the mapping operation).

Listing 28 Feedback for adding a Performance record (warning).

```
Person 'Angelique' performed a meaningful activity ('exam').
WARNING: of ~30 words (incl. ~0 potential stopwords) only
9 were retained. The rest was possibly not part of the
domain vocabulary?
about: name, schnittstelle, deklariert, definiert, abstrakt,
definiere, anzuwenden, ruckgabewerttyp, implementier,
string, parameterliste.
```

Table 9.6 German example source text versus key activated terms

Source text	Activated terms
hauptaufgaben => anforderungen klar und widerspruchsfrei zu definieren spezifikation => einholen der anforderungen im globalen sinne repraesentation => aufbereitung und kommunizieren der anforderungen fuer management, mitarbeiter, entwickler verhandlung => detailiertes ausarbeiten eines ergebnisses subprozesse der spezifikation—anforderungsdefinition: genaue definition der anforderungen in verschiedenen abteilungen, prozessen und erwartete ergebnisse anforderungsspezifikation softwarespezifikation arten von anforderungen: funktionale: welche funktionen und prozesse werden benoetigt und erwartet qualitative: welches qualitaetsniveau wird erwartet => kostenfaktor (programmiersprache, erweiterbarkeit, ...) zusaetzliche: zeitliche, budgetaere, schnittstellen, ...	spezifikation, erwartet, anforderungsspezifikation, verhandlung, prozess, ergebnis, anforderungsdefinition, repraesentation, definition, anforderung, kommuniziert, mitarbeiter, softwarespezifikation, entwickler, benoetigt, klar

Any person and the according performance record can now be further inspected using the *mpia* package routines, see Listing 29.

Listing 29 Inspecting a person and performance record.

```
lorin=ppl$people[[474]]
lorin$name
performances(lorin)
lorin[1]$getSourceText()
terms(lorin[1])
```

Therefore, the person object is first loaded again into the variable *lorin* so it can be more conveniently referenced. The last two lines then compare the source text with the most strongly activated terms (above proximity threshold). This shows, for example, that the most strongly activated terms clearly capture the key terms of the text.

The German source text for this example is listed against the key activated terms in Table 9.6. Its translation to English is provided in Table 9.7. Note that for the translation, several of the German compounds had to be translated with multiword expressions.

To see, however, which additional terms are weakly activated in the underlying social semantic space, the proximity threshold can temporarily be set to a lower value of 0 (see Listing 30).

Listing 30 Inspecting all (also weakly!) activated terms.

```
d$proximityThreshold=0
activated=lorin[1]$getActivatedTerms()
```

Table 9.7 Translation of source text and activated terms

main aims => define requirements clearly and without contradiction specification => gathering of global requirements representation => preparation and communication of the requirements for management, employees, developers negotiation => detailed elaboration of results sub-processes of specification: requirements definition: clear definition of requirements in different departments, processes, and expected results requirements specification software specification types of requirements: functional: which functions and processes are required and expected qualitative: what is the quality level expected => costs (programming language, extensibility, ...) Additional: temporal, budgetary, interfaces, ...	specification, expected, requirements specification, negotiation, process, result, requirements definition, representation, definition, requirement, communicated employee software specification developer needed clear

```
length(which(activated$values < 0.3))
d$proximityThreshold = 0.3
```

This shows, that there are 693 terms that are weakly activated in the space. 677 of which, however, are activated with an activation-strength lower than the threshold of 0.3. This shows the effect of the mapping to the Eigenbasis.

Using the human resource manager, learners can be compared: in Listing 31, three more students are inspected and compared with Lorin. The student called ‘Brandise’ was scored by the human raters with 0 points, but there is not a single strongly activated term overlapping to Lorin. Linnell’s essay was scored by the raters with 1 point, but there are only four terms overlapping, key terms central to the exam question posed are missing (such as any of the compounds containing ‘anforderung’ = ‘requirement’).

Listing 31 Comparing students and their performances.

```
lorin$scores
brandise = ppl$people[[440]]
brandise$scores
# 0
overlap(lorin[1], brandise[1])
# "anforderung"
linnell = ppl$people[[449]]
linnell$scores
# 1
overlap(lorin[1], linnell[1])
```

Table 9.8 Students' essay length

Name	# Chars	# Key terms
Lorin	673	16
Brandise	630	12
Linnell	506	17
Dall	698	29

```
# "kommuniziert" "repraesentation"
# "spezifikation" "verhandlung"
dall=ppl$people[[473]]
dall$scores
# 4
overlap(lorin[1], dall[1])
# "anforderung" "anforderungsdefinition"
# "anforderungsspezifikation" "entwickler"
# "repraesentation" "softwarespezifikation" "spezifikation"
# "verhandlung"
```

Clearly different, however, is the essay by Dall, which was—same as Lorin's—scored with the maximum number of points (4): here, the overlap lists eight terms, including three compounds with 'requirements'. This is even more evident, as all three students submitted essays with roughly a similar length, see Table 9.8.

Listing 32 plots the positions of all learners into the map projection that was created above in Listing 23. Thereby, both position and component term labels as well as arrows are suppressed.

Figure 9.14 shows the visualization generated: learner positions are depicted by 'dots' in the display, with the fill colour of the dot indicating to which essay collection the underlying performance record belongs. Figure 9.15 lists the corresponding legend.

Listing 32 Plotting the positions of all students.

```
for ( p in 1:length(ppl$people) ) {
   plot( ppl$people[[p]],
      label=FALSE, component.labels=FALSE,
      component.arrows=F, dot.cex=1
   )
}
```

The resulting plot shows how the different collections are separated by the analysis into different regions of the map, with few outliers outside of each collection cluster. Collections 7 and 1 (bottom right corner) seem to overlap which—looking at their overlapping topics ('define interface' and 'define information hiding')—is not very astonishing.

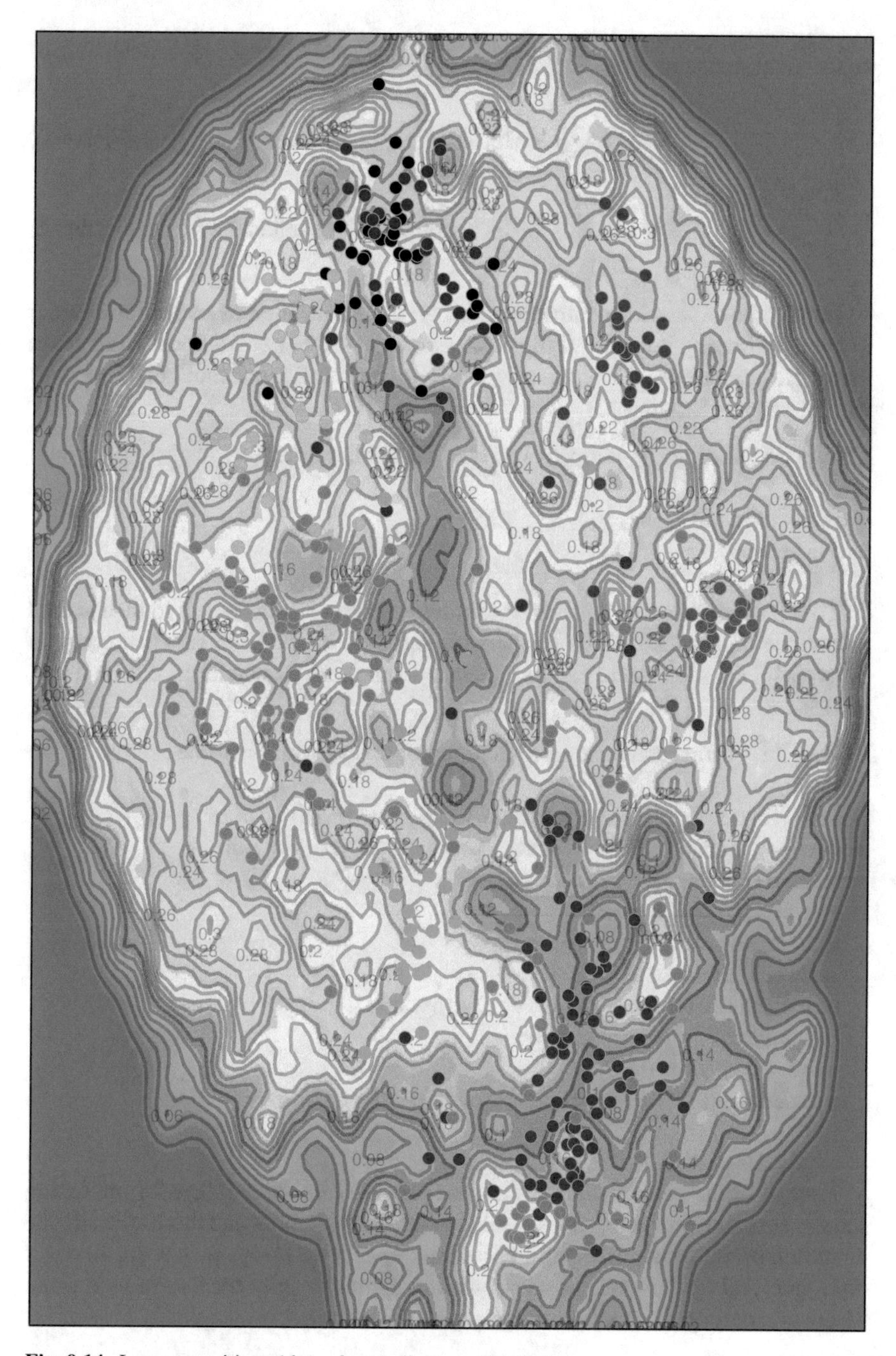

Fig. 9.14 Learner positions (*dot colour* indicates collection)

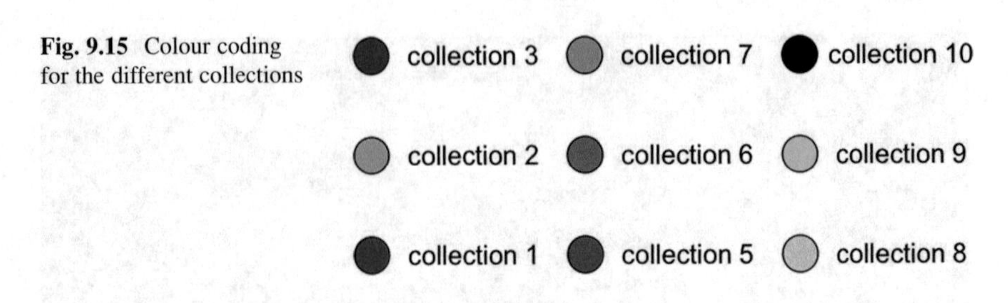

Fig. 9.15 Colour coding for the different collections

Inspect the underlying competences is best done collection by collection in order to avoid clustering artefacts at the cluster boundaries. For collection 1,[6] this gives the single competence depicted in Fig. 9.16 (at an identity threshold of 1), as done with the code of Listing 33.

Listing 33 Extracting competences (and describing them).

```
comps = competences(ppl)
terms(comps)
```

To see the differences between the individual essays with respect to their scored points, the following analysis can provide insights: the position for each group of essays with the same score can be calculated and the activated terms can be listed, see Listing 34.

Listing 34 Terms activated in each group of essays with the same scores.

```
scores = unlist( lapply(ppl$people, function(e) e$scores) )
scterms = matrix(ncol = 2, nrow = length(unique(scores)))
colnames(scterms) = c("scoregroup", "terms")
for (i in unique(scores)) {
  gstuds = which(i == scores)
  scterms[i/5+1,1] = i
  pfs = unlist( lapply(ppl$people[gstuds], performances) )
  class(pfs) = "Performance"
  pfs2 = position(pfs)
  scterms[i/5+1,2] = paste( terms(pfs2), collapse = ", " )
}
```

Executing the code results in the data provided in Table 9.9: with increasing scores, the number of terms rises, while the terms cover more and more facets of the definition of ‘information hiding’. For example, only the essays with score 40 cover both ‘get’ and ‘set’ aspects of encapsulating methods. Only the essays with score 35 cover solely ‘get’, in all other essay groups this detail is completely absent.

[6] Excluding essay 41, which had ‘no activated terms’ after being mapped to the space.

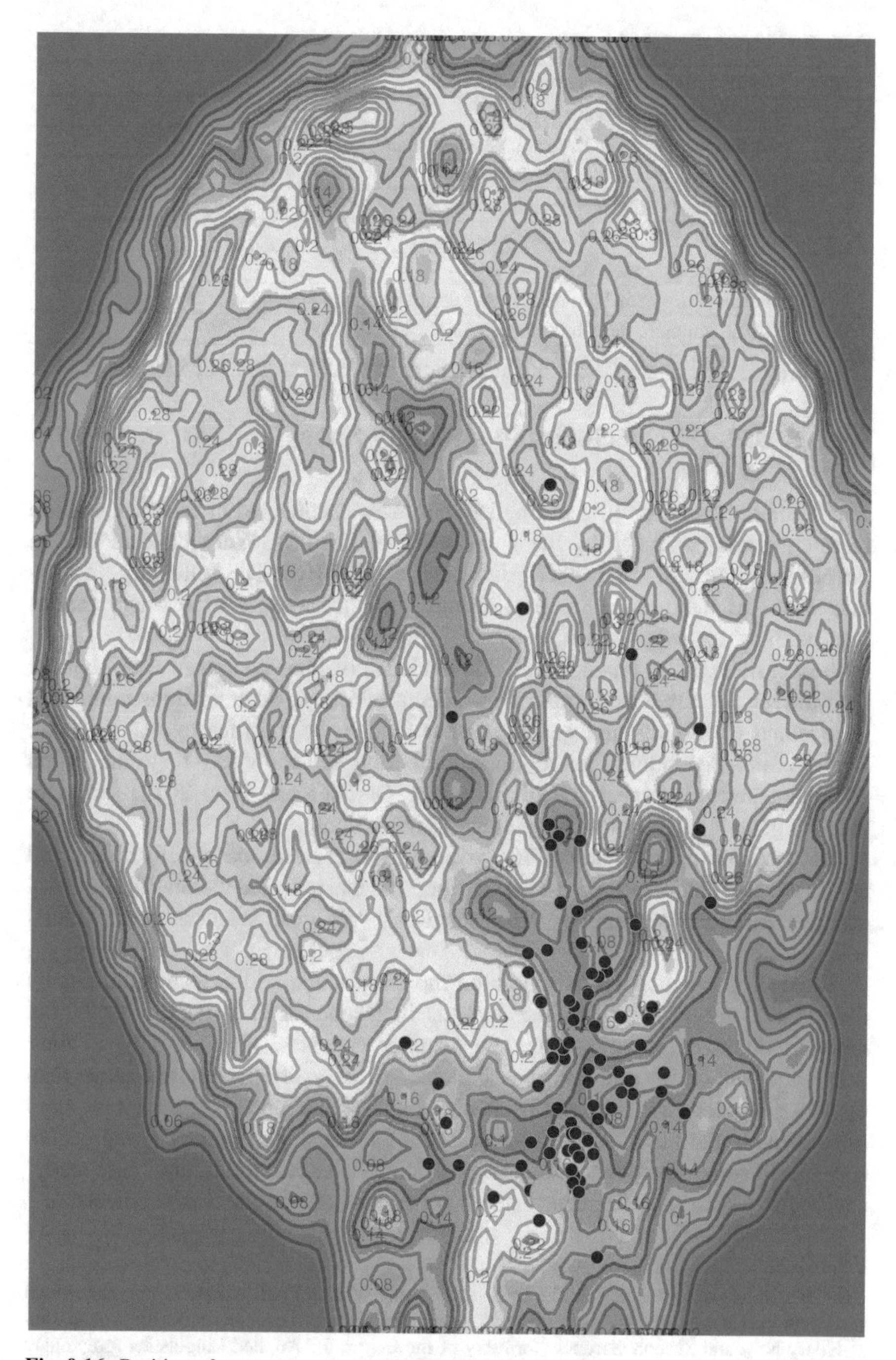

Fig. 9.16 Position of competence extracted for collection 1 (*green*)

Table 9.9 Terms activated for each group with same score

Score	Terms activated at this group position
0	nachname, angemeldet
5	nachname, angemeldet, string, package, name, deklariert, student
10	nachname, angemeldet, zugriff
15	zugriff, nachname, angemeldet, privat, hiding, beispiel
20	nachname, deklariert, angemeldet, zugriff
25	privat, zugriff, nachname
30	nachname, angemeldet, deklariert, zugriff, public, privat, note, hiding
35	privat, definiert, dass, hiding, nachname, public, get
40	note, zugriff, public, deklariert, wert, int, beispiel, direkt, set, hiding, karl, privat, get

9.4 Learner Trajectories in an Essay Space

The 'British Academic Written English' corpus was collected in an ESRC-funded research project operated by the universities of Reading, Warwick, and Oxford Brooks between 2004 and 2007 (Heuboeck et al. 2010). The following analysis makes use of this corpus[7] to demonstrate, how trajectories of learners can be inspected.

Academic learners typically provide evidence of their competence throughout their university career with multiple academic writings. Such writings come in a wide range of formats and genres. Essays tend to be the most prevalent genre family, but—depending on the discipline and study level—they are often complemented with specific variations (see Gardner and Nesi 2012, for more detailed genre analysis of writings in the Bawe corpus).

The total Bawe corpus holds 2761 writings of students almost equally distributed across the four study levels (year one to three and masters) and the four disciplinary groups 'arts and humanities', 'life sciences', 'physical sciences', and 'social sciences'.

Table 9.10 provides an overview on the distribution of the genre families by disciplinary group, following the analysis in Heuboeck et al. (2010, p. 7), but ignoring duplicate classifications. As Gardner and Nesi (2012, p. 16) in their genre analysis of the same corpus indicate, while essays "represent more than 80 % of assignments in Arts and Humanities, a far wider range of genres is required of students in the Physical and Life Sciences": In these disciplinary groups, more specialized forms such as 'methodology recounts', 'explanation', and 'case study' can be found in higher frequency (the latter as well among the Social Sciences).

[7] The data in this study come from the British Academic Written English (BAWE) corpus, which was developed at the Universities of Warwick, Reading and Oxford Brookes under the directorship of Hilary Nesi and Sheena Gardner (formerly of the Centre for Applied Linguistics [previously called CELTE], Warwick), Paul Thompson (Department of Applied Linguistics, Reading) and Paul Wickens (Westminster Institute of Education, Oxford Brookes), with funding from the ESRC (RES-000-23-0800).

Table 9.10 Documents per genre family and disciplinary group

	Arts and humanities	Life sciences	Physical sciences	Social sciences	Total
Case study	0	89	37	66	192
Critique	48	84	73	114	319
Design specification	1	2	87	3	93
Empathy writing	1	19	9	3	32
Essay	591	127	63	444	1225
Exercise	14	33	49	18	114
Explanation	9	114	56	17	196
Literature survey	7	14	4	10	35
Methodology recount	18	145	170	16	349
Narrative recount	8	25	21	18	72
Problem question	0	2	6	32	40
Proposal	2	26	19	29	76
Research report	9	22	16	14	61
	708	702	610	784	2804

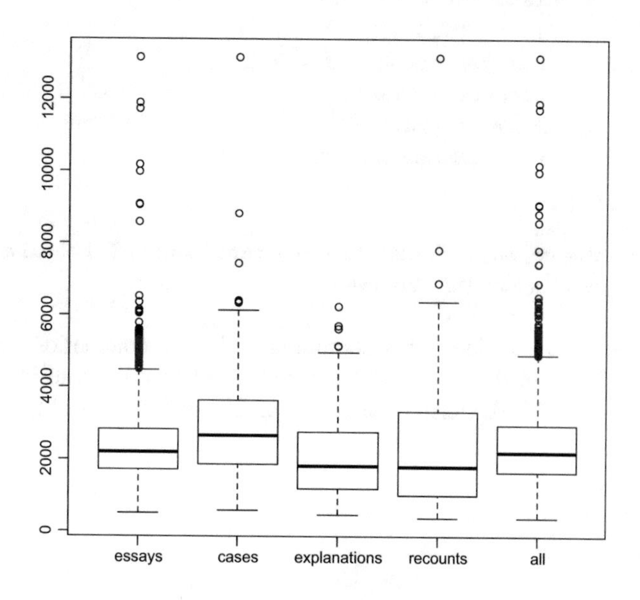

Fig. 9.17 Word length distribution across genre families

The subsequent demonstration therefore concentrates on these four genre families, yielding a total of 1676 texts.

The texts differ in length, as depicted in Fig. 9.17, with a mean length of 2466 words at a standard deviation of 1274 words. The mean length of the texts per genre family is in a similar range, case studies with 2907 words being the largest mean length and explanations (with 2094 words) being the smallest. The boxplot depicted in Fig. 9.17 identifies several outliers (marked with unfilled dots in the chart). The texts of these outliers are removed from the data set for further analysis (removing 97 texts, reducing the selection to 1579 texts). This slightly influences the mean word length to become 2253 words at a standard deviation of now 886 words.

Moreover, since the focus of this demonstration is on trajectories of learners, texts of learners are removed: learners who have written less than three and more than seven texts, thus further reducing the selection to 742[8] texts. This hardly changes the mean word length, which is now 2210 words at a standard deviation of 908 words.

Subsequently, the texts are read into a Corpus object, see Listing 35. Thereby, punctuation, numbers and English stop words[9] are removed, and only terms with a minimum length of three characters retained.

Listing 35 Filing in the corpus.

```
tm=Corpus(
  DirSource("selection/", recursive=FALSE),
  readerControl=list(
    reader=readPlain,
    language="english",
    load=TRUE,
    removePunctuation=TRUE,
    stopwords=TRUE,
    minWordLength=3,
    removeNumbers=TRUE
  )
)
tm=tm_map(tm, function(e) return(gsub("[^[:alnum:]]", " ", e)))
tm=tm_map(tm, tolower)
```

To—again—apply stemming and stem completion for improved readability, first, the dictionary with the raw word forms has to be saved into *dict* (Listing 36). This dictionary holds now 43,804 different word forms.

[8] The analysis with the full data set produces more clutter in the overlay of the paths onto the map projection: Single text positions don't have paths; and long paths sometimes produce 'lassoing' xsplines. The 'lassoing' problem could be prevented by adding a 'travelling salesman' optimization that changes the order of the path, so as to favour geodesics from one performance location to the next. This is, however, not trivial and is (yet) not implemented in the package. As a nice side effect of this filtering, the analysis also runs much faster.

[9] Using *the* tm packages stop word list for English, which originated in the author's *lsa* package.

Listing 36 Creating the dictionary required for stem completion.

```
dict = Terms(DocumentTermMatrix(
   tm,
   control = list(
      removePunctuation = TRUE,
      stopwords = TRUE,
      minWordLength = 3,
      removeNumbers = TRUE
   )
))
```

Next, stemming is applied to the vocabulary of the corpus, using the code of Listing 37. Additionally, only those terms will be retained that appear more often than twice (using the control setting 'bounds'). This reduces the vocabulary to 10,181 word stems.

Listing 37 Creating the dictionary required for stem completion.

```
tm = tm_map(tm, stemDocument, language = "english")
dtm = TermDocumentMatrix(
   tm,
   control = list(
      removePunctuation = TRUE,
      removeNumbers = TRUE,
      stopwords = TRUE,
      minWordLength = 3,
      bounds = list(global = c(3,Inf))
))
```

Next, Listing 38 applies stem completion reusing the dictionary *dict* from above. This reduces the vocabulary to 10,018 terms.

Listing 38 Stem completion.

```
sc = as.character(
   stemCompletion(rownames(dtm), dictionary = dict,
   type = "shortest")
)
sc[which(is.na(sc))] = rownames(dtm)[which(is.na(sc))]
rownames(dtm) = sc
if (any(duplicated(rownames(dtm)))) {
   dupes = which(duplicated(rownames(dtm)))
   for (i in dupes) {
      hits = which(sc == sc[i])
      target = hits[ which(! hits %in%
```

```
        which(duplicated(sc))) ]
      replvec=t(as.matrix(
        colSums(as.matrix(dtm[ hits, ])) ))
      rownames(replvec)=sc[target]
      dtm[ target,1:length(replvec) ]=replvec
    }
    dtm=dtm[!duplicated(rownames(dtm)),]
}
class(dtm)=c("TermDocumentMatrix", class(dtm))
```

The resulting term frequency distribution is depicted in Fig. 9.18: The gray rectangle indicates the window of analysis. The terms that appear more often than in 10 % of all documents and less often than 10 times are removed from analysis to concentrate on the most discriminative terms and to keep memory consumption small.

With the resulting text matrix *dtm*, the threshold for the Eigenvalue stretch truncation (at 80 % stretch) can be determined. For an example of how to calculate this manually, see Listing 39. This stretch truncation is calculated by the package routines automatically, when adding a corpus and executing its spacification.

Listing 39 Threshold determination for the stretch truncation.

```
tr=sum(dtm*dtm)
tr * 0.8
```

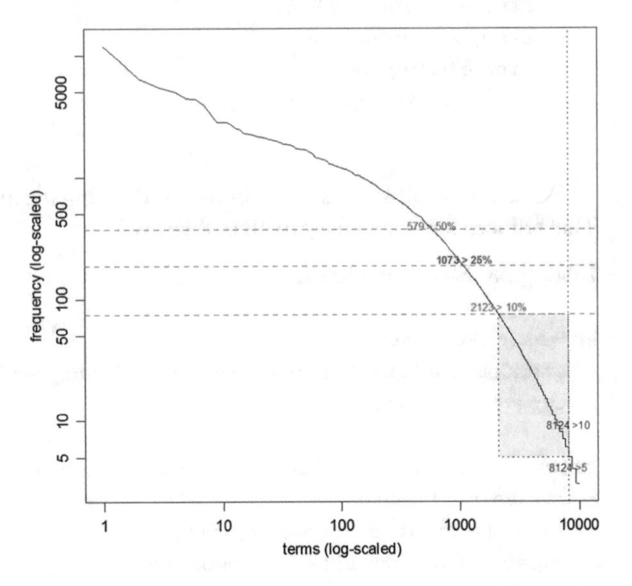

Fig. 9.18 Frequency filtering by boundary thresholds

After preparing the clean and sanitized text matrix, the package routines encapsulate the complexity of the further analysis. Listing 40 instantiates a DomainManager and adds a new Domain called 'bawe'.

Listing 40 Invoking the DomainManager and attaching new Domain.

```
dmgr = DomainManager()
d = Domain(name = "bawe")
dmgr$add(d)
```

Next, the text matrix is added via the domain's *corpus* routine. This routine accepts many different inputs formats, e.g., if the parameter were a list of filenames or a directory, it would try to create a quick and dirty full text matrix representation of the corpus by itself.

Listing 41 Eigenspace calculation using stretch truncation.

```
d$corpus(dtm)
d$spacify()
```

The truncated Eigenspace is now calculated from the cleaned text matrix *dtm* and added to the domain (to be persisted together with the proximity and visualization data for future use).

Subsequently, the map visualization data is generated in Listing 42. This uses the standard proximity threshold of 0.3 and the standard identity threshold of 1.

Listing 42 Calculating the map visualisation data.

```
d$calculateTermProximities()
d$visualiser$calculateNetCoords()
d$visualiser$calculateReliefContour()
```

After the proximity and visualization data has been calculated, the map projection can be visualized, for example, using the topographic visualization as indicated in Listing 43.

Listing 43 Visualisation of the planar projection.

```
plot(d, method = "topographic")
```

This results in the projection depicted in Fig. 9.19.

To highlight the learner positions, their writings have to mapped to the projection space. Therefore (see Listing 44), a human resource manager is instantiated into *ppl*. Using *ppl$add*, all students (of 'Sociology') are added and their performance records (the essays) assigned. The final line of the *for* loop in the listing plots the position into the map visualization. This results in the updated visualization depicted in Fig. 9.20.

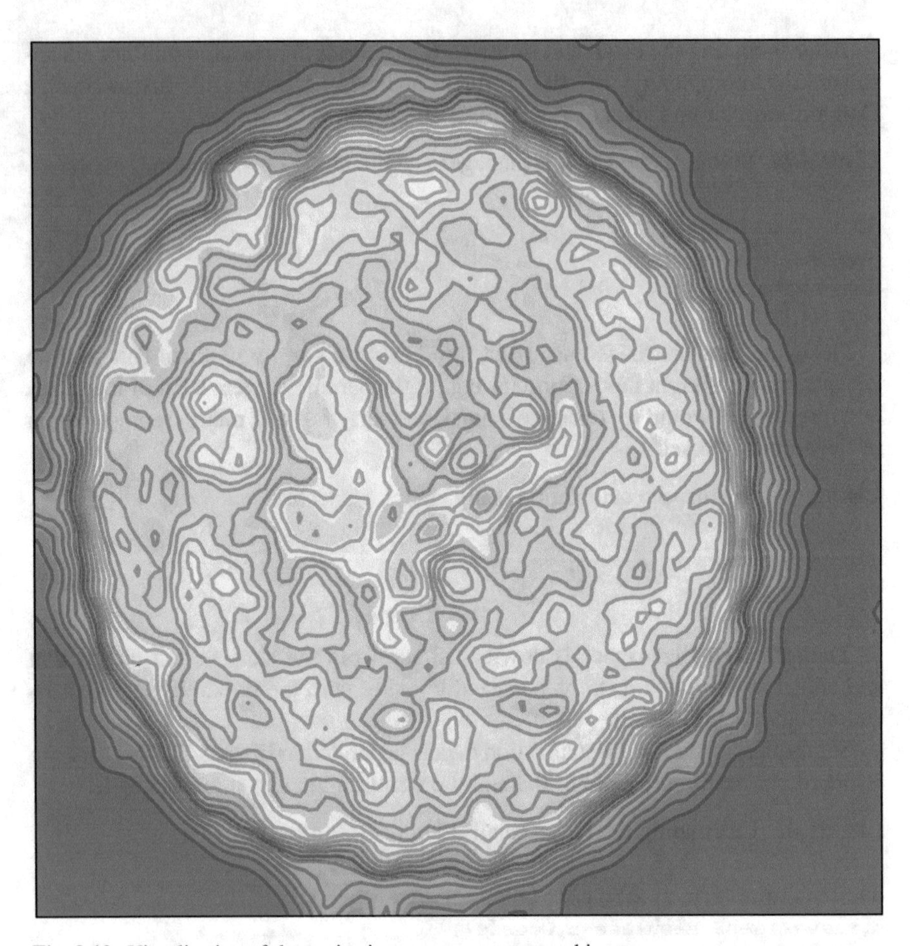

Fig. 9.19 Visualization of the projection space as a topographic map

Some basic statistics can be gathered showing *ppl* in the console ('a HumanResourceManager caring for 10 people.') and calling *performances(ppl)*, which shows 43 elements.

Listing 44 Adding learners and assigning their performance records.

```
ppl = HumanResourceManager(domainmanager = dmgr, domain = d)
for (i in 1:length(students)) {
  p = ppl$add( name = students.names[i] )
  sel = which(essays$student_id==students[i])
  for (n in sel) {
    p$perform( essay[n], purpose="Sociology" )
  }
  plot(p, dot.cex = 1.5)
}
```

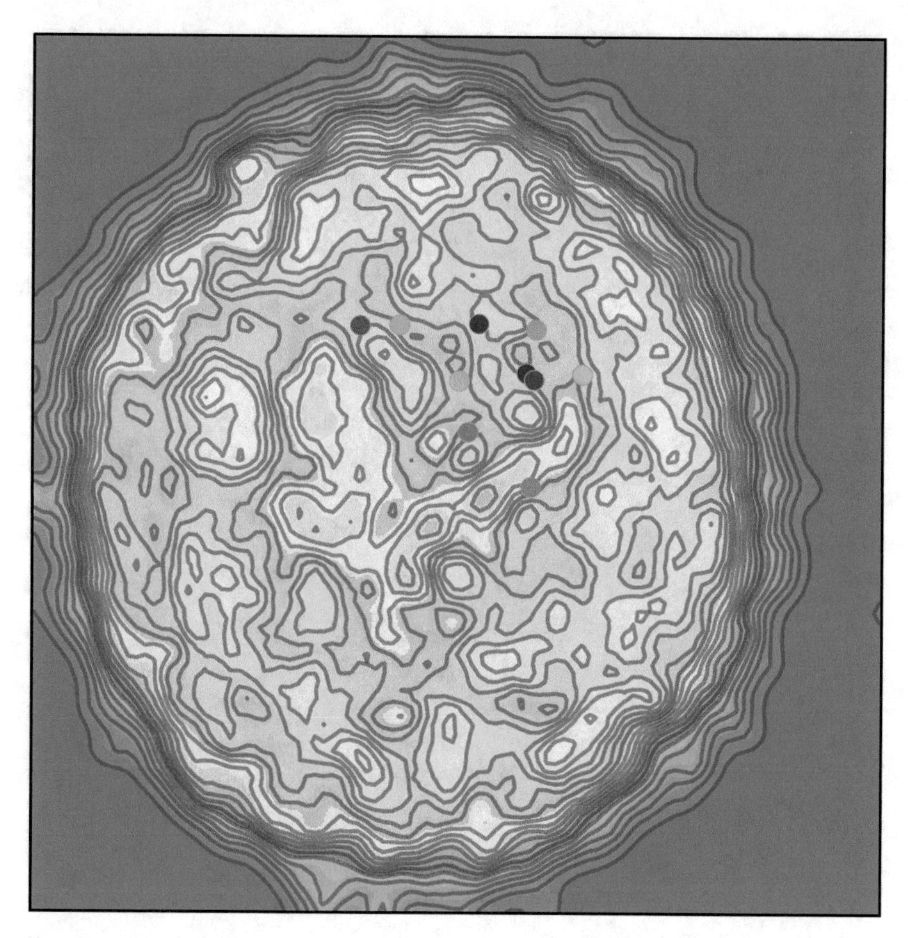

Fig. 9.20 Topographic map including learner positions (Sociology students)

Turning now to the paths, the learners expose, the following can be made visible (Fig. 9.21), when running the code of Listing 45. Since the essays of the BAWE collection originate from different semester levels and different universities, not much similarity in the pathways is directly evident (as expected). The paths, do cross, however, and underlying competences demonstrated might be shared.

Listing 45 Plotting the learning paths of the learners (Sociology).

```
plot(d, method="topographic")
for (n in 1:length(ppl$all())) {
  p=ppl$people[[n]]
  plot( path(p) )
}
```

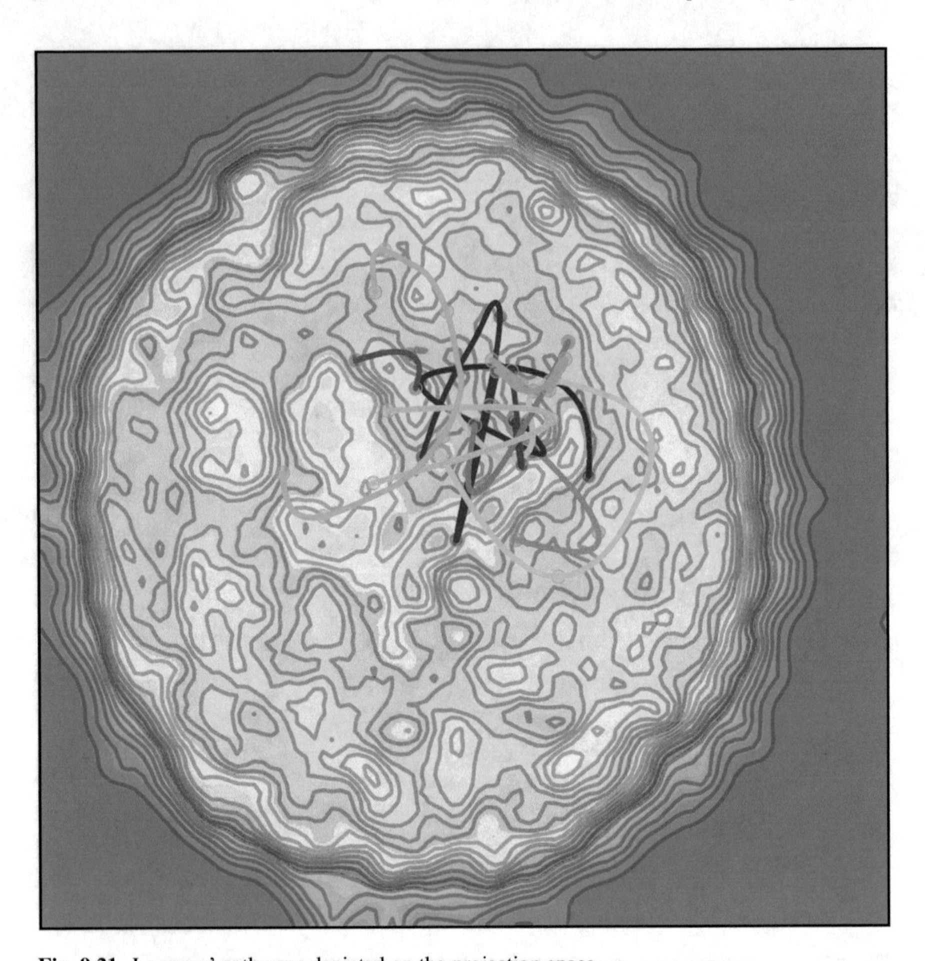

Fig. 9.21 Learners' pathways depicted on the projection space

Now, which learners are close to each other? Listing 46 picks the last person added ('Melonie') and checks which other learners are close by using *near()*. It turns out that 'Uriel' and 'Constantino' are close above proximity threshold.

This can be used (Listing 47) to re-visualize the pathways and positions of the three learners—for further visual inspection, see Fig. 9.22: it shows that there each of the other two learners has performances close to Melonie's.

Listing 46 Other learners near 'Melonie'.

```
melonie=ppl$last()
near(ppl, melonie)
# A person with name 'Uriel' and 5 textual traces.
# A person with name 'Constantino' and 3 textual traces.
# A person with name 'Melonie' and 6 textual traces.
uriel=ppl$people[[1]]
constantino=ppl$people[[7]]
```

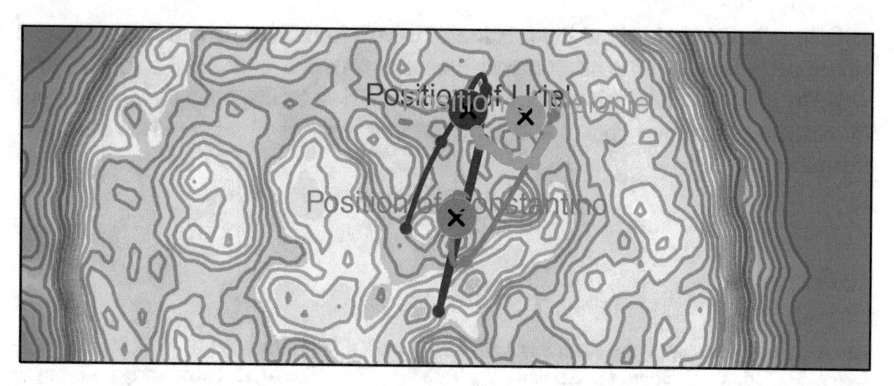

Fig. 9.22 Three learner positions (*big dots*) and paths (*splines*)

Listing 47 Revisualising paths and positions for the three learners.

```
plot(d, method="topographic")
plot(path(uriel))
plot(position(uriel))
plot(path(constantino))
plot(position(constantino))
plot(path(melonie))
plot(position(melonie))
```

9.5 Summary

This chapter turned to the third root of this work, i.e., its application for learning analytics. It briefly reviewed the state of the art for the subareas in focus—content analysis and social network analysis. Related work can be found, however, typically dealing with either of the two aspects, not their interplay.

The chapter picked up the recurring, guiding foundational example introduced originally to illustrate the working principles (and shortcomings) of social network analysis and then—again—to do the same for latent semantic analysis. It showed with the help of this example, how an integrated solution can shed light on competence development in a more holistic way and while assisting in its implementation with improved functionality for analysis (compared to the functionality provided by any of the two methods stand-alone).

Two application examples with real-life data rounded up the application demos. The first one tended again to essay scoring (as already introduced in Sect. 4.5), but this time demonstrated the additional analysis possibilities (and insights) provided by MPIA. In particular, this example illustrated, how positional information can be utilised to better understand the relation of essays to each other and to their model solutions. Moreover, the second real-life case revealed additional facilities provided by MPIA for analysis of learning paths. It did so by inspecting a larger corpus of academic writings spread out over several subject areas, study years, and institutions.

References

Arnold, K.: Signals: applying academic analytics. Educ. Q. **33**(1), (2010)

Berry, M., Do, T., O'Brien, G., Krishna, V., Varadhan, S.: SVDPACK, Particularly las2. http://www.netlib.org/svdpack/ (1992). Accessed 31 Jan 2014

Crespo García, R.M., Pardo, A., Delgado Kloos, C., Niemann, K., Scheffel, M., Wolpers, M.: Peeking into the black box: visualising learning activities. Int. J. Technol. Enhanc. Learn. **4** (1/2), 99–120 (2012)

Dawson, S., Bakharia, A., Heathcote, E.: SNAPP: realising the affordances of real-time SNA within networked learning environments. In: Dirckinck-Holmfeld, L., Hodgson, V., Jones, C., de Laat, M., McConnell, D., Ryberg, T. (eds.) Proceedings of the 7th International Conference on Networked Learning 2010. Aalborg, Denmark (2010)

Gardner, S., Nesi, H.: A classification of genre families in university student writing. Appl. Linguist. **34**(1), 1–29 (2012)

Gerdemann, D., Burek, G., Trausan-Matu, S., Rebedea, T., Loiseau, M., Dessus, P., Lemaire, B., Wild, F., Haley, D., Anastasiou, L., Hoisl, B., Markus, T., Westerhout, E., Monachesi, P.: LTfLL Roadmap, Deliverable d2.5, LTfLL Consortium (2011)

Govaerts, S., Verbert, K., Duval, E.: Evaluating the student activity meter: two case studies. In: Proceedings of the 9th International Conference on Web-based Learning 2011. Springer, Berlin (2011)

Heuboeck, A., Holmes, J., Nesi, H.: The Bawe Corpus Manual, An Investigation of Genres of Assessed Writing in British Higher Education, Version III. http://www.coventry.ac.uk/research/research-directory/art-design/british-academic-written-english-corpus-bawe/research-/ (2010)

Rebedea, T., Dascalu, M., Trausan-Matu, S., Armitt, G., Chiru, C.: Automatic assessment of collaborative chat conversations with PolyCAFe. In: Proceedings of EC-TEL 2011, pp. 299–312. (2011)

Retalis, S., Papasalouros, A., Psaromiligkos, Y., Siscos, S., Kargidis, T.: Towards networked learning analytics—a concept and a tool. In: Proceedings of Networked Learning 2006. http://citeseerx.ist.psu.edu/viewdoc/summary?doi=10.1.1.106.1764 (2006)

Rhode, D.: SVDLIBC: A C Library for Computing Singular Value Decompositions, Version 1.4. http://tedlab.mit.edu/~dr/SVDLIBC/ (2014). Accessed 31 Jan 2014

Scheffel, M., Niemann, K., Pardo, A., Leony, D., Friedrich, M., Schmidt, K., Wolpers, M., Delgado Kloos, C.: Usage pattern recognition in student activities. In: Delgado Kloos, C., Gillet, D., Crespo Garcia, R.M., Wild, F., Wolpers, M. (eds.) Towards Ubiquitous Learning. Lecture Notes in Computer Science, vol. 6964. Springer, Berlin (2012)

Trausan-Matu, S., Dessus, P., Rebedea, R., Loiseau, M., Dascalu, M., Mihaila, D., Braidman, I., Armitt, G., Smithies, A., Regan, M., Lemaire, B., Stahl, J., Villiot-Leclercq, E., Zampa, V., Chiru, C., Pasov, I., Dulceanu, A.: Support and Feedback Services Version 1.5, Deliverable d5.3, LTfLL Consortium (2010)

Chapter 10
Evaluation

Evaluation "[m]ethods cannot be differentiated in good and bad, and if a particular method fails to provide results (or even more often: results beyond tautologies), then this probably says more about their competent handling, rather than their validity or reliability", as Law and Wild (2015, p. 24) point out.

The appropriate choice of methods depends on *who* the evaluation addresses, *why* it is needed, and *what* goal it pursues (ibid). Moreover, the phase of research in general tends to favour qualitative methods for early and more exploratory stages, but quantitative methods for (cross-case) follow up (ibid, p. 24f).

The problem statement introduced in Sect. 1.2 lists three main objectives (see also Sect. 1.7 for their refinement): to represent learning, to provide analysis instruments, and to visually re-represent it back to the user.

While the reason can hence be identified as to provide evidence of efficiency and effectiveness of the (re-)representations generated to support assessment of its credibility, and while the target group of evaluation is clearly the analyst (see use cases in Sects. 3, 4.3, and 7.1), the goals are more heterogeneous, therefore calling for a mix of methods and approaches for evaluation.

Traditionally, goals of evaluation of socio-technical systems can be distinguished in—on the one side—*verification* of the implementation of a logically and analytically derived model qualification and—on the other side—*validation*, comparing computational with experimental results of an external reality (Fig. 10.1; see Oberkampf and Roy 2010, pp. 21–32, for an in depth analysis of the historic discourse).

More precisely, verification refers to "the process of assessing software correctness and numerical accuracy of the solution to a given mathematical model" (Oberkampf and Roy 2010, p. 13).

Validation on the other hand deals with assessing the "accuracy of a mathematical model" against reality, tested using "comparisons between computational results and experimental data" (ibid, p. 13).

F. Wild, *Learning Analytics in R with SNA, LSA, and MPIA*,
DOI 10.1007/978-3-319-28791-1_10

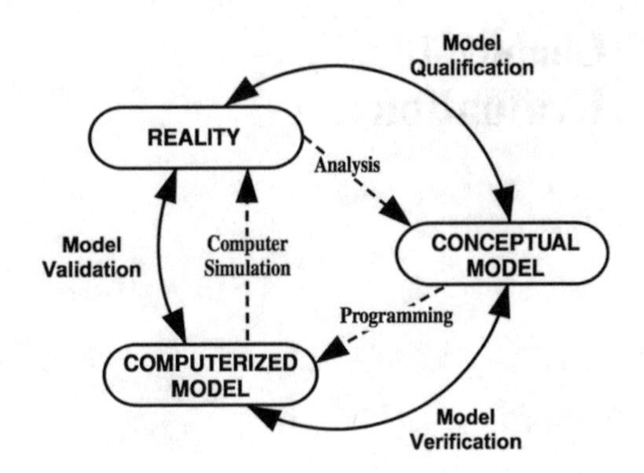

Fig. 10.1 The role of verification and validation (Schlesinger 1979, as cited in Oberkampf and Roy 2010, p. 23)

The research conducted for this book is accompanied by a series of evaluation studies, each of them contributing to both conceptual as well as algorithmic refinement and—finally—to their concluding assessment.

In this chapter, a precise summary of these studies is synthesised. Looking back, it is particularly valuable to focus not only on positive findings, but also on errors made and blind alleys turned into. The studies in sum, extended by a final accuracy study, provide evidence of the validity and utility of the approach developed.

The rest of this chapter is organised as follows. First, insights from earlier prototypes (and their evaluation studies) are briefly reported (Sect. 10.1), as they have lead to the research behind this book. The development of theoretically sound meaningful, purposive interaction analysis and samples of its application did not happen out of the blue as an answer to a Grand Challenge Problem. They were also a reaction to demands uncovered through earlier research of the author.

Then, verification results (Sect. 10.2) are reported, briefly reviewing the test-driven development methodology applied in the implementation and summarising final check results of the package.

Finally, validation is performed (Sect. 10.3), conducted in several studies. Scoring accuracy is investigated in Sect. 10.3.1, while the structural integrity of spaces (assessing convergent and divergent validity) follows in Sect. 10.3.2. Annotation accuracy is challenged and results of the according experiment are reported in Sect. 10.3.3.

Accuracy of the visualisations is probed in Sect. 10.3.4, while computational efficiency of the new factor determination technology using the trace of the original text matrix based on stretch-truncation is measured via performance gains in Sect. 10.3.5. A summary (Sect. 10.4) concludes the chapter.

10.1 Insights from Earlier Prototypes

Step by step, earlier research of the author lead from open exploration to the initial ideas that—then more focused—were elaborated for this work in iterations. It is important to see, how the ideas evolve over time as a result of findings of these earlier works.

Initially, in a phase of open exploration, latent semantic analysis was at the core of the focus of interest, leading to a number of effectiveness studies including investigating influencing parameters and configuration settings (Wild et al. 2005a, b).

Several application studies followed. Wild and Stahl (2007a) reported on term similarity measurements and essay scoring. Feinerer and Wild (2007) applied and evaluated it for coding of qualitative interviews. Wild and Stahl (2007b) looked into assessing social competence, whereas Wild (2007) provided first thoughts on a conceptual model and its implementation developed for the package.

In Wild et al. (2007), a first web prototype application follows: ESA is an essay scoring application developed for. LRN, the learning management system (Fig. 10.2). Its front-end is written in tcl, paired with stored procedures written in R for the postgreSQL database.

Refinements over the years have lead to a multitude of releases and the current *lsa* package (Wild 2014, version 0.73, see ChangeLog for more details), which includes explicit support and language resources for English, German, French, Dutch, Polish, and Arabic.[1]

In Wild et al. (2008), positioning for conceptual development is further investigated, here for the first time introducing network plots to visualise proximity relationships, see Fig. 10.3: The black vertices indicate those essays scoring equal to our higher than 50 % of the possible number of points.

While high precision and recall results could be obtained (with differences found in the way how machine scores are generated, see Wild et al. 2008), one of the key shortcomings of this visualisation method became painstakingly evident: every time, the visualisation is generated, the positions in the plot area change.

Concept visualisation studies follow in Wild et al. (2010c), applied to a study of the lexical semantics extracted from a corpus of papers in the field of technology-enhanced learning, thereby using divisive clustering and manual mounting of sub component force-directed layouts on top of a to explore the lexical semantics of spaces.

Further experiments in visualisation and monitoring conceptual development follow, leading to the development of first ideas of MPIA (then called 'meaningful interaction analysis (MIA') and its implementation in the Conspect (a blend of 'concept' and 'inspection') prototype as a widget for Moodle, Blackboard, and other compatible environments.

[1] Both UTF-8 native and Buckwalter transliterations.

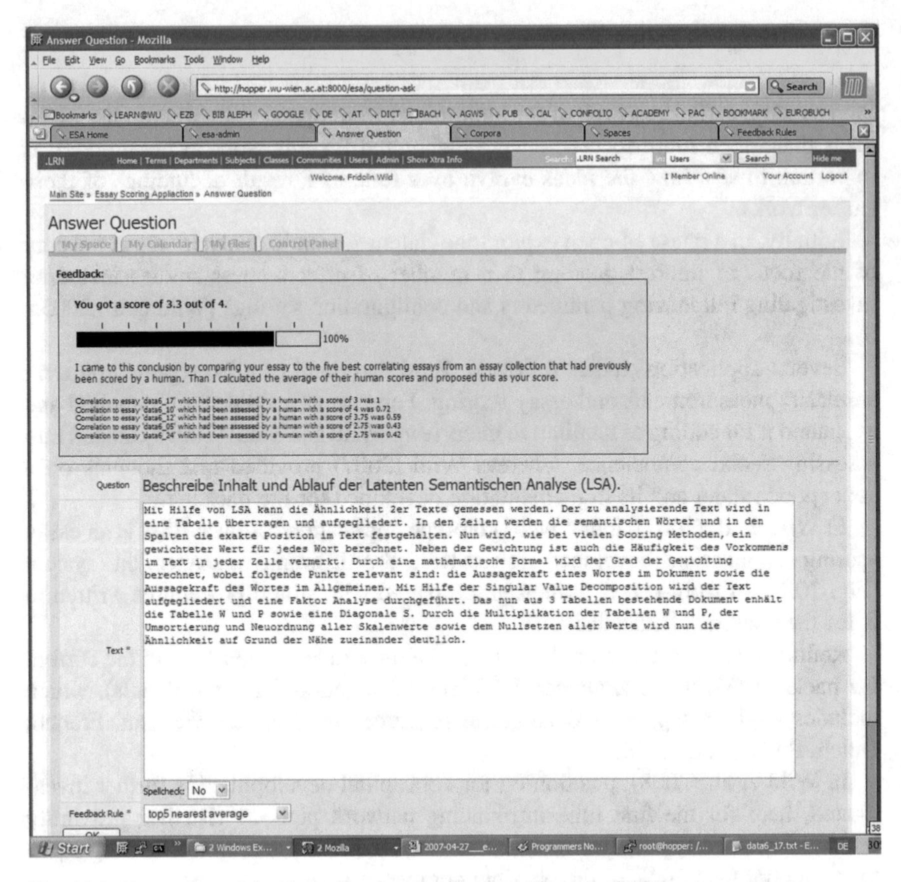

Fig. 10.2 Screenshot of the Essay Scoring Application (ESA) for LRN

For Conspect, two types of evaluations were conducted iteratively over the three versions developed in order to assess validity. Focus groups and a streamlined cognitive walkthrough, as qualitative methods, were used to—among others—evaluate usability, pedagogical effectiveness, relevance, efficiency, and satisfaction (Armitt et al. 2009, 2010, 2011).

Several complementary quantitative studies were applied in order to measure accuracy of the system (Wild et al. 2010a, b, 2011).

While the idea of Conspect was very positively evaluated (and so did its representation accuracy), its implementation was not. The main problems of the final version 3 were to be found to be in the "granularity of output ('too low level') and the "cognitively demanding visualisation" (Fig. 10.4 and Fig. 10.5).

Qualitative insights ultimately leading to the inception and proposal of MPIA were largely gained from this web-app implementation Conspect and its evaluation trials. Conspect was subjected to the LTfLL project's evaluations in three rounds.

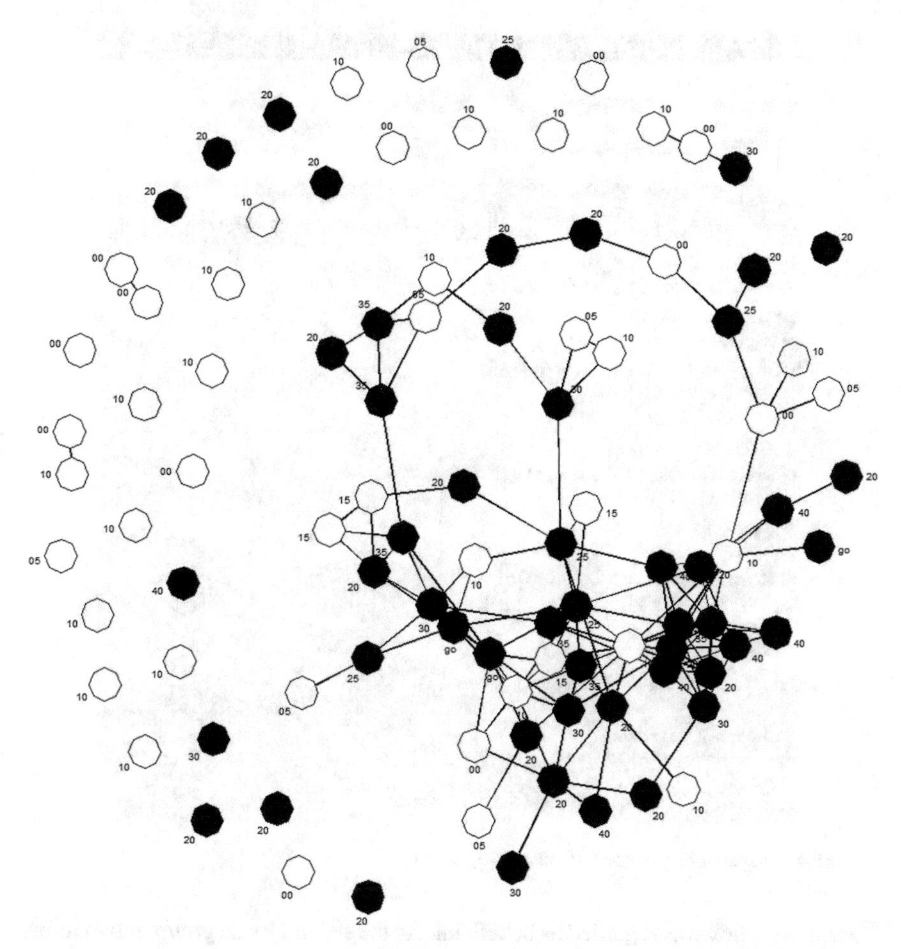

Fig. 10.3 Proximity network visualization of essays in a space (Wild et al. 2008, p. 7)

Each round fed back into development, leading to release of a new version, including an additional one following the final round.

Six medical students were interviewed in a focus group in trials of the first prototype version of Conspect. They appreciated particularly that they are able to compare the representation of their conceptual development with those of their peers (Fig. 10.6). One student stated, for example, that "I find it useful [...] especially the comparing part. You can see what's been missed out on. I think it's more useful to see what you've missed out on than to see overlap—[but] it's always good to know if you've covered extra".

This qualitative evaluation of the first prototype found that the service concept is regarded valuable, "subject to the clarity with which the service could represent the results" (Armitt et al. 2009). The key strength was seen in the possibilities for the identification of conceptual coverage and in its assistance for tailoring feedback to support needs of individuals as well as groups. Particularly the opportunity to get

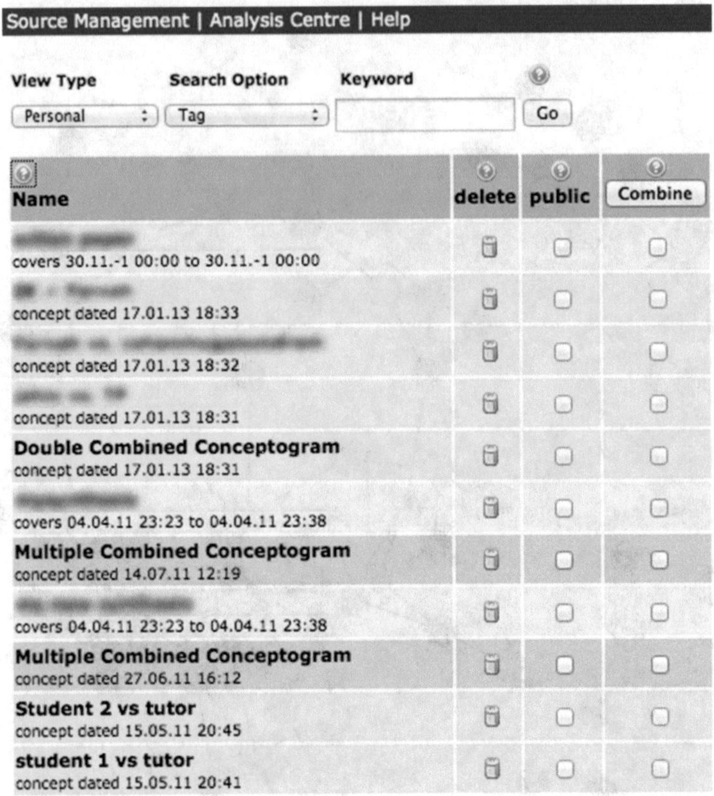

Source Management | Analysis Centre | Help

View Type: Personal | Search Option: Tag | Keyword | Go

Name	delete	public	Combine
covers 30.11.-1 00:00 to 30.11.-1 00:00			
concept dated 17.01.13 18:33			
concept dated 17.01.13 18:32			
concept dated 17.01.13 18:31			
Double Combined Conceptogram concept dated 17.01.13 18:31			
covers 04.04.11 23:23 to 04.04.11 23:38			
Multiple Combined Conceptogram concept dated 14.07.11 12:19			
covers 04.04.11 23:23 to 04.04.11 23:38			
Multiple Combined Conceptogram concept dated 27.06.11 16:12			
Student 2 vs tutor concept dated 15.05.11 20:45			
student 1 vs tutor concept dated 15.05.11 20:41			

Fig. 10.4 Snapshot of Conspect (Conspect 3, 2011)

instant feedback was regarded as beneficial. As the second focus group of two tutors identified missing facilities to aggregate information about groups of learners, the qualitative evaluation report concluded that the major value of Conspect is "to help self-directed learners find their own route through the curriculum, by showing the ground remaining to be covered and helping them achieve similar coverage to their peers" (Armitt et al. 2010). For the second version of the prototype, this shortcoming was resolved and functionality to merge multiple conceptual graphs was added.

Additionally, a usability evaluation of Conspect was conducted to identify usability problems of the first prototype as input for the implementation plan of the second prototype. Methodologically this was done as a usability inspection in form of a streamlined cognitive walkthrough (Spencer 2000). Since the aim of the analysis lay on the identification of potential problems rather than measuring the degree of usability, the results, however, are not relevant for here.

The third validation round concluded with a number of open issues: "single word responses being inappropriate in a complex domain, stemming leading to ambiguity, issues with the granularity of the output ('too low level'), all compounded by a cognitively demanding visualisation." Moreover, "stemming of words is a major

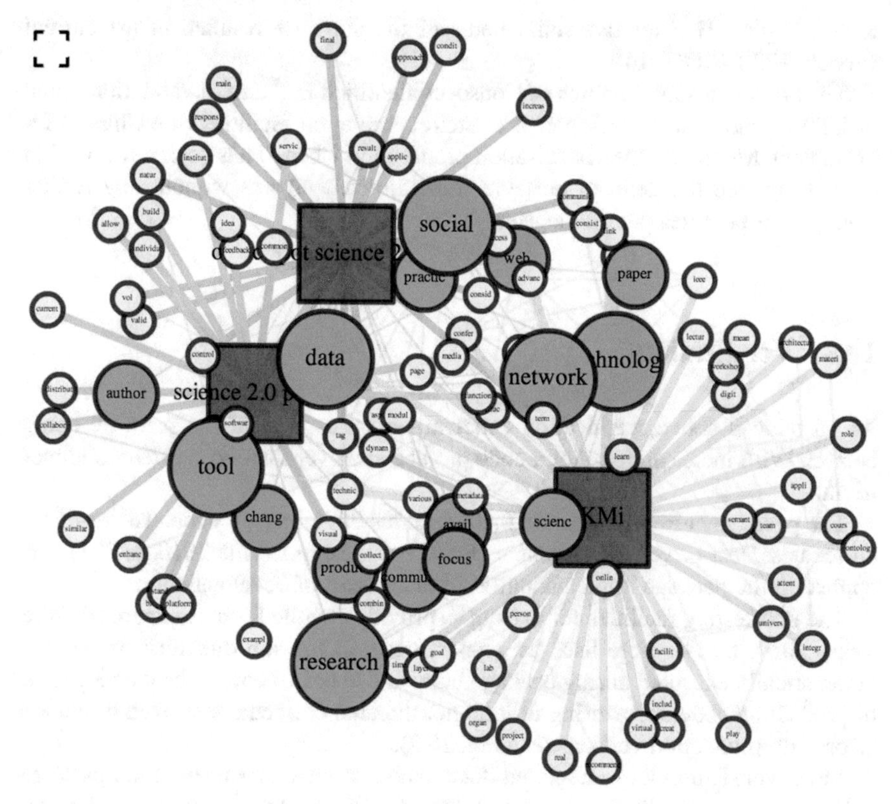

Fig. 10.5 Conspect's 'multiple merge' functionality (Conspect 3, 2011)

Graph 1	Overlap		Graph 2
abl	enhanc	enhanc	data
project	metadata	metadata	practic
workshop	paper	paper	scienc
abstract	research	research	servic
base	approach	approach	allow
call	build	build	applic
contribut	softwar	softwar	author
develop	technolog	technolog	blog
exchang			collabor

Fig. 10.6 Conspect's comparison table

shortfall of CONSPECT when applied in complex domains, where the lemma takes meaning specifically from its suffix" (Armitt et al. 2011).

Finally, starting summer 2013 and building on this earlier work, the mathematical formulation of MPIA and its implementation in the software package *mpia* for

R was begun. The iterative refinements of the software resulted in the current version 0.73 (Wild 2014).

The two main shortcomings of Conspect identified in the third validation round of LTfLL lead to the development of stem completion facilities (see Chap. 9 for examples). Moreover, the visualisation proposed in Chap. 6 has been developed in order to reduce the demand unstable force-directed layouts without any further aggregation facilities put on the user.

10.2 Verification

As introduced above, verification concentrates on aspects of the software being correct. With increasing size of a code project, such verification is often no longer trivial.

The package currently (version 0.60) features 9671 lines of code. To ensure its correctness, test-driven development (Beck 2003; Sommerville 2010, p. 221) was applied in the package implementation from the start of development on.

The unit testing facilities for packages, provided by the R environment (R Core Team 2014, p. 14), were used as support: tests in the sub directory 'tests/' are automatically executed during package build and check to ensure the development of production code or rewriting of it do not threaten correctness or even introduce errors into production versions (see Fig. 10.7).

Moreover, R provides integrated documentation check routines in the package build and check facilities. They test for identity of documentation and code. Moreover, the check routine also runs the 22 examples included in the documentation to test if they pass without errors and warnings.

Listing 1 Routine to catch errors in testing

```
tryCatch(
  {
    dmgr=new("DomainManager")
    dmgr$get("madeupname")
  },
  error=cat("passed."),
  finally=function(e) stop ("test failed: no error")
)
```

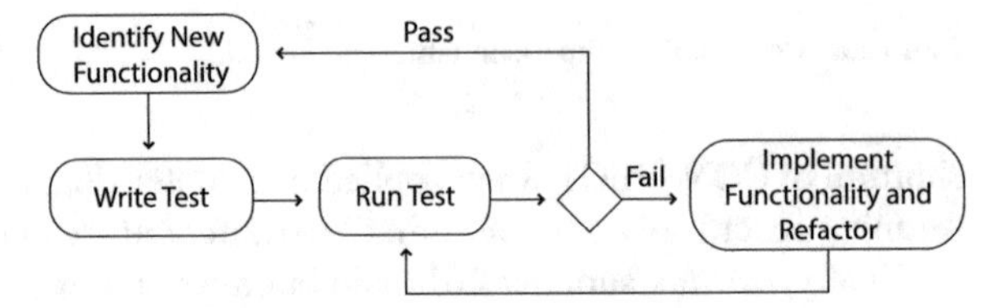

Fig. 10.7 Test-driven development (redrawn from Sommerville 2010, p. 222)

Table 10.1 Unit tests in the mpia package

Class	Tests	Scope
DomainManager, Domain	7	Create, flush cache, domain not found, add, retrieval by name, materialisation, upgrading of Domain objects to newer class versions
HumanResourceManager	4	Create, Adding persons, listing persons, adding Performances, flushing positions, terms of Performances, names, path by timestamp
Performances	2	Create, Position, Summary of position, Create
Person	11	Create, Generate traces, Read, Write, Add without domain, getActivatedTerms, Multiple domains per person, Path by creation time, Proximity, Proximity by interval path
Visualiser	10	Create, Upgrade visualiser objects to new class, pdf file with toponymy only, test materialisation of netcoords, plotMap (all methods), toponymy (all methods), position plot, path plot
	34	*Total*

Fig. 10.8 Extract of the package check output

```
[...]
* this is package 'mpia' version '0.60'
[...]
* checking examples ... OK
* checking for unstated dependencies in tests ... OK
* checking tests ...
 Running 'tests.R'
OK
[...]
```

The package uses 34 unit tests, where required they are encapsulated in a tryCatch call as listed in Listing 1. An overview on the tests executed with each package build and check is provided in Table 10.1. For more details on the examples, see Annex 1: documentation of the *mpia* package.

The 34 unit tests and the 22 examples contained in the documentation together ensure that—throughout development—changes in production code do not affect connected functionality, causing the package to break.

Figure 10.8 provides an extract of the package build and check output, showing that all tests and examples run without error.

10.3 Validation

As introduced above, validation refers to assessing the "accuracy of a mathematical model" against reality, tested using "comparisons between computational results and experimental data" (Oberkampf and Roy 2010, p. 13).

This section reports on several experiments and evaluation studies that—together—work towards assessing the validity of both the mathematical model developed in Chap. 5 as well as its visualisation using the planar projection proposed in Chap. 6.

While some of these results have been conducted along the way, leading up and feeding into the development of MPIA, findings have been updated and extended where possible to reflect changes introduced.

In particular, this relates to the essay scoring experiment of Sect. 10.3.1, the cardsort and annotation experiment of Sect. 10.3.2 (reanalysed from Wild et al. 2011) and Sect. 10.3.3.

Accuracy thereby is investigated from several angles. The first section reports on accuracy in an application, i.e. for automated scoring of essays.

While the *essay scoring experiments* (Sect. 10.3.1, introduced already in Chap. 7) were conducted initially without the application of the new stretch truncation method and while they use a rather simple scoring method of the average proximity to three model solutions, they show that accurate analytics with MPIA can be created. Moreover, it is described, how a more precise scoring algorithm can be set up using MPIA's functionality for inspection, which solves the significant shortcoming of lacking inspection facilities.

Both *convergent and divergent validity* of term clustering in the Eigenspace is investigated (Sect. 10.3.2) in order to establish that the social semantic models calculated are performing well in establishing conceptual relations similar to how humans would. The research retold here roots in the earlier publication in Wild et al. (2011), but is extended.

Annotation accuracy (Sect. 10.3.3) is tested to see if the quality of the mapping of texts to their Eigenspace representation serves the successful extraction of relevant term descriptors. Again, this work roots in Wild et al. (2011) and is therefore only briefly summarised here.

Since the projection of the underlying graph onto a plane in the visualisation of MPIA looses accuracy for the sake of better visual readability, the *degree of loss* is critically assessed in Sect. 10.3.4.

10.3.1 Scoring Accuracy

For the experiment presented here, the nine essay collections described already above in Chap. 7 were utilised to calculate separate MPIA spaces. Chapter 7 provides more detail on the set up of the experiment, so here only the maximal correlation achieved with human ratings is reported.

Table 10.2 lists the collections, the number of essays they contain each, and the maximum Spearman rank correlation Rho of the machine scores assigned and the scores given by the human raters. In average, a correlation of 0.60 could be achieved, with results ranging as low as 0.42 for collection one to as high as 0.71 for collections five and ten.

Since collection one ('define information hiding') was evaluated with the lowest correlation of machine scores to human ratings, an additional scoring method was

Table 10.2 Spearman's Rho of human scores to machine scores

#	Topic: Assignment	Essays	Rho
1	*Programming: define 'information hiding'*	102	0.42***
2	*E-Commerce: define 'electronic catalogue'*	69	0.60***
3	*Marketing: argue for one of two wine marketing plans*	40	0.64***
5	*Information systems: define 'meta-data'*	23	0.71***
6	*Information systems: define 'LSA'*	23	0.61***
7	*Programming: define 'interface'*	94	0.58***
8	*Procurement: argue reengineering of legacy system versus procurement of standard software*	45	0.64***
9	*Information systems: explain 'requirements engineering'*	46	0.56***
10	*Information systems: explain component/integration/system test*	39	0.71***
		481	0.60 ∅

***p-value < 0.001

tested. This new method was now developed in an 80 % stretch truncated Eigenspace.

To implement this scoring method, the three model solutions were added to a Person object called 'tutor' (see Listing 2). Then, for each student essay, the term overlap of the learner's performance record ('the essay') with the terms at the position of the tutor was counted and the number of overlapping terms assigned as the machine score.

With this scoring method in the stretch truncated space, the Spearman rank correlation with Rho 0.48 was achieved (at a p-value < 0.001), thus slightly improving the correlation with human scores, but now providing instruments for feedback generation. For example, Listing 3 shows how to retrieve the most highly activated terms for the tutor position and Table 10.3 illustrates this with an example of a student essay, scored by the human raters with 15 out of 40 points.

Listing 2 Scoring by term overlap.

```
tutor=ppl$add( name="tutor" )
tutor$perform( solutions[1], activity="gold 1" )
tutor$perform( solutions[2], activity="gold 2" )
tutor$perform( solutions[3], activity="gold 3" )
p=ppl$add( name="student" )
p$perform( essays[1], activity="essay (1)" )
length( overlap( p[1], position(goldpers) ) )
```

Listing 3 Retrieving the terms at the position of the tutor (three model solutions).

```
needed=terms(position(tutor))
has=terms(p[1])
missing=names(needed)[! needed %in% has]
```

Table 10.3 Term overlap between tutor position and a 15 points essay

Needed	Has	Missed
daten, implementiert, deklariert, public, variable, private, angemeldet, nachname, zugegriffen, information, hiding, auszulesen, vorteil, zugriff, zugreifen	implementiert, variable, information, daten	deklariert, public, private, angemeldet, nachname, zugegriffen, hiding, auszulesen, vorteil, zugriff, zugreifen

A further example of scoring is provided in Sect. 9.3, with additional inspection facilities outlined in e.g. Table 9.8.

10.3.2 Structural Integrity of Spaces

To assess the validity of truncated Eigenspaces, two further experiments were conducted: a cluster analysis evaluation and an annotation accuracy evaluation. The experiment investigating clustering capabilities is summarised in this section and is reported in more detail in Wild et al. (2011), while the annotation accuracy evaluation follows in the next.

The experiment aims to compare the ability of MPIA spaces to cluster together what belongs together, while separating unrelated constructs. Sometimes this is also called assessing the convergent and discriminant (or divergent) validity.

The Research Methods Knowledge Base (2014) requires for the establishment of convergent validity to show that "measures of constructs that theoretically *should* be related to each other are, in fact, observed to be related to each other".

Moreover, to establish discriminant validity, it needs to be shown that "measures of constructs that theoretically should *not* be related to each other are, in fact, observed to not be related to each other" (ibid).

For the experiment, postings from four case reports (A1, M1, M2, M3) posted by real learners in a university discussion forum about the topic 'safe prescribing' in Medicine were used to extract a list of about 50[2] top-loading descriptor terms from their resulting space representation. Top-loading hereby refers to those concepts that were activated the most by projecting the posting into the space described above. This was done to ensure that it was *possible at all* to cluster the terms extracted from the space of 21,091 distinct terms after stemming. The space was calculated from a corpus of 24,346 PubMed abstracts using 300 dimensions.

The extracted descriptors of the texts were printed on cards and the participants of the study were instructed to arrange them in at least two piles, thereby grouping the concepts according to their closeness. This procedure was performed for each of the four postings.

[2] The number of concepts extracted varied slightly for each of the four postings depending on their content.

As participants, 18 first-year medical students were recruited from the University of Manchester Medical School, half of them being male, the other half female. Ages of the students ranged from 19 to 21. The students were compensated with £10 Amazon vouchers for the 2 h of their time they spent on this experiment (and the one summarised in the next section).

The data collected are analysed in three steps, summarising what was already reported in Wild et al. (2011). First, the frequency of co-occurrence of term pairs across participants is analysed in order to measure how accurate the Eigenspace term clustering (generated with kmeans) compared to the humans' clusterings. Then, the edit distances required to change one card sort into another is inspected, yielding more insight into the actual distance between humans and between humans and the machine clustering. Finally, silhouette widths are inspected in order to further inspect discriminance of clusters.

With respect to the co-occurrence frequency of the pairs of terms across participants, the participants were found to vary largely in the way they clustered the provided concept cards, as the co-occurrence analysis of term-pairs in the card piles shows.

Figure 10.9 provides the overview on the full distribution of co-occurrence frequency of term pairs amongst the human subjects' card sorts: from left to right it lists each term and how often it was put into a pile together with each other term. The mean across all terms is 3 with a standard deviation of 3. There are term pairs

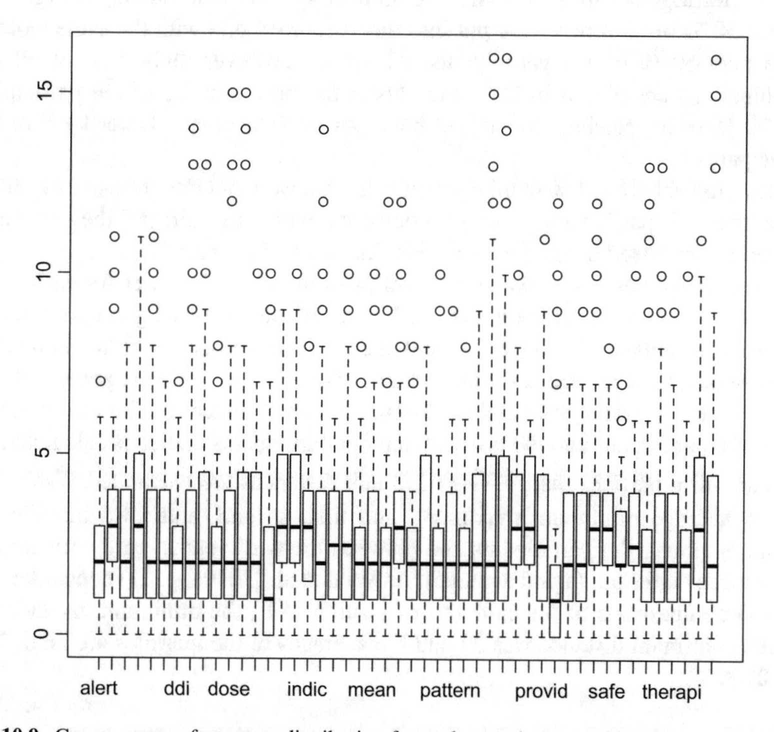

Fig. 10.9 Co-occurrence frequency distribution for each term (own graphic)

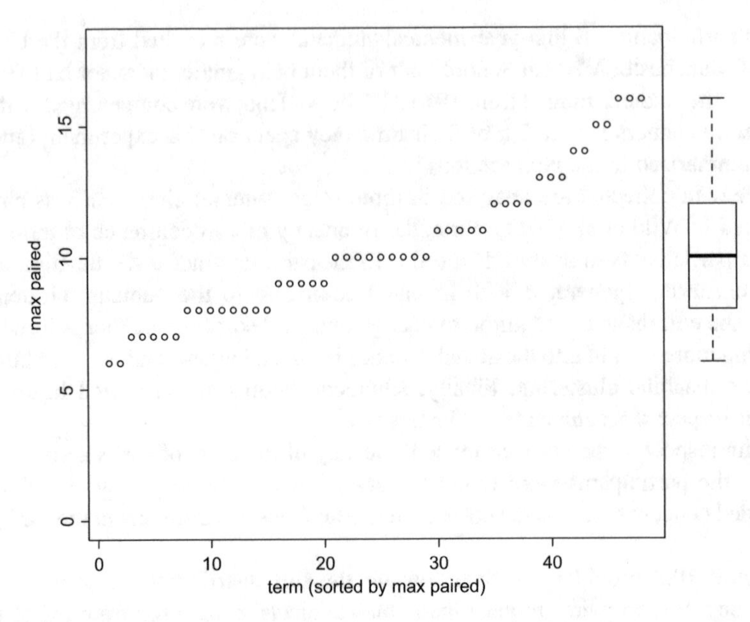

Fig. 10.10 Maximum co-occurrence of each term pair (own graphic)

that cluster together into the same pile more frequently than others, as Fig. 10.10 shows: 58 % of all terms were put into the same card pile with the same word by more than 50 % of the participants. All in all, however, only 1 % of all pair combinations are placed in the same cluster by more than 12 of the participants and 90 % of all pairings found had been put into the same cluster by 6 or less participants.

Next, the cohort of 1 % of all concept pairs that were put into the same cluster by more than 12 participants will be compared with the pairs of the automated clustering generated using the cosine similarities in the space.

The average cosines between the term pairs in the top 1 % tier for each of the four postings is shown above in Fig. 10.11: two of them showed good proximity of 0.7, one was remarkably high at 0.9, and one unveiled a moderate correlation of 0.4.

Subsequently, the card sort edit distance (Deibel et al. 2005) is generated with the UW card sort analyzer (Deibel 2010): the edit distance required to resort the cards of a participant into the order of another participants choice is calculated. To compare all n participants, $\frac{n \cdot (n-1)}{2} = 153$ edit distances are calculated. Table 10.4 reports average minimum, maximum, and first, second, and third quartile edit distances among the participants and between the participants' card sorts and the Eigenspace card sorts (again average). As Wild et al. (2010a, p. 21) conclude: "For the 153 comparisons of the card sort of posting 'A1', the minimum distance was 21, the maximum distance was 34 and the averages of the quartiles are 26.6, 28.7, and 29.8."

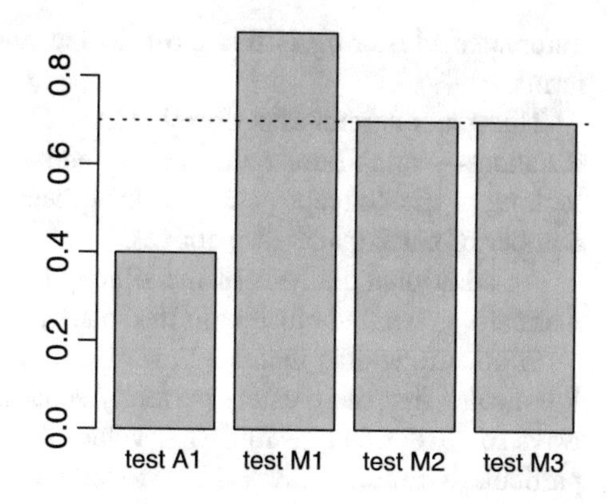

Fig. 10.11 Correlation between top 1 % pairs and Eigenspace pairs (own graphic)

Table 10.4 Card sort edit distances (Wild et al. 2011, p. 15)

Card sorting results in terms of edit distance							
		Min	Quartile 1	Quartile 2	Quartile 3	Max	#Comparisons
Test A1	Avg.	21	26.6	28.7	29.8	34	153
MPIA		30	31.25	33	34	35	18
Test M1	Avg.	18	22.6	24.7	26.4	31	153
MPIA		24	27	28	29	31	18
Test M2	Avg.	21	30.9	32.7	34.5	40	153
MPIA		31	33.3	35.5	36	37	18
Test M3	Avg.	20	28.7	31.0	33.1	38	153
MPIA		30	31	32.5	35	37	18

Table 10.5 Differences of card sort distances (Wild et al. 2011, p. 15)

Test	diff	%diff	#Cards	%Cards	#Words
3	2.26	6.5	52	67	1117
4	2.33	7.0	51	65	533
2	3.2	11.5	43	65	557
1	4.5	13.6	48	68	228
Average		9.7			

As can be read from Table 10.4, the edit distances between the automatically generated clusters and the human clusters were larger than those between the humans in each case.

Table 10.5 further investigates this difference: The subtracted average of the space edit distances and the human edit distances is listed in 'diff', whereas '%diff' reflects this as percentages. The percentage of cards that have to be moved (i.e., the edit distance) is reported in '%cards'.

The percentage difference of edit distances for the automatically generated clusters in comparison to the intelligently created clusters in the third column suggests, with a mean of 9.7 %, that the automatically created clusters have an approximately 10 % larger edit distance than the human study subjects. The automated clustering is this close to the human agreement on clustering related terms.

This general trend of performing in an equal range as humans with respect to edit distances—which here rather means equally bad—is much more positive, when looking at the concept pairs that have been put into the same cluster by a larger number of participants (see above).

An additional analysis of the silhouette widths (Rousseeuw 1987) of the automatically generated clusters in the space provides an explanation.

Silhouette widths indicate how discriminant the clusters chosen are: a value of 1 indicates that the clusters perfectly separate the data and no better (competing) ways to cluster can be found. A value of -1 signals that the clusters chosen are particularly unfortunate. Values around 0 mean that the clusters are not necessarily very discriminant and many cluster members would in fact also fit into another cluster provided.

The silhouette widths for the chosen clusters range between -0.1 and -0.03, meaning that the clusters chosen were not necessarily very discriminant, although also not very bad: there are term pairs with very high proximity in them, but the average silhoutte width per cluster is significantly reduced by outliers in the clusters. Using a higher number of iterations instead of the rather small number used for the calculation of the kmeans clusters should improve the clustering. Moreover, using the top-loading terms of a projected text may not have been such a good idea, as the activated terms not necessarily are the ones in closest, semantic proximity to each other in the space.

Alternative clustering methods (hclust, method 'complete linkage'; agnes) in an 80 % stretch-truncated space calculated from 505 case study forum posts in the same area was tried to see if clustering can be improved, leading to silhouette widths between –0.009 and 0.5 (hclust) and 0–0.21 (agnes). This finding has lead to replacing the clustering methods in MPIA used for groups and competences.

It is astonishing, but not uncommon that the human participants of the experiment had a rather small overlap in their judgements. Still, the high correlation in the top cohort with highest agreement in pair-wise cluster memberships indicates that the algorithm as such is performing well.

10.3.3 Annotation Accuracy

The next experiment looked at the accuracy of the concept representations generated to annotate the five postings on 'safe prescribing' with concept descriptors. The experiment is summarised here, more detail can be found in Wild et al. (2011).

To ensure that not only convergent, but also discriminant validity can be established (and to assess the baseline), a set of ten top-loading descriptors for each of the five postings was generated and complemented with an additional set of five random terms ('distracters') that were randomly chosen from the vocabulary covered by the space. Participants (same as in the previous experiment) were asked

Table 10.6 Interrater agreement between humans and MIA

	Free marginal kappa	Without distracters	Distracters only
Text 1	0.4	0.2	0.7
Text 2	0.4	0.3	0.5
Text 3	0.4	0.3	0.5
Text 4	0.4	0.3	0.8
Text 5	0.3	0.2	0.5
Average	0.4	0.3	0.6

Table 10.7 Interrater agreement with conflated categories

	Free marginal kappa	Without distracters	Distracters only
Text 1	0.5	0.4	0.8
Text 2	0.5	0.4	0.7
Text 3	0.6	0.4	0.8
Text 4	0.6	0.4	1.0
Text 5	0.5	0.4	0.7
Average	0.5	0.4	0.8

to rate on a Likert scale of 1–5 how good each of these 15 concepts described the posting.

The rating data obtained is analysed below using the free marginal kappa measure (Randolph 2005a, b), "an interrater reliability statistic that is applicable when raters are not constrained by the number of entries per category" (Wild et al. 2011, p. 17).

Table 10.6 shows the interrater agreement for the full metric of the Likert scale used. The first column provides the results for all 15 terms (the 10 MPIA-chosen descriptors and the 5 random distracter terms), whereas the second and third list the kappas for the set without distracters and the set with only the distracters respectively.

Table 10.7 shows the same results for conflated categories: ratings of 4 and 5 are conflated into a 'descriptive' class, ratings of 1 and 2 are conflated to indicate 'not descriptive', and ratings of 3 are kept as 'neutral'. This provides a clearer picture, when the strength of the rating is neglected and the analysis focuses on the binary selection of concepts as descriptors.

As expected, the interrater agreement is high when 'distracters only' are analysed. In the case of non-conflated categories, the agreement ranges from 0.5 to 0.8 with an average of 0.6. For conflated categories, agreement is higher and ranges from 0.7 to 1.0 with an average of 0.8.

The results on the full set of 15 descriptors and the results on the 10 descriptors selected by the deployed algorithm, however, indicate that humans again do not have very high interrater agreement: the average is only 0.4 and 0.3, where 0.7 is considered good (non-conflated categories). This is slightly better, when conflated categories are considered: Humans judge that the terms chosen by the deployed algorithm describe a text with a kappa of either 0.5 (all 15) and 0.4 (without distracters) when conflated categories are considered.

It remains to be tested, whether higher agreement can be found among human test subjects, when advanced learners of later years are tested that have developed a more common frame of reference.

These results have been fed back into the calibration of the cut-offs used to improve the selection of representative concepts, leading to more rigorous cut-offs to avoid noise.

10.3.4 Visual (in-)Accuracy

The degree of loss from the proximity in the MPIA space to its projection onto the cartographic map plane can be measured. Listing 4 documents the formulas used to calculate the measurements. In the formula for the loss through the outlier reattachment, the difference in the proximity driven link erosion and outlier reattachment is measured by subtracting the *termProximities* stored in a Domain from their raw cosine proximities (divided by the number of values). Since this sum removes all ego values, the length of the diagonal has to be subtracted. As the raw proximity matrix typically has a mean floating around 0, the low degree of loss is not astonishing.

Listing 4 Measuring the map projection accuracy.

```
lsavecs=d$space$tk %*% diag(d$space$sk)
t2t=cosine(t(lsavecs))
# loss introduced in the outlier reattachment
sum(t2t - d$termProximities) /
(nrow(t2t)*ncol(t2t)-length(diag(t2t)))
# calculate distances in the map projection
map=d$visualiser$netcoords
distances=matrix(ncol=nrow(map), nrow=nrow(map))
for (i in 1:nrow(map)) {
   for (n in 1:nrow(map)) {
      distances[i,n]=sqrt(
         (map[i,1]-map[n,1])^2 +
         (map[i,2]-map[n,2])^2
      )
      }
}
# normalize cosines
t2tnorm=( t2t-min(t2t) ) / ( max(t2t) - min(t2t) )
# measure map projection loss
( sum((1-t2tnorm)-distances) ) /
   ( nrow(t2t)*ncol(t2t)-length(diag(t2t)) )
```

Table 10.8 Visual (in-)accuracy (% loss)

Domain	# Terms	Stretch truncation (%)	Link erosion (% loss)	Map Projection (% loss)
Essays	1447	80	1.6	25.4
Medical	2341	80	4.4	28.2
Bawe	4096	80	2.2	38.7

This is slightly different for the subtraction of the calculated Pythagorean distances: here, the normalised proximity values *t2tnorm* have to be subtracted from 1 to convert them to distances.

Table 10.8 then provides the results. It is positive, that the link erosion (and outlier reattachment) introduces almost no loss compared to the raw proximity values, while—at the same time—only with it, the planar projection is possible. The significant loss in accuracy of the projection surface is clear: projecting a multidimensional graph onto a two-dimensional plane is otherwise not possible. Since the visualisation serves the purpose to guide exploration and help in forming hypotheses, while the actual measurements and further analysis takes place in the high dimensional Eigenspace, this can accepted.

10.3.5 Performance Gains

One of the significant advantages of the mathematically sound Eigenspace stretch truncation proposed in Chap. 5 is the ability to stop calculation at the desired number of Eigendimensions without requiring resolving of all fundamental equations calculation of the full Eigenspace via calculation of the trace of the original text matrix.

To verify this claim, the following performance evaluation was conducted. With rising text matrix size, the gain is expected to increase significantly, whereas for small text matrices it will not make a big difference.

The performance tests were conducted on R (3.1.0, 64bit) for OsX (10.9.2) on a Maxbook Pro with 8GB main memory (1333 Mhz DDR3) and a 2.3 Ghz Intel Core i7 processor.

Calculations were performed using svdlibc (Rhode 2014) and the (unpublished[3]) svdlibc native interface for the lsa package.

As can be seen in Table 10.9, with rising corpus size the effective performance gain through the stretch truncation rises as well: while the ‘demo’ (used in Sect. 9.2) is too small to have any impact on performance, the gain in ‘business’ (Sect. 9.3) and ‘Bawe’ (Sect. 9.4) is significant. Moreover, a similarly sized corpus ‘medical’, with case study discussion forum contributions of medical students about ‘safe prescribing’ shows a similar gain.

[3] There currently is no svdlibc implementation available for Windows, which prevents releasing this interface across platforms.

Table 10.9 Performance measurement of the stretch truncation

Name	Raw t	Raw n	Sanitising	# Terms	# Docs	Full	MPIA	Gain (%)
Demo	56	14	0.35 s	17	14	0.51 s	0.50 s	0
Business	4154	481	16.10 s	1447	479	3.43 s	1.35 s	61
Medical	6091	505	3.70 s	2341	505	5.23 s	1.78 s	64
Bawe	43,804	742	432.96 s	4096	742	13.62 s	6.97 s	49

While the demo uses a stretch-truncation of 59 % (and only three dimensions), the other three were calculated with an Eigenspace stretch truncation of 80 %.

It should not go without mention that the *mpia* package provides routines for materialisation, allowing exchanging of spaces and the according visualisation data. Once the network coordinates are calculated, for example, they can be materialised and from then on will be served from the cache, whenever the domain is re-instantiated to memory. These persistence facilities thus help to prevent any recalculation of results, if not explicitly requested—in practice often saving additional time.

10.4 Summary and Limitations

The evaluation studies and experiments presented here—together—provide evidence of verification and validation of meaningful purposive interaction analysis as a method and software package.

The test-driven development verified stability and consistency of the implementation. Moreover, it ensured that the functional requirements were satisfied with progressing development.

The validation studies reported in (Sect. 10.3) looked into accuracy of an application of MPIA, namely essay scoring. Nine collections in the wider domain of business and economics were utilised to assess that—for these heterogeneous cases—evidence is provided that MPIA works. Moreover, the inspection capabilities in an 80 % stretch-truncated Eigenspace were tested rounding up Sect. 10.3.1 to show, how the analysis instruments provided by MPIA can be used to build more enhanced scoring methods.

Two experiments (see Sects. 10.3.2 and 10.3.3) were used to establish whether convergent and discriminant validity may be given. For four cases studied in the medical domain, this can be concluded both for structural accuracy as well as annotation accuracy.

The visual accuracy and the connected degree of loss was measured in Sect. 10.3.4, thereby finding that the link erosion and outlier reattachment has low influence, while the planar projection onto a two dimensional surface—as expected—has. Since the visualisation serves exploration and since actual measurements are conducted in the analysis workflow in the actual Eigenspace, this vagueness is at least tolerable, if not desired. Jurafsky and Martin (2004, p. 506), for example, postulate a certain degree of vagueness as a computational desideratum for representations (though they stay equally vague in to what degree).

The performance tests presented in Sect. 10.3.5 towards the end of Sect. 10.3 complete the validation studies, assessing performance gains for the real-life cases studied of 49–64 %, and indicating a further rise with growing matrix size.

Repeatability and reproducability are constructs often evaluated to assess reliability, when proposing a new evaluation methodology. Since an algorithm, however, produces always the same results, repeatability was not further investigated. Multi-angulation guided the validation studies, using different topics, different languages, and different applications to triangulate reproducibility. As mentioned several times throughout this work, examples strive to support the re-execution (same code, same data) of research and foster reproducibility (same code, different data). Moreover, the domain manager class's interoperability facilities allow sharing domains and the data they include.

The lessons learnt from the experiments and their set up include the following. Re-use of data collected is extremely difficult, often not possible—other than intended. For example, the card sort data collected is bound to the text matrix the space was calculated from. Moreover, the method for selection of terms used by projecting a text into the space and selecting its most highly activated terms is problematic: it tends to select terms that not necessarily have in high proximity in the underlying space semantics. While this as such is not a problem, as there are many terms in several tests and with many subjects, this may be the reason for the rather low interrater reliability of the human subjects: the setup of the experiment asked them to group concepts by similarity—and not many of the subjects dared to discard terms they did not find very close to any of the other. Moreover, the selection of terms is dependant on the text matrix used: for example, where spaces with stem-completed vocabulary are introduced, it drops not only those terms that have the same stem form, but further affects the compositional structure of the Eigendimensions—rendering it virtually impossible to recreate the selection.

There are notable limitations of the evaluation work conducted. The evaluation did not empirically investigate usability and utility, most particularly it did not investigate its practicability in application in and for instructional design.

Moreover, the usual limitations typical to any text-mining-related research apply: it would have been nicer to have more studies in more domains, with more participants, and with more languages.

References

Armitt, G., Stoyanov, S., Braidman, I., Regan, M., Smithies, A., Dornan, T., Rusman, E., van Bruggen, J., Berlanga, A., van Rosmalen, P., Hensgens, J., Wild, F., Hoisl, B., Trausan-Matu, S., Posea, V., Rebedea, T., Dessus, P., Mandin, S., Zampa, V., Simov, K., Osenova, P., Monachesi, P., Mossel, E., Markus, T., Mauerhofer, C.: Validation Report II, Deliverable d7.2, LTfLL Consortium (2009)

Armitt, G., Stoyanov, S., Hensgens, J., Smithies, A., Braidman, I., van Bruggen, J., Berlanga, A., Greller, W., Wild, F., Haley, D., Hoisl, B., Koblischke, R., Trausan-Matu, S., Posea, V., Rebedea, T., Mandin, S., Dessus, P., Simov, K., Osenova, P., Westerhout, E., Monachesi, P., Mauerhofer, C.: Validation Report III, Deliverable d7.3, LTfLL Consortium (2010)

Armitt, G., Stoyanov, S., Hensgens, J., Smithies, A., Braidman, I., Mauerhofer, C., Osenova, P., Simov, K., Berlanga, A., van Bruggen, J., Greller, W., Rebedea, T., Posea, V., Trausan-Matu, S., Dupre, D., Salem, H., Dessus, P., Loiseau, M., Westerhout, E., Monachesi, P., Koblischke, R., Hoisl, B., Haley, D., Wild, F.: Validation Report IV, Deliverable d7.4, LTfLL Consortium (2011)

Beck, K.: Test Driven Development: By Example. Addison-Wesley, Boston, MA (2003)

Deibel, K.: UW Card Sort Analyzer: Version 1.1. University of Washington, Seattle, WA (2010)

Deibel, K., Anderson, R., Anderson, R.: Using edit distance to analyze card sorts. Expert. Syst. **22**, 129–138 (2005)

Feinerer, I., Wild, F.: Automated coding of qualitative interviews with latent semantic analysis. In: Mayr, Karagiannis (eds.) Proceedings of the 6th International Conference on Information Systems Technology and its Applications (ISTA'07). Lecture Notes in Informatics, vol. 107, pp. 66–77, Gesellschaft fuer Informatik e.V., Bonn, Germany (2007)

Jurafsky, D., Martin, J.H.: Speech and Language Processing. Pearson Education, Delhi (2004). Third Indian Reprint

Law, E., Wild, F.: A multidimensional evaluation framework for personal learning environments. In: Kroop, S., Mikroyannidis, A., Wolpers, M. (eds.) Responsive Open Learning Environments. Berlin, Springer (2015)

Oberkampf, W., Roy, C.: Verification and Validation in Scientific Computing. Cambridge University Press, New York (2010)

R Core Team: R: A Language and Environment for Statistical Computing, R Foundation for Statistical Computing, Vienna, Austria. ISBN 3-900051-07-0. http://www.R-project.org/ (2014)

Randolph, J.J.: Free-marginal multirater kappa: an alternative to Fleiss' fixed-marginal multirater kappa. In: Joensuu University Learning and Instruction Symposium. Joensuu, Finland (2005a)

Randolph, J.J.: Online Kappa Calculator. http://justusrandolph.net/kappa/ (2005b). Accessed 6 March 2016

Research Methods Knowledge Base: Convergent & Discriminant Validity. http://www.socialresearchmethods.net/kb/convdisc.php (2014)

Rhode, D.: SVDLIBC: A C Library for Computing Singular Value Decompositions, version 1.4. http://tedlab.mit.edu/~dr/SVDLIBC/ (2014). Accessed 6 March 2016

Rousseeuw, P.J.: Silhouettes: a graphical aid to the interpretation and validation of cluster analysis. J. Comput. Appl. Math. **20**(1), 53–65 (1987)

Schlesinger, S.: Terminology for model credibility. Simulation **32**(3), 103–104 (1979)

Sommerville, I.: Software Engineering, 9th edn. Addison-Wesley, Boston, MA (2010)

Spencer, R.: The streamlined cognitive walkthrough method: working around social constraints encountered in a software development company. In: Proceedings of the CHI 2000, pp. 353–359. ACM Press (2000)

Wild, F.: An LSA package for R. In: Mini-Proceedings of the 1st European Workshop on Latent Semantic Analysis in Technology-Enhanced Learning. Open University of the Netherlands: Herleen (2007)

Wild, F.: lsa: Latent Semantic Analysis: R Package Version 0.73. http://CRAN.R-project.org/package=lsa (2014)

Wild, F., Stahl, C.: Investigating unstructured texts with latent semantic analysis. In: Lenz, H.J., Decker, R. (eds.) Advances in Data Analysis, pp. 383–390. Springer, Berlin (2007a)

Wild, F., Stahl, C., Stermsek, G., Neumann, G.: Parameters driving effectiveness of automated essay scoring with LSA. In: Proceedings of the 9th International Computer Assisted Assessment Conference (CAA), Loughborough, pp. 485–494 (2005a)

Wild, F., Stahl, C., Stermsek, G., Penya, Y., Neumann, G.: Factors influencing effectiveness in automated essay scoring with LSA. In: Proceedings of the 12th International Conference on Artificial Intelligence in Education (AIED), Amsterdam, The Netherlands (2005b)

Wild, F., Koblischke, R., Neumann, G.: A research prototype for an automated essay scoring application in LRN. In: OpenACS and LRN Spring Conference, Vienna (2007)

Wild, F., Stahl, C.: Using latent semantic analysis to assess social competence. In: Mini-Proceedings of the 1st European Workshop on Latent Semantic Analysis in Technology-Enhanced Learning. Open University of the Netherlands: Herleen (2007b)

Wild, F., Hoisl, B., Burek, G.: Positioning for conceptual development using latent semantic analysis. In: Proceedings of the EACL 2009 Workshop on GEMS: GEometical Models of Natural Language Semantics, pp. 41–48. Association for Computational Linguistics, Athens, Greece (2008)

Wild, F., Haley, D., Buelow, K.: Monitoring conceptual development with text mining technologies: CONSPECT. In: Proceedings of eChallenges 2010, Warsaw (2010a)

Wild, F., Haley, D., Buelow, K.: CONSPECT: monitoring conceptual development. In: Proceedings of the ICWL'2010, pp. 299–308. Shanghai (2010b)

Wild, F., Valentine, C., Scott, P.: Shifting interests: changes in the lexical semantics of ED-MEDIA. Int. J. E-Learning. **9**(4), 549–562 (2010c)

Wild, F., Haley, D., Buelow, K.: Using latent-semantic analysis and network analysis for monitoring conceptual development. JLCL **26**(1), 9–21 (2011)

Chapter 11
Conclusion and Outlook

This book addresses three current Grand Challenges for research and development in technology enhanced learning, challenges on which wide consensus within the research community was formed with the help of the EC-funded Network of Excellence STELLAR (Gillet et al. 2009; Fischer et al. 2014). Together, the selected top ten challenges as well as their 22 complements on the extended short list promise to drive innovation in the three main areas 'awareness', 'engagement', and 'collaboration'. These key areas of innovation are not independent, but slightly overlapping, as depicted in Fig. 11.1, and achievements on one may very well affect the others.

Engagement is about "the study of cognitive affect and motivation, helping build and sustain passion in learning" (Wild et al. 2013a, b, c), while in massive collaboration "the new capabilities [..] gained through advances in infrastructure meet open practices spanning across big crowds of learners and teachers in both formal and informal learning" (ibid).

The three interconnected Grand Challenge this book helps to address (see also Sect. 1.2) are rooted in the awareness strand aimed at "fostering awareness with the help of learning analytics relates to the study of digital traces left behind when (co-) constructing knowledge and developing competence using predictive models that allow for advising on performance improvement of individuals" (Wild et al. 2013a, p. 25).

The three challenges addressed are about 'new forms of assessment for social TEL environments' (Whitelock 2014a), 'assessment and automated feedback' (Whitelock 2014b), and about 'making use and sense of data for improving teaching and learning' (Plesch et al. 2012). Their achievement creates more awareness. They are, however, not without impact on engagement and collaboration (for a further discussion, see Sect. 11.2).

These three interlocked challenges have been summarized in Sect. 1.2 into one overarching problem statement for this work, namely "*to automatically represent conceptual development evident from interaction of learners with more knowledgeable others and resourceful content artefacts; to provide the instruments required*

F. Wild, *Learning Analytics in R with SNA, LSA, and MPIA*,
DOI 10.1007/978-3-319-28791-1_11

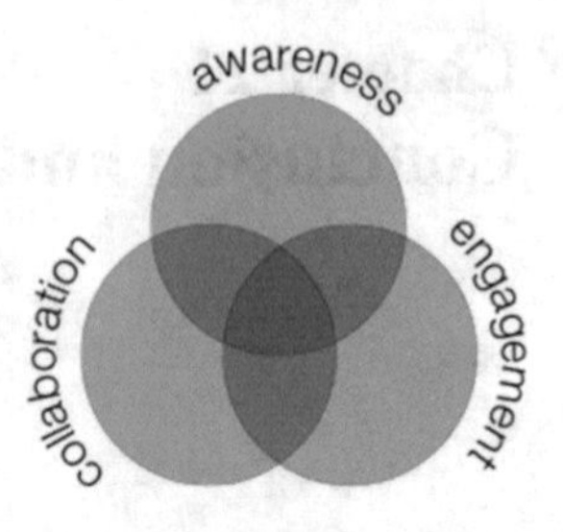

Fig. 11.1 Areas of innovation in TEL (Wild et al. 2013a, p. 25)

for further analysis; and to re-represent this back to the users in order to provide guidance and support decision-making about and during learning."

In this final chapter, the achievements against this problem statement will be summarised and the contributions along the derived three objectives will be critically examined.

The remainder of this chapter is organised as follows. First, the three derived research objectives are revisited and the contributions brought forward by this book are related to them (Sect. 11.1). Then, connections to other research areas are pointed out (Sect. 11.2) and open points for future research defined (Sect. 11.3). Concluding remarks wrap up the book (Sect. 11.4).

11.1 Achievement of Research Objectives

The problem statement guiding this work consists of three parts, each of them setting an objective on its own, as defined in the introduction to this book in Sect. 1.7.

The first objective is about the theory and model "*to represent learning*" that takes place in interaction with others and with content artefacts. The second is for the software implementation "*to provide the instruments required for further analysis*". The final one then turns to the application area in learning analytics "*to re-represent learning*" to provide guidance and support decision-making in and on learning.

The progress achieved against each of these three key objectives will be summarized in the next sections, each of them dedicated to one of the three objectives, concluding with a final synthesis of the contributions.

As will be seen, each of the objectives has been met and thus the overall problem statement guiding this research can be reassessed as 'problem overcome'. This, however, does not preclude further research in this area. On the contrary, as the theory chapter already emphasizes, *mpia* is only one possible model that fits the theory and there may be other algorithms that offer more (or less) effectiveness and efficiency.

As the research front in learning analytics moves forward, it would be very interesting to review and compare with additional other algorithms, especially with more remote ones.

Table 11.1 The three objectives

	Epistemology	Algorithms	Application area
Objective	• O1: represent	• O2: analysis instruments	• O3: re-represent
Main aim	• Define key concepts and their relations	• Derive and develop computational model for representation and analysis	• Adapt, deploy, and subject to testing in application area

The three objectives (see Sect. 11.1.1) and their main aim are listed in Table 11.1. Objective one about representing learning means to define the key concepts and their relations in coherent theory, fit for deriving an actual implementation of an algorithm from it.

Objective two (see Sect. 11.1.2) is about providing the analysis instruments required, focusing on the computational model to be derived and developed for representation and interactive analysis.

The final objective three (see Sect. 11.1.3) then looks at re-representing learning back to the user to adapt, deploy and test the *mpia* model and implementation in the application area of learning analytics.

11.1.1 Objective 1: Represent Learning

With respect to epistemology, the study of knowledge, this work contributes a theory of *learning from a methodo-culturalist perspective*.

In this tradition, information is a logical abstractor, i.e. an equivalence class of sentences for which equality in meaning as well as purpose can be attested invariant of speaker, listener, and formulation.

Consequently, competence is defined as the potential that capacitates a person to perform an information act, when challenged. In learning, people strive to enlarge this potential. Competence development is the information purpose of learning.

The potential as such, however, is not accessible for observation. It cannot be measured. Only when put to action, this potential becomes visible in performance.

Evidence of competence reflected in performance (such as, e.g., a corpus of student essays) can be collected and—when the shared purpose of developing a particular competence is known and the set representative—such collection can be used to investigate the textual cues that separate success from failure.

The set being 'representative' thereby is the fact introduced into theory that is wrong to a certain degree: any abstractor holds *all* possible sentences, in which the piece of information of concern can be expressed, whereas a 'representative set' only holds enough elements to cover all (or 'most') of its variations. At the same time, this reduction allows the introduction of an actual computational model that sacrifices completeness for the ability to predict based on heuristics. While controlling purpose, it trades accuracy in the 'long tail' in representation of meaning for automation and improved oversight.

These theoretical foundations are logically derived through analysis, clearly stating the axioms it rests on such as the principle of methodical order. All key concepts information, purpose, meaning, competence, performance, learning, performance collection, disambiguation, proximity, expertise cluster, introspection, etc. are clearly defined and for the two core concepts information and competence a brief idea history is provided. Practical implications (such as model quality characteristics) as well as limitations of the theory are outlined.

Any theory is only as good as its application. Evaluation hence focuses on two main aspects: model verification is conducted using test-driven development and a range of unit tests (Sect. 10.2); model validation compares prediction results with the performance of human experts in carefully chosen experiments (Sect. 10.3).

11.1.2 Objective 2: Provide Instruments for Analysis

Three algorithmic models are investigated for their applicability to provide the sought instruments for analysis: SNA, LSA, and the novel MPIA. Whereas the first two in isolation fall short of meeting the requirements of theory to be able to analyse both meaning and purpose equivalence at the same time, the newly introduced MPIA does not.

Social network analysis is described extensively in Chap. 3, thereby utilising two examples to first introduce into the method to then exemplify its use on a real-life case. The deliberations clearly elaborated where the advantages and the main weaknesses of the method lie: though the structural measures and filtering operations proposed over time are extensive and powerful, it lacks facilities to represent and inspect meaning of the content of conversations, while tending to the social relations and other similar incidents.

Latent semantic analysis (Chap. 4), on the other hand, does not share this lack in representing and analysing meaning through its facilities to construct a lower-dimensional, latent semantic vector space from a corpus, but it falls short in providing the complex measurement and filtering instruments SNA is popular for, in particular for inspecting purposes and relational context. Moreover, LSA fall short in providing a sound justification for how many dimensions to keep in the Eigensystem truncation. The chapter additionally describes the package *lsa* implemented by the author. Two application cases first foster understanding to then illustrate with a comprehensive application example the power of the method.

In the Chaps. 5–8, a novel fusion algorithm is elaborated. Meaningful, purposive interaction analysis (MPIA) brings together the graph analysis and graph visualisation methods known from SNA with a further developed Eigensystem-based semantic representation method (Chap. 5) rooting in LSA. The method resolves the truncation conundrum and proposes a novel, stretch-based truncation method, which can be calculated ex ante using the trace of the originating text matrix, thus offering a computational shortcut. The differentiation of proximity and identity introduces an important conceptual distinction that motivates inter alia the spring-embedder-based planar projection used for visualisation, as presented in the subsequent chapter.

Moreover, this subsequent Chap. 6 adds a visual graph representation to the set of analysis instruments that offers projection stability while enabling step-by-step investigation of content and relational structure expressed in social semantic performance networks. In particular, this contributes novel means for analysis of locations, positions, and pathways. The visualisation itself is composed using a multistep procedure from link erosion, to planar projection, application of kernel smoothening, and tile colouring using hyposometric tints. The resulting cartographic representations can be utilised to inspect positional information of the textual evidence held in performance collections, for individuals, or of whole groups. The analytical instruments introduced in Chap. 5 are extended with visual instruments in this Chap. 6.

Chapter 7 elaborates further recommendations on corpus sanitisation and calibration for specific domains. The experiment conducted using a test collection and variations of calibration parameter settings sheds light on tuning and corpus size required to build effective spaces. It finds that the size is only a weak driving factor and also small spaces can already be very effective. Smaller corpora, however, tend to rely more on the representativity of evidence collected than bigger ones, allowing the latter to supplement actual evidence for training with incidences of general language. Similarly, recommendations for vocabulary filtering (and matrix pruning) could be obtained.

The theoretical foundations in matrix algebra, the visualisation method proposed, and the recommendations on tuning are all picked up in the implementation presented in Chap. 8, which reports on the software package *mpia* with its class system and interfaces. The package provides all functionality needed to create and analyse representations of social semantic performance networks, including visualisation analytics.

The model of meaningful, purposive interaction analysis implements the representation theory with a practical software package.

Table 11.2 provides an overview on the difference in analysis use cases of MPIA in comparison to its predecessors SNA and LSA. As can be seen from the table, while use cases of SNA (see also Sect. 3.1: Fig. 3.2) and LSA (see also Sect. 4.2: Fig. 4.5) focus mainly on lower level operations, the MPIA use cases support analysis of learning (Sect. 8.1: Fig. 8.1). The use cases necessary for LSA are largely encapsulated in the higher level routines provided by MPIA: for example, while it is possible to apply sanitising or weighting operations, the package chooses a useful default setting, following the recommendations of the experiments presented in Chap. 8.

Similarly, the spacify routine provided chooses a stretch-based, mathematically sound truncation of the Eigenspace (per default set to 80 %), which can be calculated in advance, before entering the computationally intense singular value decomposition.

The differences are most evident, when looking at these instruments for analysis provided. While SNA differentiates only into nodes and edges and while LSA

Table 11.2 Comparison of the use cases

SNA use cases	LSA use cases	MPIA use cases
Filter • Component, • Nodes, • Edges *Measure* • Component-level, • Node-level, • Graph-level	*Text matrix* • Filter (by index, by boundary) • Weighting (global vs. local) • Sanitise (stemming, stopword removal) *LSA space* • Factor reduce • Fold-in *Similarity measurement* • Pearson's r, Spearman's Rho, Cosine	*Domain management* • Retrieve, materialise, upgrade, get, spacify *HR management* • Add Person, remove, Person, groups, performances, position *Evidence management* • Perform (read, write) • Competences • Path *Inspection* • Near • Proximity • Identity • '+' • Overlap • Terms *Visualisation* • Plot • Toponymy • Path, Position, Location

knows only texts and terms, MPIA provides various interfaces for managing evidence, human resources, and domains. Moreover, additional interfaces are included for discovery of groups, underlying competences, and extended relational measures (from 'proximity' over 'identity' to higher level functions such as 'near'). Instruments for further inspection of raw term and document vectors are provided, allowing adding, investigating overlap, or inspecting highly activated terms.

The visualisation methods complement this. They contribute to the mix the cartographic plotting methods, automated toponymy determination, and interfaces for interacting with the visualisation stage through adding locations, positions, and pathways.

11.1.3 Objective 3: Re-represent to the User

Learning analytics is the chosen application area for the representation algorithms and theory brought forward in this book. To re-represent back to the user means to *adapt*, *deploy*, and *subject to testing* in this very application area of learning analytics.

Only in its application, the means for representation provided by the algorithms and the facilities for analysis become contested with respect to its utility for the users. Moreover, only with precise application cases, validation studies become possible—providing the required context for external validation against real-life data.

Manifold example cases of learning analytics are provided throughout the book in Chaps. 3, 4, and 9. They are complemented with the validation studies selected and presented in Chap. 10.

Turning to first part of this objective, the following can be said about *'adapting and deploying'* MPIA as well as the predecessor algorithms SNA and LSA to the application area of learning analytics.

There is one foundational application case that guides through the root algorithms SNA (Sect. 3.2) and LSA (Sect. 4.3), to then be picked up again in the introduction of MPIA (Sect. 9.2). The purpose of this shared example case is to work out clearly, where the advantages and limitations of each of the methods lie—and how MPIA can help to overcome the latter. At the hand of this foundational example, the key differences in application between the predecessor algorithms SNA and LSA to MPIA are made visible.

Though entirely made up, the example is situated in a usage scenario very typical for small and large enterprises: often human resource management is underequipped, company wide learning management solutions are missing, and procurement of trainings and education anyway drafts from a whole ecosystem of vendors and providers.

In such situation, often little is known about the actual competence development of employees, as there is no shared catalogue or centralised system for it. The closest thing to a training register is the financial ledger of the human resource development unit, merely listing people, price, and the name of the training measure conducted.

The social network analysis case uses this knowledge about purposes to derive adjacency information from the incidences of attendance of nine employees in twelve training measures to then identify a worthy replacement for the person on sick leave, who can fill in for the scheduled work with a client that requires particular competence. Moreover, it demonstrates filtering and aggregation facilities to figure out who could be the next best replacement working in a related area.

A second real-life case takes this into a world of bigger data. Among the tens of thousands of postings in the discussion boards of a full university over several years, a lot of valuable information is hidden. For example, learning analytics about the engagement of both tutors and learners can be created with SNA by looking into their purposive interaction along topical discussion boards of courses—uncovering in prestige scores how successful such engagement is.

Both SNA example cases show the benefits of using social network analysis: it helps leverage the power of purpose. Through the graph representation, a powerful instrument becomes available to inspect context, adjacency, affiliation, etc.—on node, component, and graph level. At the same time it also clearly exposes the symbol grounding problem of SNA: data not entered are not available and, thus, SNA remains largely blind to content. Even in cases where information about content is explicitly available (such as through tags), handling and maintenance become ever so complicated, as SNA lacks the interfaces to separate concerns as required for analysing learning.

The foundational example is revisited and complemented in Sect. 4.3, there focusing on the advantages and disadvantages of latent semantic analysis. The example shows how inspecting the short training memos participants wrote about them can help implicitly derive relational data about the courses.

The example shows, how LSA helps to map the memos describing 14 different training opportunities in a factor-reduced vector space that is capable of clustering better together key descriptor terms and memo documents in clusters reflecting the different subject areas. It thereby demonstrates the advantages LSA provides for generating relational data from natural language descriptions of the training opportunities, but it also at the same time demonstrates its shortcoming in handling of the relational data SNA is so good with: the purposes, i.e. the mapping of persons and their attendances, gets lost. Moreover, the graph manipulation and analysis facilities SNA is popular for are not utilised in LSA.

The second real-life example in Sect. 4.5 shows, how LSA can be utilised to create learning analytics for automated scoring of student essays, using a scoring method of comparing each student essay in the latent semantic space to three 'gold standard' model solutions and assigning the average correlation as score. Through comparing the machine assigned scores in a collection of 74 essays to human expert scores, a Spearman's rank correlation of Rho = 0.69 can be achieved, showing that results of machine scoring can reach a performance level similar to human agreement.

The foundational example is picked up again in Chap. 9, when turning to describing the application of *MPIA for learning analytics*. These application examples show, how MPIA extends the state of the art by combining the advantages of its predecessor methods.

With the help of this foundational example, the chapter shows, how the *mpia* package functionality is exploited to first instantiate a human resource manager under whose umbrella the required person objects are added. The course memos are then added utilising the purposive context available from the SNA incidence matrix, so that for each learner the according memo is logged as performance. The subsequently evolving analysis demonstrates many of the analytical instruments available in *mpia*: it shows the projection plot and subsequent visual analysis using paths and positions; it demonstrates inspection of term activation, overlaps, competence extraction, and diverse measurement operations using atomic identity and proximity to identify persons in close vicinity with near.

Two further learning analytics examples substantiate what *mpia* has to offer. The second example revisits the essay-scoring example already introduced before in the LSA chapter, but—here—extended by eight further essay collections from different subject areas of one university. In the analysis, the strength of MPIA to investigate positional information of learners to each other and to model solutions becomes evident—including showcasing its visual facilities with the cartographic plots. Moreover, introspection facilities to look into shared meaning across groups of same-scored essays are explained.

The final example of learning analytics with MPIA rounds up the demonstration of capabilities with a case study of investigating learning paths in the various genres of student writings collected in the 'British Academic Written English' corpus. It particularly demonstrates, how partially overlapping learner trajectories can be analysed to identify locational proximity of the learning journeys conducted.

The three application examples of learning analytics together provide evidence of the types of learning analytics made possible with MPIA and its implementation into a software package. They also demonstrate, how the interfaces meet demands of analysis, thus allowing rapid prototyping of novel learning analytics applications.

Together they show, how an integrated solution can foster analysis in a more holistic way, looking at the same time at purposive context and conveyed meaning.

With respect to the second part of the definition of this objective, i.e. '*subject to testing*' of the algorithms and implementation, the following can be said.

Verification through test-driven development ensured that the actual implementation presented in Chap. 8 is inline with the constructs of the theory developed in Chap. 2. The 22 examples included in the documentation of the package as well as the 34 unit tests written check whether functionality still works, whenever extending the package with additional routines. This was already valuable during development, but it will also be an asset in further development and maintenance of the software package *mpia*, as this will help to determine whether any changes packages *mpia* is depending on affect its functionality.

Several studies contested the validity of the approach (Sect. 10.3) by probing its accuracy in annotation, its ability to assure structural integrity, and its visual accuracy. Moreover, the computational performance of *mpia* was measured.

The first validation study focused on the *accuracy in application* of MPIA in learning analytics for essay scoring. The test of the capabilities of MPIA to produce near human expert results with automated scoring of nine collections of essays from the wider area of business and economics studies provide evidence that the method can be adapted and deployed with success. Additionally, this Sect. 10.3.1 also demonstrated the innovation possible regarding improved inspection facilities with *mpia*, using an overlap scoring method.

Two further experiments were presented in Sects. 10.3.2 and 10.3.3 to challenge *convergent and divergent validity* of the representation algorithm, thereby using student learners as subjects to gather empirical data to compare machine judgements with. The studies found evidence for both structural as well as annotation accuracy.

Visual accuracy was evaluated in Sect. 10.3.4, thereby assessing the degree of loss taken into account in the visualisations through link erosion, outlier removal, and planar projection. While link erosion and outlier reattachment was found to be of low influence, the planar projection onto a two dimensional surface—as expected—has bigger, but tolerable impact. Since the visualisation serves exploration and since actual measurements are conducted in the analysis workflow in the actual Eigenspace, this vagueness is at least tolerable, if not desired.

The *performance tests* presented in Sect. 10.3.5 towards the end of Sect. 10.3 complete the validation studies, attesting significant performance improvements of 49–64 %, with expectation to rise with growing matrix size.

11.1.4 Summary

As Plesch et al. (2012) postulate for the measurement of achieving the *grand challenge to TEL research and development* of 'making use and sense of data for improving teaching and learning', success can be measured via two milestones: "The first milestone in solving this GCP is to reach a better understanding of how real time and outcome data has to be collected and presented so that teachers can react more precisely towards their students' needs." Moreover, they add "the successful development and evaluation of technologies that support teachers in their efforts of monitoring their students learning progress and success is the second milestone that has to be reached."

As the analysis of achievements against objectives in Sects. 11.1.1–11.1.3 underline, both these milestones have been reached and a novel theoretical foundation, a derived computational model, and the package release of *mpia* contribute one possible and validated solution to the challenge.

Whitelock (2014a, p. 54) postulates for the challenge of 'new forms of assessment for social TEL environments' progress in "learning network analysis—assessing networks and driving the development of groups and networks that provide effective support for learners", analysis of "learning dialogue—assessing the quality of dialogue, and using this formative assessment to guide the development of learning dialogue", "learning behaviour analysis—assessing the activity of individuals and groups, and using this formative assessment to guide the development of skills and behaviours associated with learning", "learning content analysis—assessing the resources available to learners, and using this information to recommend appropriate materials, groups and experts", "summative analysis of networks, dialogue, behaviour and content that is valued by learners, teachers and society", as well as "development of recommendation engines that use these analytics to provide personalised recommendations that support learning and that are valued by learners, teachers and society".

Progress against each of them was delivered in this book, providing one validated possible technique and a sound theoretical fundament, while at the same time—of course—not excluding the development of alternatives.

Among the measures with which to evaluate success, Whitelock (2014a, p. 54) lists the "engagement with learning—supported by directed feedback", "quality of online learning dialogue" and "engagement with online learning networks" as key measures.

The examples presented and evaluation studies reported provide evidence of success along these key measures, see summary in Table 11.3.

Table 11.3 Summary of contributions against objectives

	Epistemology	Algorithms	Application area
Objective	• O1: represent	• O2: analysis instruments	• O3: re-represent
Main aim	*Define* • Define key concepts and their relations	• Derive and develop computational model for representation and analysis	• Adapt, deploy, and subject to testing in application area
Contribution	• Learning from a methodo-culturalist perspective: semantic appropriation theory	• Meaningful, purposive interaction analysis (MPIA)	• Learning Analytics with MPIA
Chapters	• Learning theory and requirements (Chap. 2)	• Algorithmic roots in SNA (Chap. 3) and LSA (Chap. 4); • MPIA (Chaps. 5 and 6) • Calibration for specific domains (Chap. 7) • Package implementation (Chap. 8)	• SNA and LSA application cases (Chaps. 3 and 4) • MPIA Learning Analytics Applications (Chap. 9) • Evaluation (Chap. 10)

Whitelock (2014b, p. 23) postulates for addressing the challenge of 'assessment and automated feedback' the main activity of "wide-scale development, evaluation and implementation of new formative assessment scenarios including the development and evaluation of technologies that make for example intensive use of text-/data mining or natural language processing approaches".

Clearly, a novel approach using text- and data mining techniques and natural language processing was proposed and a rich set of formative assessment scenarios was provided in the learning analytics application examples spread out over the book (see Sect. 11.1.3 for a summary).

Whitelock (2012, p. 1) lists the points of "releasing teachers as the sole assessor and source of feedback/feed forward" and "visualising student competence and changing competence" as key measures for evaluating success. Both have been achieved.

The three challenges towards which the problem statement, objectives, and—consequently—this work are tuned came out in the top ten in the ranking exercise conducted in STELLAR, assessing their likelihood to leave lasting educational, technological, social, and economic impact.

Moreover, all five *algorithmic quality characteristics* postulated in Sect. 2.15 are satisfied. Evidence of divergent and convergent validity was provided and so are the requested capabilities for introspection and visual inspection. Moreover, computational performance is significantly improved for all but small matrices compared to predecessor Eigenspace-based methods through the trace-based short cut proposed.

While a lot of research fails to "cross the 'chasm'" (Dede 2006; Moore 2002; Wild et al. 2013c), the software developed for this book is already successful and the *lsa* package is already used around the globe in the Americas, Europe, Africa, and Asia.

The *mpia* package is too young to judge, as the first release is taking place in parallel to the submission of this book. The problems it fixes should, however, ensure its uptake—and so should the commitment to re-executable and reproducible research followed throughout this book.

Several of the code examples provided in this book, however, were made available by the author via his public tutorial folder on cRunch, the infrastructure for computationally intense learning analytics hosted by the author,[1] thereby contributing to the success of the infrastructure attracting over 150 developers so far (27.2.2014) and therein contributing to lower barriers for uptake. Where such code examples were released, it is indicated within the chapter, where to find it.

11.2 Open Points for Future Research

As outlined in the introduction, this work has three roots: in epistemology, in algorithms, and in the application for learning analytics. Along the way of achieving the objectives, interesting new questions became possible and new routes of investigation popped up in each of these rooting areas.

With respect to *epistemology*, the connection to dispositions, especially epistemic beliefs and commitments would be very valuable to further explore, as they shape learning and other information seeking behaviour (Knight et al. 2014).

With respect to *algorithms*, several new questions arise. The research front on distributional semantic models (Sahlgren 2008) already started turning to *pathways in spaces*, to see if a full semantic representation theory can be derived, with the thought of using ordered pathways and composition of vector units (Turney 2012) for representation of sentence- or unit-level semantics not being far fetched. The main driving factor for this research seems to determine the size of the bag-of-words or ordered-words units, their investigation also promising deeper insights for sanitising, hopefully reducing the amount of pruning still required today.

Though the computational shortcut provided by the *stretch-truncation* based prediction on dimensionality reduction of the Eigensystem is significant, more research on the ideal amount of stretch is needed. Now that there is a clear explanation connecting the trace and Eigenvalues to variability and providing a shortcut to their prediction via calculation in the trace of the original text matrix, it remains to be seen, whether a common default value can be found that is stable across languages or language families (including more remote ones such as Chinese) and that is stable across different, more heterogeneous application cases (branching out to other areas of media informatics such as information retrieval). Investigating the relation between bag size, number of words loading on each Eigendimension, and the amount of stretch expressed in the Eigenvalue promises new insights.

[1] http://crunch.kmi.open.ac.uk/

There are other mathematical facilities that could be used to extend the Eigenspace factorisation models into a *multi-layered model* (e.g., the Kronecker product, see Langville and Stewart 2004; van Loan 2000).

Comparison with further competing algorithms is needed, for example, benchmarking effectiveness and studying efficiency with ontology-based inference algorithms. Similarly, how do parsing algorithms compare that take word order and sentence structure into account?

Since spaces are valuable resources, sharing and trading them could help lower barriers to uptake. The *mpia* package already provides facilities for persistence and materialisation. Data shop routines (such as the limited ones already available on cRunch) could provide a common meeting place for sharing (or selling) such space data, leveraging community network effects. Moreover, implementation of upload and remoting features into *mpia* could be very convenient to the users, especially, when desktop machines do not provide the memory or calculatory power required for the endeavour in mind.

With respect to the visualisation methods proposed, two notable extensions would be possible. First, the force-directed placement algorithm at the core of the projection method proposed in Chap. 6 is engineered in a way that it prefers vaguely circular layouts (through its built in preference for symmetry). Alternative methods could make better use of typically rectangular displays prevalent today. In such case it remains to be shown that the accuracy is not significantly reduced or—better—in fact improved. Similarly, web-app map interfaces such as the ones made available by Google can be used to provide panning and zooming mechanisms of own maps. A quick prototype developed by the author shows that such procedure is feasible.[2]

Other processing modes for the visualisation are possible, as already indicated in Chap. 6. For example, as an alternative to terminology focused analysis, it would be possible to emphasise evidence, i.e. using the left-singular Eigenvectors over the $A^T A$ pairwise document incidences. In such visualisation, it would be the document proximity in the stretch-truncated Eigenspace that would govern the layout of the planar projection. It is also possible to use a combination of both, terminology and evidence, to establish the projection surface, then emphasising relations of both performances and their term descriptors. Both these alternatives come with a limitation: they provide less stable and less readable projection surfaces, as the relations are no longer clear and the evidence may continuously change the picture.

With respect to the application roots in *learning analytics*, the following can be said. Several learning analytics applications were proposed in this book—all utilising the improved facilities for holistic inspection of meaning and purpose and the extended interfaces provided in the package implementation. This does, however, not exclude innovating further and different learning analytics applications. On the contrary, since the foundations have been laid and examples have been provided, innovating new learning analytics instruments now becomes easier—

[2] See online demo: http://people.kmi.open.ac.uk/fridolin/mia-map/

hopefully yielding in the predicted growth and impact the STELLAR challenges promise.

11.3 Connections to Other Areas in TEL Research

Though this work is deeply rooted in the awareness strand currently pursued in the research community around technology-enhanced learning, it is not unrelated to the other two main directions of research, see Fig. 11.2.

When extended with capabilities to inspect affect, the representation technologies presented and showcased here could help understand and capture *engagement* in a better way. Building up passion in learning, preventing attrition and increasing retention, and engaging the disengaged are key challenges for this area (Wild et al. 2013a).

With respect to the other area, *collaboration*, the awareness and monitoring mechanisms provided here could form a key instrument to study new approaches to instructional design, in fields such as learning design, activity design, and even curricular design, uncovering valuable insights to teaching and educational practice.

Turning to the first area, engagement, in more depth, the following endeavour can be outlined. As studies show, cognition and affect are to inseparable sides of the same coin. They are intertwined and successful and engaging learning and cognition seeks oscillation between particular affective states, while trying to avoid others.

D'Mello and Graeser (2012) contributed a first model for which they were able to validate the key transitions postulated. In their model (see Fig. 11.3), learners are seen to typically oscillate between the affective state of equilibrium, in which everything flows, and disequilibrium, a state of confusion, entered whenever an impasse (a problem, conflicting facts, etc.) is encountered. If this impasse can be directly resolved, learners fall back into flow. In many cases, this may require going through a state of delight and achievement, when pro-actively working towards resolving this impasse and achieving set goals.

Is the state of confusion, however, not resolved, blocked goals and failure lead to frustration and learners are stuck. If getting stuck happens too often, learners ultimately get bored and disengage.

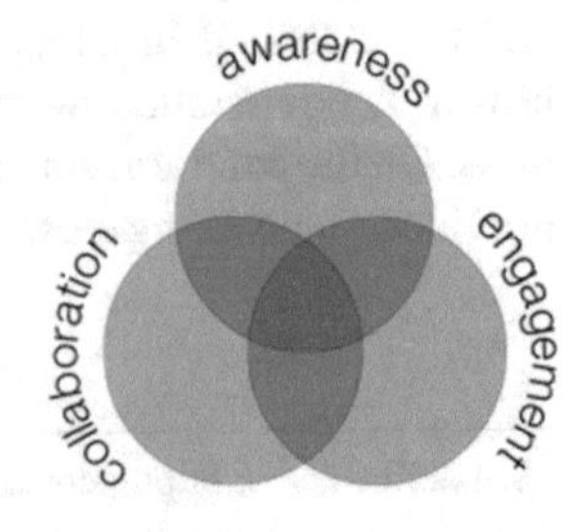

Fig. 11.2 Key research strands in TEL (Wild et al. 2013a, p. 25)

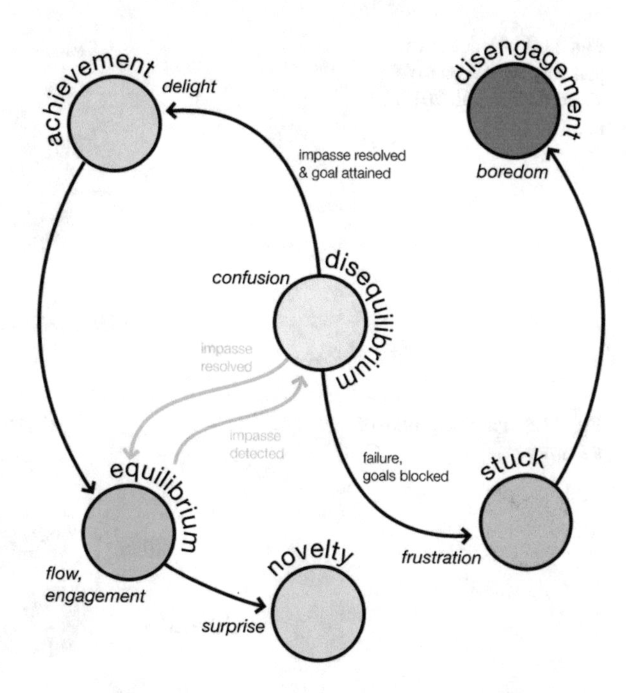

Fig. 11.3 Cognitive-affect model (redrawn from D'Mello and Graeser 2012; reproduced from Wild et al. 2013a, b, c)

This model can be deployed to create computational cognitive-affect models, branching off from MPIA to utilise for example sentiment detection algorithms to fit the representation of social semantic performance networks with information on (aggregate) affect states of learners. Figure 11.4 (reproduced from Wild et al. 2013a, p. 33) illustrates this concept: there are two distinct pathways leading from one cognitive-affective state on the left to the other on the right. Both pathways start off in a situation of inhibition (also indicated by the yellow node colour) to get back to flow and equilibrium on the right (indicated by the green node fill colour). The difference between the two is that the lower pathway does it 'the hard way', taking the learner through a state of frustration (orange) and then back to inhibition (yellow), while the top pathway moves—more engagingly—through achievement (dark green) back to flow.

Taking this idea a bit further, it becomes possible to create a passion profile for each learner, showing which areas the learner currently is passionate about (or disengaged in: from left to right, from red to green, as indicated in Fig. 11.5). Moreover, which exact topics or assignments this relates to can be—maybe interactively—be derived through mapping of topics to nodes (and the according labels on the axis on the bottom of the display). The third dimension of such a passion profile could account for the difference between developing domain-specific competences—and the other areas such as creativity and innovation as well as social competence, as indicated by the axis label to the left of Fig. 11.5. This way, new and expressive interfaces to understand and promote engagement in learning could be developed.

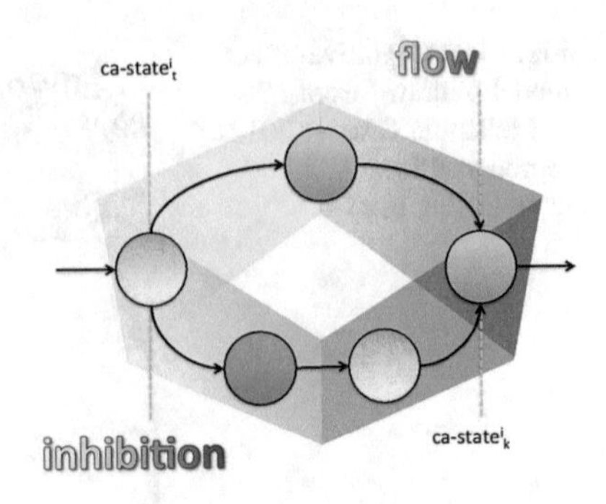

Fig. 11.4 Affect-laden pathways in a cognitive space (Wild et al. 2013b, p. 33)

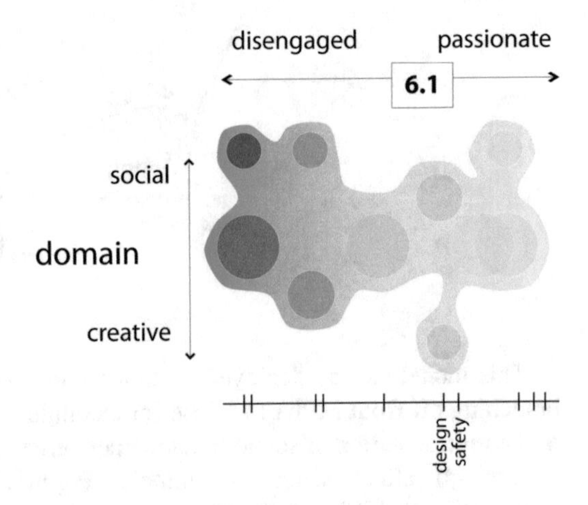

Fig. 11.5 Passion profile of a learner

As for the other research strand, *collaboration*, MPIA can form a key instrument to study optimal instructional design, as it provides the analysis instruments required to evaluate impact of activity design and curricular planning on learning. It could very well be used to inform orchestration and especially for self-regulated learning is has much to offer for allowing building of instant feedback mechanisms as demonstrated several of the application examples.

With the advent of a third generation of augmented reality technology and the Internet of Things (cf. Wild et al. 2014a, b; Megliola et al. 2014), the boundaries between interaction in the real world and digital interaction become even more blurred than it already are. Through fusion tracking of objects registered to a single workplace reference space, for example, interaction with physical entities such as persons, places, and tools becomes accessible.

Besides being a promising (and challenging) new user interface model, this offers also novel facilities for study, suddenly allowing for bridging the formerly often separated observation and analysis of digital and real world traces.

Exciting new APIs for tracking, such as the experience API (ADL 2013; Megliola et al. 2014), already use sentence structure for tracking interaction with the real world in subject-predicate-object triples. Adapting MPIA to such data might very well turn out to provide the analysis instruments needed to deal with the big data expected from automated tracking with sensors and computer vision.

11.4 Concluding Remarks

Within this final chapter, a final conclusion was elaborated, turning back to the grand challenges selected that this work addresses and assessing the achievement of the problem statement and derived objectives. Section 11.1 provides a summary of the contributions along each of the three objectives to represent, provide instruments for further analysis, and re-represent learning back to the user. Open and new problems identified are listed in Sect. 11.2 for each of the three areas this work contributes. Finally, connections to other relevant areas in technology-enhanced learning are made in Sect. 11.3, introducing several new research challenges that would bring the three strands of research and study of awareness, engagement, and collaboration closer together.

References

ADL: Experience API, version 1.0.1. Advanced Distributed Learning (ADL) Initiative, U.S. Department of Defense (2013)

D'Mello, S., Graeser, A.: Dynamics of affective states during complex learning. Learn. Instr. **22**, 145–157 (2012)

Dede, C.: Scaling up: evolving innovations beyond ideal settings to challenging contexts of practice, Chapter 33. In: Sawyer (ed.) Cambridge Handbook of the Learning Sciences, pp. 551–566. Cambridge University Press, Cambridge (2006)

Fischer, F., Wild, F., Sutherland, R., Zirn, L.: Grand Challenges in Technology Enhanced Learning. Springer, New York (2014)

Gillet, D., Scott, P., Sutherland, R.: STELLAR European research network of excellence in technology enhanced learning. In: International Conference on Engineering Education & Research, August 23–28, 2009, Seoul, Korea (2009)

Knight, S., Buckingham Shum, S., Littleton, K.: Epistemology, assessment, pedagogy: where learning meets analytics in the middle space. J. Learn. Anal. **1**(2), 23–47 (2014)

Langville, A., Stewart, W.J.: The Kronecker product and stochastic automata networks. J. Comput. Appl. Math. **167**, 429–447 (2004)

Megliola, M., De Vito, G., Sanguini, R., Wild, F., Lefrere, P.: Creating awareness of kinaesthetic learning using the experience API: Current practices, emerging challenges, possible solutions. In: Proceedings of the ARTEL'14 Workshop, CEUR Workshop Proceedings, to Appear (2014)

Moore, G.: Crossing the Chasm: Marketing and Selling Disruptive Products to Mainstream Customers. Revised edition. HarperBusiness, first published 1991 (2002)

Plesch, C., Wiedmann, M., Spada, H.: Making Use and Sense of Data for Improving Teaching and Learning (DELPHI GCP 10). http://www.teleurope.eu/pg/challenges/view/147663 (2012). Accessed 23 Jul 2012

Sahlgren, M.: The distributional hypothesis. Rivista di Linguistica **20**(1), 33–53 (2008)

Turney, P.: Domain and function: a dual-space model of semantic relations and compositions. J. Artif. Intell. Res. **44**, 533–585 (2012)

Van Loan, C.F.: The ubiquitous Kronecker product. J. Comput. Appl. Math. **123**, 85–100 (2000)

Whitelock, D.: New forms of assessment of learning for social TEL environments (ARV GCP 18). In: Fischer, F., Wild, F., Zirn, L., Sutherland, R. (eds.) Grand Challenge Problems in Technology Enhanced Learning: Conclusions from the STELLAR Alpine Rendez-Vous. Springer, New York (2014a)

Whitelock, D.: Assessment and automated feedback (ARV GCP 7). In: Fischer, F., Wild, F., Zirn, L., Sutherland, R. (eds.) Grand Challenge Problems in Technology Enhanced Learning: Conclusions from the STELLAR Alpine Rendez-Vous. Springer, New York (2014b)

Whitelock, D.: Assessment and Automated Feedback (ARV GCP 7). http://www.teleurope.eu/pg/challenges/view/147638 (2012). Accessed 07 Aug 2014

Wild, F., Lefrere, P., Scott, P., Naeve, A.: The grand challenges, Chapter 1. In: Wild, F., Lefrere, P., Scott P.J. (eds.) Advances in Technology Enhanced Learning: iPad Edition. The Open University. ISBN 978-1-78007-903-5 (2013a)

Wild, F., Scott, P., Da Bormida, G., Lefere, P.: Learning by Experience, Deliverable d1.2. TELLME consortium (2013b)

Wild, F., De Liddo, A., Lefrere, P., Scott, P.: Introduction. In: Wild, F., Lefrere, P., Scott P.J. (eds.) Advances in Technology Enhanced Learning: iPad Edition. The Open University. ISBN 978-1-78007-903-5 (2013c)

Wild, F., Scott, P., Karjalainen, J., Helin, K., Lind-Kohvakka, S., Naeve, A.: An augmented reality job performance aid for kinaesthetic learning in manufacturing work places. In: Rensing, C., et al. (eds.) EC-TEL 2014. LNCS 8719, pp. 470–475. Springer, Berlin (2014a)

Wild, F., Scott, P., Karjalainen, J., Helin, K., Naeve, A., Isaksson, E.: Towards data exchange formats for learning experiences in manufacturing workplaces. In: Proceedings of the ARTEL'14 Workshop, CEUR Workshop Proceedings, http://ceur-ws.org/Vol-1238/ (2014b)

Erratum to: Learning Analytics in R with SNA, LSA, and MPIA

Fridolin Wild

Erratum to:

F. Wild, *Learning Analytics in R with SNA, LSA, and MPIA*,
DOI 10.1007/978-3-319-28791-1

In the copyright page (p. iv) of the Front Matter, the sentence

"Cover photo by Sonia Bernac (http://www.bernac.org)" has been included.

The updated original online version for this book can be found at
DOI 10.1007/978-3-319-28791-1

F. Wild, *Learning Analytics in R with SNA, LSA, and MPIA*,
DOI 10.1007/978-3-319-28791-1_12

Annex A: Classes and Methods of the *mpia* Package

A.1 Class 'DomainManager'

See Fig. 1.

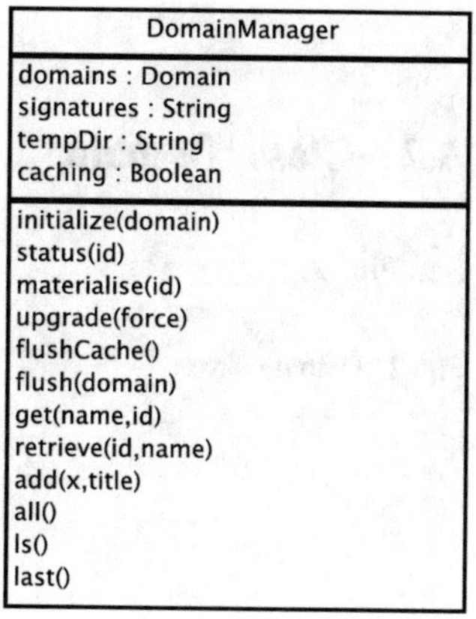

Fig. 1 Class DomainManager

DomainManager

domains : Domain
signatures : String
tempDir : String
caching : Boolean

initialize(domain)
status(id)
materialise(id)
upgrade(force)
flushCache()
flush(domain)
get(name,id)
retrieve(id,name)
add(x,title)
all()
ls()
last()

Fields

domains	Reference pointers to the domain objects currently held *in memory*.
signatures	The unique identifiers of the domain objects held in memory. Each signature is the MD5 hash value of the serialised space field of the domain.
tempdir	Path to the local cache directory. Per default, this is the cache subfolder of the directory the mpia package has been installed to.
caching	Logical flag indicating whether caching is enabled (true) or not (false).

F. Wild, *Learning Analytics in R with SNA, LSA, and MPIA*,
DOI 10.1007/978-3-319-28791-1

Methods

initialize (domain)	Constructor, optionally calls *get(domain)* to retrieve domain object from cache to memory. Returns reference pointing to the *DomainManager* object.
add (x, title)	Create a new domain object and add it to the list of domains available in memory.
retrieve (id, name)	Load a materialised domain object from the local cache directory.
get (name, id)	Return the reference pointer to the domain object requested, load it from the cache directory if needed.
materialise (id)	Save the domain object to local cache directory.
status(id)	Check for availability of domain with signature *id*.
flush(domain)	Remove all (or specified) domain objects from memory.
flushcache()	Delete all domain object cache files from the local cache directory.
upgrade(force)	Update all domain objects currently in memory to the latest class code version.
last()	Return the reference pointer to the last domain added.
ls()	Return the names of all domains currently in memory.
all()	Return reference pointers to all domains currently in memory.
print()	Pretty printing: listing the number of domains in memory.
show()	Display the object by printing its key characteristics.

A.2 Class 'Domain'

See Fig. 2.

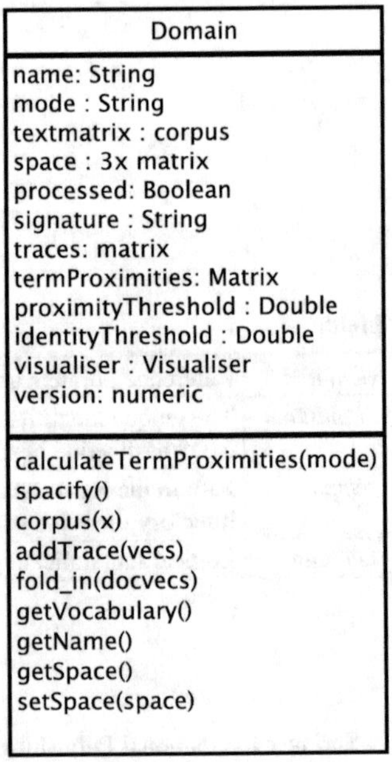

Fig. 2 Domain class

Fields

name	The name of the domain (character string).
mode	Mode of analysis: currently only 'terminology' is supported, though other views are theoretically plausible (such as focusing on the 'incidences' provided as documents or 'both').
textmatrix	The *TermDocumentMatrix* holding the raw data from which an MPIA space is constructed with spacify.
space	The Eigensystem: holds the three truncated matrices resulting from the singular value decomposition.
processed	Logical flag, denotes whether the *space* was already calculated from *textmatrix*.
signature	A unique identifier of the domain (MD5 hash value of the serialised space); automatically assigned by the *spacify* method.
traces	Internally used to store temporary fold-in data (of positions).
termproximities	The symmetric matrix of cosine proximities for all term pairs.
proximitythreshold	The threshold for associative closeness to be considered *near*.
identitythreshold	The threshold for associative closeness to be considered same.
visualiser	Reference pointer to its *Visualiser* object.
version	The version number of the *Domain* class.

Implemented generics

plot	Visualise the projection surface of the domain as plain or perspective plot.
toponymy	Analyse the places in the visualisation and label landmarks accordingly
summary	Print basic descriptive statistics about the data held.

Methods

initialize (name, ...)	Constructor; name should preferably be a unique identifier.
calculateTermProximities (mode)	Determine the associative closeness of all term pairs in a given domain, defined as their cosine proximity in the Eigenspace.
spacify()	Determine optimal number of dimensions for the conceptual space and convert the source vectors to a *space* in its Eigenbasis.
corpus(x)	Create a document-term matrix from corpus *x* (either a list of files or directory, a *Source* object, or a *TermDocumentMatrix*). Store it internally in field *textmatrix*.
addTrace(vecs)	Interface for adding query document-term matrix vectors using fold-ins: project new texts into an existing Eigenspace.
fold_in(docvecs)	Internally used fold-in routine: returns a context vector appendable to the right singular Eigenvectors (*not* a document-term matrix vector such as provided by *lsa::fold_in*).
getVocabulary()	Returns the list of terms used in the conceptual vector space.
getName()	Returns the (manually assigned) label of the domain.
getSpace()	Returns the space object.
setSpace(x)	Set the space object.
print()	Pretty printing of the domain object.
show()	Display the object by printing its key characteristics.
copy(shallow)	Internal routines required for upgrading objects to newer versions of the *Domain* class.

A.3 Class 'Visualiser'

See Fig. 3.

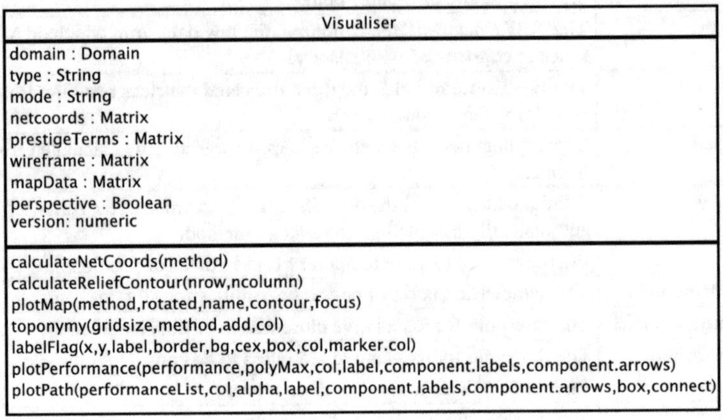

Visualiser
domain : Domain type : String mode : String netcoords : Matrix prestigeTerms : Matrix wireframe : Matrix mapData : Matrix perspective : Boolean version: numeric
calculateNetCoords(method) calculateReliefContour(nrow,ncolumn) plotMap(method,rotated,name,contour,focus) toponymy(gridsize,method,add,col) labelFlag(x,y,label,border,bg,cex,box,col,marker.col) plotPerformance(performance,polyMax,col,label,component.labels,component.arrows) plotPath(performanceList,col,alpha,label,component.labels,component.arrows,box,connect)

Fig. 3 The Visualiser class

Fields

Domain	Back reference from the *Visualiser* to its *Domain*.
Type	The type of the current map plot: one of 'topographic', 'persp', 'wireframe', 'contour'.
Mode	The focus of the current map plot: currently only 'terminology' is supported ('incidences' or 'both' would be alternatives).
Netcoords	The exact coordinates of the planar projection (before grid-tiling aggregation for surface elevation).
prestigeTerms	The prestige scores for the terminology: symmetric matrix with scores for all term pairs.
Wireframe	The relief contour: a wireframe of elevation levels (resulting from the grid-tiling aggregation of the *netcoords*).
mapData	Viewing transformation matrix required for projecting additional visual data into the display area (such as returned by e.g. persp).
Perspective	Logical flag, indicating whether the current plot is using perspective (or whether its plain), default is *true*.
Version	Current version number of the *Visualiser* class (used for updating cached objects).

Implemented generics

summary	Print basic descriptive statistics about the visualisation data held.

Methods

initialize(domain, ...)	The constructor, no actual calculation is done in this step
calculateNetCoords (method)	Method used internally to calculate a position through planar projection.
calculateReliefContour (nrow, ncol)	Method used internally to calculate a smoothened wireframe with the elevation levels reflecting the grid square density of the network coordinates (see *calculateNetCoords*).
plotMap(method, rotated, name, contour, focus)	Display a perspective or plain plot with the knowledge cartographic map.
toponymy(gridsize, method, add, col)	Find interesting places on the map and label them.
topo.colors.pastel(n)	Helper method to create a colour palette of hyposometric tints.
labelFlag(x, y, label, border, bg, cex, box, col, marker.col)	Helper method to plot a label flag onto a particular location on the map.
plotPath(performanceList, col, alpha, label, component.labels, component.arrows, box, connect)	Plot a sequence of markers on the map, each indicating the exact location of a performance. If *connect* is set to *true*, an x-spline will be used to connect the locations.
plotPerformance(performance, polyMax, col, label, component.labels, component.arrows)	Plot a marker onto the exact location of a performance (and add markers for the top loading terms activated by it). If *component.arrows* is *true*, arrows will be plotted starting from the location of the performance and pointing towards the position of the constituting key terms. *polyMax* sets the number of component terms to be highlighted, per default 3.
summary()	Display summary statistics about the visualisation data held (use generic *summary* instead).
print()	Display short info about the object.
show()	Display the object by printing its key characteristics.
copy(shallow)	This method is required for updating the *Visualiser* object to newer versions of the class without loosing its data.
newDevice (name, pdf, filename)	Standardised interface to open a new plot on the device of choice regardless of operating system.
closeDevice()	Standardised interface to close the plot on the device of choice regardless of operating system.

A.4 Class 'HumanResourceManager'

See Fig. 4.

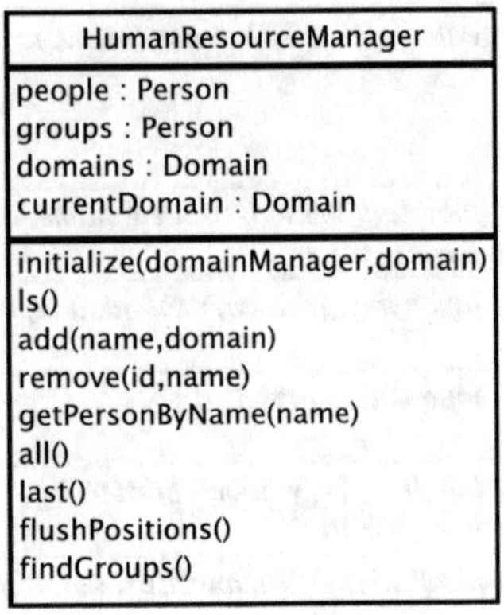

Fig. 4 *Human Resource Manager* class

HumanResourceManager

people : Person
groups : Person
domains : Domain
currentDomain : Domain

initialize(domainManager,domain)
ls()
add(name,domain)
remove(id,name)
getPersonByName(name)
all()
last()
flushPositions()
findGroups()

Fields

people	A vector of references pointing to the individuals' *Person* objects.
groups	A vector containing reference pointing to a group of persons (itself being an object of class *Person*).
domains	Reference pointing to the *DomainManager* object.
currentdomain	Reference pointing to the current *Domain* object.

Implemented generics

competences	Calculate competence positions amongst all performances of all persons.
groups	Identify groups of persons that have a position closer to each other than the minimal identity threshold.
names	List the names of all persons cared for.
near	Identify all persons that are competent in the area of the given performance or that have a similar competence profile as the given person.
performances	Return the list of all performances of all persons.
proximity	Returns the pairwise proximity of all persons to each other (in matrix form).

Methods

initialize (domainmanager, domain, environment)	Constructor: requires a *DomainManager* to be handed over.
findGroups()	Internally used to identify groups with similar competence positions.
flushPositions (domain)	Calculating positions creates temporary traces in the domain; if memory is a scarce resource, flushing these traces every once in a while may be helpful.
add (name, domain)	Create a new *Person* object and add it to the list of people.
remove (id, name)	Remove a person.
last()	Return reference pointing to the last person added.
all()	Return a vector of references pointing to all person objects cared for.
getPersonByName (name)	Return reference pointing to the person object of the person with the name 'name'.
ls(environment)	Return a character vector containing the names of all persons.
print()	Pretty print basic data about the personnel cared for.
show()	Display the object by printing its basic stats.

A.5 Class 'Person'

See Fig. 5.

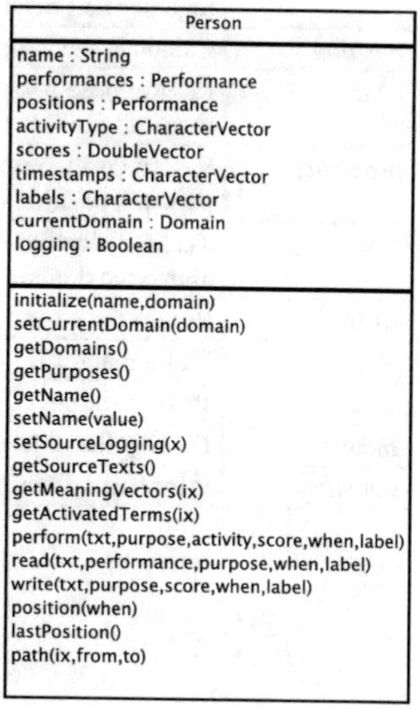

Person

name : String
performances : Performance
positions : Performance
activityType : CharacterVector
scores : DoubleVector
timestamps : CharacterVector
labels : CharacterVector
currentDomain : Domain
logging : Boolean

initialize(name,domain)
setCurrentDomain(domain)
getDomains()
getPurposes()
getName()
setName(value)
setSourceLogging(x)
getSourceTexts()
getMeaningVectors(ix)
getActivatedTerms(ix)
perform(txt,purpose,activity,score,when,label)
read(txt,performance,purpose,when,label)
write(txt,purpose,score,when,label)
position(when)
lastPosition()
path(ix,from,to)

Fig. 5 The *Person* class

Fields

name	The real name of the person (character string).
performances	Vector with references pointing to the person's performance objects.
positions	Vector temporarily holding references pointing to positions calculated from selections or all of the performances. May be cleared any time.
activityType	Character string describing the type of activity (e.g. 'read' or 'write').
scores	Double holding a grade of the performance (if available) or \code{NULL} if not.
timestamps	Vector holding timestamps for each performance about when it was enacted (character string).
labels	Vector holding character strings with labels of each performance (e.g. 'text written for exam').
currentDomain	Reference pointing to a domain object.
logging	Logical flag indicating on whether the person's performance objects shall internally store the raw source text.

Implemented generics

names	Return character string with the realname of the person.
names<-	Set real name of the person (to given character string).
[	Accessor to the performances (e.g. *fridolin[1]* returns the first performance).
performances	Return vector with references pointing to the person's performances.
path	Return vector with references pointing to the person's performances, sorted chronologically or by given index positions.
terms	Return list with one entry for each performance containg the terms (as a vector with the character strings of the terms).
competences	Calculate competences held by the person (clustering those performance together that belong together).
position	Calculate the competence position held by the person.
near	Return those persons cared for by a given *HumanResoureManager* that are in proximity to this person.
proximity	Return cosine proximity value of the person's competence position to another person's position or a performance.
==	Test for identity: return \code{TRUE} if two persons are identical (i.e. close above the domain's identity threshold).
cosine	Return the cosine value between the person object and other given objects.
plot	Plot a marker for the person's current competence position.
print	Pretty printing of the object.
show	Display the object by printing its key characteristics.
summary	Print basic descriptive statistics about the data held.

Methods

initialize(name, domain)	Constructor: create a new *Person* object with given *name* and in a given *Domain*.
perform (txt, purpose, activity, score, when, label)	This method is called from *read* and *write* to instantiate a *Performance* object and fill it with the data handed over.
write (txt, purpose, score, when, label)	This method is used when learners write a text. It instantiates a *Performance* object and fill it with data (calling *perform*). The parameters contain the raw text (*txt*), a human-readable character string for its *purpose*, if available a score, a character string *when* holding the timestamp, and a human-readable *label* to be used when displaying the performance in visualisations. If no label is handed over, the method constructs one. If no timestamp is handed over, it picks the current date and time via *Sys.time()*.
read (txt, performance, purpose, when, label)	his method is used when learners read a text. It instantiates a *Performance* object and fill it with data (calling *perform*). The parameters contain the raw text (*txt*), a human-readable character string for its *purpose*, a character string *when* holding the timestamp, and a human-readable *label* to be used when displaying the performance in visualisations. If no label is handed over, the method constructs one. If no timestamp is handed over, it picks the current date and time via *Sys.time()*.
path (ix, from, to)	Return a vector with references pointing to the persons *Performance* objects, ordered chronologically or by index *ix*, possibly restricted to the interval *from* to *to*.
position (when)	Return a newly instantiated *Performance* object for a given time and date via *when* or up to the current date and time.
lastPosition()	Return the latest competence position occupied by the person. If no position was previously calculated, the current position is calculated and returned. This method thus caches performance positions.
setName(value)	Set the realname of the person (use the generic *names<-* instead).
getName()	Get the realname of the person (use the generic names instead).
getActivatedTerms (ix)	Returns a list holding the names of the top activated terms in a vector for each performance.
getMeaningVectors (ix)	Returns a matrix with the performances' meaning vectors in the rows.
getSourceTexts()	Return a vector of character strings containing the source texts.
setSourceLogging (x)	Change the setting for source logging: *true* will make all future performances internally store the raw source text (and not only the meaning vectors).
getPurposes()	Return a vector of character strings containing the purpose labels.
getDomains()	Return a vector of references pointing to the domain each performance resides in.

(continued)

setCurrentDomain (dom)	Return the current domain the person is active in, that is used by all future *performance*, *read*, and *write* acts.
print()	Pretty printing the object (use the generic *print* instead).
show()	Display the object by printing its key characteristics (use the generic *show* instead).

A.6 Class 'Performance'

See Fig. **6**.

Fig. 6 The *Performance* class

Performance

name : String
sourcetext : String
logging : Boolean
meaningvector : Vector
terms : CharacterVector
domain : Domain
purpose : String

initialize(text,purpose,domain,name,weighting,logging)
getPurpose()
getDomain()
getMeaningVector()
getActivatedTerms(threshold)
setMeaningVector(vec)
getSourceText()
setName(value)
getName()

Fields

name	Character string holding a human-readable label.
sourcetext	Character string holding the original source text.
logging	Logical flag indicating whether to internally store the source text (or discard it, once the meaning vector is constructed).
meaningvector	Index position of the meaning vector in the Domain's *traces* field.
terms	The top loading terms (above *threshold*).
domain	Reference pointing to the according Domain object.
purpose	Character string indicating the intended purpose of the performance.

Implemented generics

+	Add together two meaning vectors, return result vector.
==	Test two performances for identity (i.e. whether they are close to each other above the proximity threshold).
competences	Calculate the competence positions underlying the vector with references to performances. In case this is only one performance, it returns the performance.
cosine	Calculate the cosine between performances.
names	Return character string with the human-readable label of the performance.

(continued)

names<-	Set the human-readable label for the performance.
near	Return cosine closeness value of performances (or performances and persons) tested.
overlap	Return a vector containing those terms that are shared by all performance objects under investigation.
plot	Plot a marker indicating the location of the performance on the projection surface created by the map visualisation of the domain.
position	Return the competence position held by a vector of performances (calculated as the centroid of the performances).
proximity	Return *true* if the performances in the argument(s) are in proximity to each other (above the *proximityThreshold* of the *Domain*).
terms	Return a vector of character strings containing the top loading terms of the performance (activated above threshold).
print	Pretty printing the performance object.
show	Display the object by printing its key characteristics.
summary	Describe the performance (top loading terms, name, source text).

Methods

initialize(text, purpose, domain, name, weighting, logging)	Constructor: requires the sourcetext *text* and reference pointing to *domain*, can optionally set *purpose* and *name*. *Logging* indicates whether the sourcetext will be stored internally. A *weighting* function can be handed over.
getName()	Get the name of the performance (better use the generic *names* instead).
setName(value)	Set the name of the performance (better use the generic *names<-* instead).
getSourceText()	Return a character string with the raw source text of the performance.
setMeaningVector(vec)	Add the meaning vector to the domain's *trace* matrix field and internally store the index position in field *meaningvector*.
getActivatedTerms(threshold)	Return a vector of character strings containing the top loading terms of the performance (activated above threshold). Use the generic *terms* instead.
getMeaningVector()	Fetch the meaning vector from the domain's *traces* matrix and return it.
getDomain()	Return reference pointing to the according domain object.
getPurpose()	Return character string with the human-readable purpose the performance aimed at achieving.
print()	Pretty printing the performance object (use the generic *print* instead).
show()	Display the object by printing its key characteristics (use the generic *show* instead).

Printed in the United States
By Bookmasters